CYCLIC NUCLEOTIDE SIGNALING

METHODS IN SIGNAL TRANSDUCTION SERIES

Joseph Eichberg, Jr., Series Editor

Published Titles

Cyclic Nucleotide Signaling, Xiaodong Cheng
TRP Channels, Michael Xi Zhu
Lipid-Mediated Signaling, Eric J. Murphy and Thad A. Rosenberger
Signaling by Toll-Like Receptors, Gregory W. Konat
Signal Transduction in the Retina, Steven J. Fliesler and Oleg G. Kisselev
Analysis of Growth Factor Signaling in Embryos, Malcolm Whitman and Amy K. Sater
Calcium Signaling, Second Edition, James W. Putney, Jr.
G Protein-Coupled Receptors: Structure, Function, and Ligand Screening, Tatsuya Haga and Shigeki Takeda
G Protein-Coupled Receptors, Tatsuya Haga and Gabriel Berstein
Signaling Through Cell Adhesion Molecules, Jun-Lin Guan
G Proteins: Techniques of Analysis, David R. Manning
Lipid Second Messengers, Suzanne G. Laychock and Ronald P. Rubin

CYCLIC NUCLEOTIDE SIGNALING

Edited by

Xiaodong Cheng

CRC Press
Taylor & Francis Group
Boca Raton London New York

CRC Press is an imprint of the
Taylor & Francis Group, an **informa** business

CRC Press
Taylor & Francis Group
6000 Broken Sound Parkway NW, Suite 300
Boca Raton, FL 33487-2742

Printed on acid-free paper
Version Date: 20150413

International Standard Book Number-13: 978-1-4822-3556-2 (Hardback)

Visit the Taylor & Francis Web site at
http://www.taylorandfrancis.com

and the CRC Press Web site at
http://www.crcpress.com

Contents

Series Preface

The concept of signal transduction is now long established as a central tenet of biological sciences. Since the inception of the field close to 50 years ago, the number and variety of signal transduction pathways, cascades, and networks have steadily increased and now constitute what is often regarded as a bewildering array of mechanisms by which cells sense and respond to extracellular and intracellular environmental stimuli. It is not an exaggeration to state that virtually every cell function is dependent on the detection, amplification, and integration of these signals. Moreover, there is increasing appreciation that in many disease states, aspects of signal transduction are critically perturbed.

Our knowledge of how information is conveyed and processed through these cellular molecular circuits and biochemical switches has increased enormously in scope and complexity since this series was initiated 15 years ago. Such advances would not have been possible without the supplementation of older technologies, drawn chiefly from cell and molecular biology, biochemistry, physiology, and pharmacology, with newer methods that make use of sophisticated genetic approaches as well as structural biology, imaging, bioinformatics, and systems biology analysis.

The overall theme of this series continues to be the presentation of the wealth of up-to-date research methods applied to the many facets of signal transduction. Each volume is assembled by one or more editors who are pre-eminent in their specialty. In turn, the guiding principle for editors is to recruit chapter authors who will describe procedures and protocols with which they are intimately familiar in a reader-friendly format. The intent is to assure that each volume will be of maximum practical value to a broad audience, including students and researchers just entering an area, as well as seasoned investigators.

It is hoped that the information contained in the books of this series will constitute a useful resource to the life sciences research community well into the future.

Joseph Eichberg
Michael Xi Zhu
Series Editors

Preface

The cyclic nucleotide second messenger systems are of major historical significance in biology. Since the original prize-winning discovery of cAMP by Earl Sutherland more than half a century ago, deciphering the cyclic nucleotide signaling pathways has led to several major scientific breakthroughs, including the discoveries of cAMP-dependent protein kinase/protein kinase A (PKA), G proteins, nitric oxide, and G protein–coupled receptors, for which Robert Lefkowitz and Brian Kobilka shared the Nobel Prize in Chemistry in 2012. In addition to the aforementioned long-lasting impact on fundamental biology and chemistry, the study of cyclic nucleotide second messengers has also revolutionized the practices of modern medicine and pharmaceutics in view of the fact that current medicines target the cyclic nucleotide signaling pathway more than any other pathway.

With the surprising discovery of the exchange proteins directly activated by cAMP (EPACs), the research field is currently undergoing a renaissance, which has generated a large amount of excellent original work across a wide range of research topics. Because of the diverse roles that cyclic nucleotides exert under physiological and pathological conditions, assorted methodologies, such as fluorescence imaging, structural biology, molecular and cell biology, genetics, proteomics, pharmacology, and drug discovery, are employed to investigate various aspects of these important signaling molecules. Without a comprehensive coverage of methodology for studying cyclic nucleotide signaling, choosing and carrying out the proper protocol among many possible approaches in these broad areas can be a daunting task. This volume is aimed to bridge this gap by bringing together novel and current techniques, spanning from x-ray crystallographic analysis at the atomic level to in vivo genetic knockout in whole organisms, employed in the cyclic nucleotide signal transduction field. The book is organized in chapters, written by experts, and focuses on describing experimental protocols and tools, as well as their applications in solving biological problems with the goal of providing clear directions for laboratory use. Because of the breadth of the cyclic nucleotide signaling field, the applications of some methods are described in multiple chapters to provide distinctive views on how the methodology can be utilized to study different physiological functions. Some chapters include discussion on the usefulness and pitfalls associated with the use of certain tools and techniques. We believe that this volume will be a useful resource for investigators, new or experienced, in the cyclic nucleotide and related fields.

Finally, we would like to thank all authors for their valuable contributions that made the book possible. Special thanks also go to Professor Michael Zhu for sharing his knowledge and experience in preparing the book, as well as to Dr. Chuck Crumly and Ms. Joselyn Banks-Kyle at CRC Press, Taylor & Francis Group.

Editor

Xiaodong Cheng is a professor in the Department of Integrative Biology and Pharmacology at the University of Texas Health Science Center at Houston. He received his bachelor's degree from Peking University in Beijing, China in 1985, and his master's degree from Shanghai Institute of Biochemistry, Chinese Academy of Science in 1988. He obtained his PhD from the University of Texas Medical Branch in Galveston in 1994. After completing his postdoctoral studies at the University of California, San Diego in 1999, he returned to the University of Texas Medical Branch to start his own laboratory as an assistant professor in the Department of Pharmacology and Toxicology, where he was promoted to associate professor in 2004 and professor in 2009. In 2013, Dr. Cheng moved to his current position. He is also a member of the Brown Foundation Institute of Molecular Medicine and the Texas Therapeutics Institute at the Texas Medical Center. Dr. Cheng's major research interests are cAMP-mediated cell signaling and drug discovery, especially those involving the exchange proteins directly activated by cAMP (EPAC). Recently, his research group has discovered novel small molecule inhibitors for EPAC proteins and established animal disease models, and is in the process of evaluating the therapeutic potential of targeting EPAC proteins. His long-term goals are to unravel the signaling intricacies of EPAC proteins and to design pathway-specific probes for these important signaling molecules so that their functions can be pharmaceutically exploited and modulated for the treatment of human diseases. Dr. Cheng is an editor of *Acta Biochimica et Biophysica Sinica* and is an editorial board member of *Molecular Pharmacology* and *PLOS ONE*.

Contributors

Madoka Akimoto
Department of Chemistry and Chemical Biology
McMaster University
Hamilton, Ontario, Canada

Zeynep Bastug
Heart Research Center Göttingen
Georg August University Medical Center
University of Göttingen
Göttingen, Germany

Daniela Bertinetti
Department of Biochemistry
University of Kassel
Kassel, Germany

Stephen Boulton
Department of Biochemistry and Biomedical Sciences
McMaster University
Hamilton, Ontario, Canada

Elke Butt
University Clinic Wuerzburg
Institute of Clinical Biochemistry and Pathobiochemistry
Wuerzburg, Germany

Xiaodong Cheng
Department of Integrative Biology and Pharmacology
The University of Texas Health Science Center
Houston, Texas

Oleg G. Chepurny
Department of Medicine
State University of New York (SUNY) Upstate Medical University
Syracuse, New York

Dar-Chone Chow
Department of Pharmacology
Baylor College of Medicine
Houston, Texas

Jennifer Deem
Department of Pharmacology
University of Washington
Seattle, Washington

Carmen W. Dessauer
Department of Integrative Biology and Pharmacology
University of Texas Health Science Center at Houston
Houston, Texas

Stein Ove Døskeland
Department of Biomedicine
University of Bergen, Norway

Niels Eijkelkamp
Laboratory of Neuroimmunology and Developmental Origins of Disease
and
Laboratory of Translational Immunology
University Medical Center Utrecht
Utrecht, the Netherlands

Rajanish Giri
Department of Chemistry and Chemical Biology
McMaster University
Hamilton, Ontario, Canada

Kirill Gorshkov
Department of Pharmacology and Molecular Sciences
The Johns Hopkins University School of Medicine
Baltimore, Maryland

Jan Haavik
Department of Biomedicine
University of Bergen, Norway

Claudia Hahnefeld
Department of Biochemistry
University of Kassel
Kassel, Germany

Cobi J. Heijnen
Laboratory of Neuroimmunology
Department of Symptom Research
The University of Texas M.D. Anderson Cancer Center
Houston, Texas

Friedrich W. Herberg
Department of Biochemistry
University of Kassel
Kassel, Germany

Lars Herfindal
Department of Biomedicine
University of Bergen, Norway

George G. Holz
Department of Medicine and Pharmacology
State University of New York (SUNY) Upstate Medical University
Syracuse, New York

Gilbert Y. Huang
Department of Pharmacology
Baylor College of Medicine
Houston, Texas

Brian W. Jones
Department of Pharmacology
University of Washington
Seattle, Washington

Annemieke Kavelaars
Laboratory of Neuroimmunology
Department of Symptom Research
The University of Texas M.D. Anderson Cancer Center
Houston, Texas

Choel Kim
Department of Pharmacology
Baylor College of Medicine
Houston, Texas

Jeong Joo Kim
Department of Pharmacology
Baylor College of Medicine
Houston, Texas

Rune Kleppe
Department of Biomedicine
University of Bergen, Norway

Antonius Koller
Department of Pharmacology
Baylor College of Medicine
Houston, Texas

Colin A. Leech
Department of Medicine
State University of New York (SUNY) Upstate Medical University
Syracuse, New York

Yong Li
Department of Integrative Biology and Pharmacology
University of Texas Health Science Center at Houston
Houston, Texas

Robin Lorenz
Department of Biochemistry
University of Kassel
Kassel, Germany

Lise Madsen
National Institute of Nutrition and Seafood Research
Bergen, Norway

G. Stanley McKnight
Department of Pharmacology
University of Washington
Seattle, Washington

Fang Mei
Department of Integrative Biology and Pharmacology
The University of Texas Health Science Center
Houston, Texas

Giuseppe Melacini
Department of Chemistry and Chemical Biology
and
Department of Biochemistry and Biomedical Sciences
McMaster University
Hamilton, Ontario, Canada

Kody Moleschi
Department of Chemistry and Chemical Biology
McMaster University
Hamilton, Ontario, Canada

Stefan Möller
Department of Biochemistry
University of Kassel
Kassel, Germany

Viacheslav O. Nikolaev
Institute of Experimental Cardiovascular Research
University Medical Center Hamburg-Eppendorf
Hamburg, Germany

Holger Rehmann
Molecular Cancer Research
Center of Molecular Medicine
and
Centre of Biomedical Genetics and Cancer Genomics Centre
University Medical Center Utrecht
Utrecht, the Netherlands

Robert Rieger
Department of Pharmacology
Baylor College of Medicine
Houston, Texas

Michael W. Roe
Departments of Medicine and Cell and Developmental Biology
State University of New York (SUNY) Upstate Medical University
Syracuse, New York

Frode Selheim
Department of Biomedicine
University of Bergen, Norway

Pooja Singhmar
Laboratory of Neuroimmunology
Department of Symptom Research
The University of Texas M.D. Anderson Cancer Center
Houston, Texas

Tamara Tsalkova
Department of Integrative Biology and Pharmacology
The University of Texas Health Science Center
Houston, Texas

Bryan VanSchouwen
Department of Chemistry and Chemical Biology
McMaster University
Hamilton, Ontario, Canada

Linghai Yang
Department of Pharmacology
University of Washington
Seattle, Washington

Jin Zhang
Department of Pharmacology and Molecular Sciences
and
The Solomon H. Snyder Department of Neuroscience and Department of Oncology
The Johns Hopkins University School of Medicine
Baltimore, Maryland

1 Discovery of Small Molecule EPAC Specific Modulators by High-Throughput Screening

Xiaodong Cheng, Tamara Tsalkova, and Fang Mei

CONTENTS

1.1 INTRODUCTION

Since the landmark discovery of the prototypical second messenger, adenosine 3′,5′-cyclic monophosphate (cAMP), by Earl Sutherland and colleagues more than half a century ago,[1,2] the studies of cyclic nucleotide-mediated signaling pathways have produced long-lasting effects on fundamental biology and chemistry, with at least seven Nobel Prizes, including the 2012 Award for Chemistry to Robert Lefkowitz and Brian Kobilka for their studies of the G-protein coupled receptors, being awarded in related research fields. The knowledge gained from these endeavors has also revolutionized the practices of modern medicine and pharmaceutics considering the fact that current therapeutics target the cyclic nucleotide signaling pathway more than any other pathway.

In humans, the effects of cAMP are mainly transduced by two ubiquitously expressed intracellular cAMP sensors, protein kinase A (PKA, also known as cAMP-dependent protein kinase) and exchange protein directly activated by cAMP (EPAC, also known as cAMP-GEF),[3,4] as well as cyclic nucleotide-activated ion channels (CNG and HCN) in certain tissues.[5] At first glance, the discovery of the EPAC family of proteins in 1998 was quite unexpected because their *cousin* PKA was initially identified more than 30 years earlier,[6] and for many years, the PKA family of proteins were considered the main intracellular effectors of cAMP. However, the existence of EPAC proteins was not a total surprise, bearing in mind that PKA-independent cAMP functions have been reported.[7–10]

The identification of EPAC proteins heralded the arrival of a new era for cAMP signaling research because many cAMP functions previously attributed exclusively to PKA were in fact also transduced by EPAC. In mammals, there are two EPAC isoforms, EPAC1 and EPAC2, encoded by two separate genes. EPAC1 and EPAC2 have discrete tissue-specific expression patterns. Although EPAC1 is more ubiquitously expressed, EPAC2 is mainly found in the central nervous system, pancreatic islets, and adrenal gland.[4] It also seems that levels of EPAC1 and EPAC2 are differentially regulated at various developmental stages.[11] At the cellular level, the two isoforms of EPAC assume distinct cellular localizations and presumably participate in the formation of different signalsomes.[12–16] These differences in cellular behaviors between EPAC1 and EPAC2 are consistent with the fact that most of the biological and cellular functions described in the literature are nonredundant.

A number of recent studies using mouse knockout models of EPAC1 and EPAC2 have provided valuable information about the physiological functions and relevance of these two important signaling molecules in disease.[17–22] However, not all reported phenotypes for the same gene knockout are consistent. Such apparent *irreproducible results* can actually be an intrinsic property of the knockout approach. Theoretically, it is possible that a particular gene knockout may produce a spectrum of apparent phenotypes depending on the unique genetic background of the mouse strain on which the knockout is based. Researchers often deliberately pick a specific genetic background from a plethora of inbred mouse strains to create a disease model because only certain mouse strains are susceptible to the induction of a specific disease. To complicate the matter further, compensation or adaptation may mask the role of the target gene in a knockout model. One potential solution is the use of pharmacological approaches to validate the genetic phenotype of a knockout mouse model. Because most current therapeutics are small molecules, it is of critical importance to develop appropriate chemical probes as complementary approaches to genetics in probing biological functions, as well as in identifying protein targets for drug development. Considering the important functions that EPAC proteins play in physiology and pathophysiology, developing EPAC-specific pharmacological probes capable of distinguishing EPAC proteins from other cyclic nucleotide receptors and between EPAC isoforms is an essential area of research for basic biology and drug discovery.

Over the years, an array of cyclic nucleotide analogs have been developed and applied successfully to selectively manipulate different effectors of the cyclic nucleotide signaling pathways[23,24] (also see Chapters 2, 4, and 12). Sequence analysis of the cAMP binding domain (CBD) of EPAC reveals that a glutamate residue conserved in other CBD-containing proteins is absent in EPAC proteins.[25] X-ray crystal structural analysis of PKA regulatory subunit demonstrates that this residue forms hydrogen bonds with the 2′-hydroxyl of the cAMP ribose group and is important for cAMP binding to PKA.[26] This unique property of EPAC CBD suggests that hydrogen bonding with the 2′-hydroxyl of the cAMP ribose group that is critical for PKA function is not required for EPAC, which lead to the identification of an EPAC-selective cAMP analog, 8-(4-chloro-phenylthio)-2′-*O*-methyladenosine-3′,5′-cyclic monophosphate (8-CPT-2′-*O*-Me-cAMP, also known as 007).[25,27] Although 007 exerts approximately 100-fold selectivity toward EPAC over PKA and has become a widely used tool in EPAC-related research,[25,27–31] there are several concerns associated with 007-based EPAC-selective agonists. Besides the fact that 007 class compounds are incapable of differentiating between EPAC1 and EPAC2, they have been shown to inhibit phosphodiesterases (PDEs).[24] Moreover, a series of recent studies have raised concerns that 8-CPT-conjugated cAMP analogs and their metabolites can exert diverse biological functions through cAMP-independent mechanisms.[32–35] These findings point out a potential common limitation of the application of cyclic nucleotide mimetics.

Here, we summarize recent efforts in identifying noncyclic nucleotide EPAC-specific modulators using high-throughput screening (HTS) efforts. We also briefly describe the applications of these EPAC-specific pharmacological modulators in probing the biological functions of EPACs under physiological and pathological conditions.

1.2 ASSAY DEVELOPMENT

1.2.1 Choice of Assay Method

The most important initial decision to make in the development of HTS assay for a defined target is the choice of assay format. Biochemical-based assays using isolated macromolecular components were more prevalent during earlier HTS campaigns. In recent years, cell-based assays have gained popularity in part due to technological advances and new instruments that have significantly expanded our ability to monitor cellular readouts that meet the requirements of HTS. One intrinsic advantage of a cell-based assay is the ability to perform HTS of the target protein in a more physiological setting within the native cellular context. With the development of genetically encoded fluorescent cyclic nucleotide reporters based on fluorescence resonance energy transfer (FRET; see Chapters 5 and 8), the activation of cyclic nucleotide sensors, such as EPAC, can be followed in real time in living cells. Furthermore, these biosensors can be stably expressed in mammalian cell lines and screened using a plate reader (see Chapter 3). However, the signal-to-noise levels of these plate-based FRET readouts are moderate at most. Accurate determination of biosensor activation requires averaging multiple sample repeats (8–16), which renders the cost of an HTS prohibitive using this approach. In contrast, such cell-based assays are quite suitable and complementary to be used as a secondary screen for hit validation and lead optimization.

Recently, an attempt to use bioluminescent resonance energy transfer (BRET)–based assays to identify and validate EPAC1 modulators has been reported.[36] This specific BRET sensor was constructed using engineered human EPAC1 (amino acids 149–881) lacking the N-terminal Disheveled, Egl-10, pleckstrin (DEP) domain with two mutations (T781A and F782A) that disrupt its nucleotide exchange activity. The engineered EPAC1 construct is sandwiched between Renilla luciferase and a modified green fluorescent protein (GFP) variant citrine to generate CAMYEL, which produces a strong BRET signal and allows the detection of cAMP in live cells.[37] With a small focused library of cyclic nucleotide analogs and 133 compounds originating from a virtual docking screening, Brown and colleagues validated that the CAMYEL-based EPAC1 BRET sensor was capable of identifying EPAC modulators. The authors performed the assay using cell lysates expressing CAMYEL instead of intact cells.[36] CAMYEL produces a higher signal-to-noise ratio than FRET-based cAMP sensors and may be more suited for HTS, but it remains to be seen if the assay is robust enough for large-scale screenings. With these considerations in mind, we proceeded to design a biochemical assay using purified EPAC proteins.

1.2.2 Assay Design

Although the minimal biochemically active EPAC construct requires only the guanine nucleotide exchange factor (GEF) region and the contiguous CBD, the use of truncate EPAC proteins may diminish the possibility of identifying isoform-specific or allosteric modulators that target regulatory components/properties in EPAC other than direct cAMP binding. Therefore, we decided to use full-length recombinant EPAC proteins for our assay. This turned out to be critically important for the identification of allosteric modulators and isoform EPAC-specific inhibitors, which will be covered in Section 1.4.2.

1.2.2.1 Assay Readout

Unlike most typical enzymes, EPAC proteins are not directly involved in catalyzing a specific chemical reaction. Instead, the binding of cAMP to EPAC induces a major conformation change, which frees the C-terminal GEF domain from the autoinhibition of the N-terminal regulatory region and allows the docking of small GTPases, Rap1 or Rap2. This interaction significantly reduces the guanine nucleotide affinity of Rap proteins, which binds guanosine-5′-diphosphate (GDP) and guanosine-5′-triphosphate (GTP) extremely tightly. As a consequence, Rap proteins switch from a GDP-bound inactive state to an active Rap–GTP complex due to the estimated 10:1 ratio of GTP over GDP present in the cellular milieu. These biochemical properties dictate the choice of readouts for designing suitable HTS assays for EPAC. One well-known in vitro guanine nucleotide exchange activity assay using a fluorescent guanine nucleotide analog with the *N*-methyl-3′-*O*-anthraniloyl (MANT) moiety[38] can be conveniently adapted to determine the activity of EPAC proteins. However, attempts to optimize the MANT-GDP/GTP assay for HTS have not been successful due to the low signal-to-noise ratio.

Therefore, we decided to explore the cAMP-binding property of EPAC for assay development. For the benefit of simplicity, sensitivity, and cost-effectiveness, our first choice was to design a direct competition assay using fluorescence as a readout. Fluorescent nucleotide analogs have been widely used as tools for structural and functional analyses.[39] Various fluorescent cAMP analogs have also been developed and applied to probe the properties of various cAMP-binding proteins, including PKA,[40] PDE,[41] and cyclic nucleotide activated channel,[42] as well as EPAC.[43] Much to our delight, when several commercially available fluorescent cAMP analogs were tested, the fluorescent intensity of one analog, 8-NBD-cAMP (8-(2-[7-nitro-4-benzofurazanyl] aminoethyl-thio) adenosine-3′,5′-cyclic monophosphate), increased dramatically when titrated with purified full-length EPAC2. As shown in Figure 1.1, binding of 8-NBD-cAMP to EPAC2 led to a dose-dependent increase in fluorescent signal with a maximal 212-fold signal enhancement under EPAC2 saturation concentrations. Fitting the titration curve led to an apparent dissociation constant of 270 nM for 8-NBD-cAMP. More importantly, the fluorescent signal change was completely reversible in the presence of excess cAMP, which competed with 8-NBD-cAMP

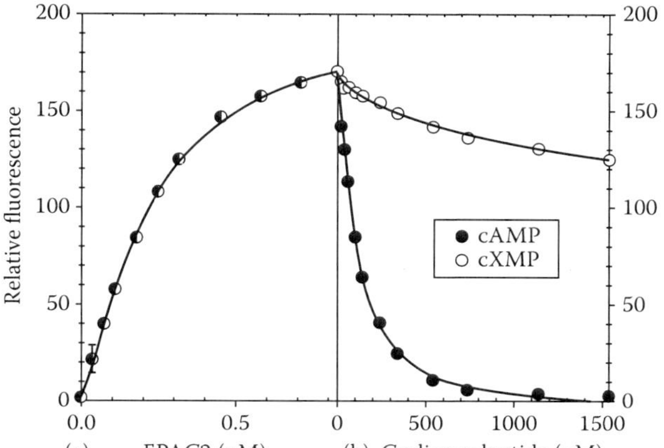

FIGURE 1.1 Changes in 8-NBD-cAMP fluorescence as a function of EPAC2 and cyclic nucleotide concentrations. (a) Change in fluorescence when 8-NBD-cAMP (0.1 μM) is titrated with EPAC2. (b) The fluorescence change can be reversed by the addition of excess cAMP but not by cXMP.

in binding to EPAC2. Interestingly, the extent of fluorescence increase associated with EPAC1 binding was much less impressive, which is consistent with the finding that binding of 8-NBD-cAMP to an isolated fragment of the CBD of EPAC1 leads to a maximal sixfold increase in fluorescence intensity.[43] This very large reversible fluorescence change of 8-NBD-cAMP after its binding to EPAC2 protein made it an excellent readout for designing a sensitive and robust HTS assay for EPAC2 antagonists.

1.2.2.2 Assay Miniaturization and Optimization

An important step in HTS assay development is assay miniaturization and optimization so the desired throughput could be achieved in a cost-effective manner without compromising assay performance quality. The key issue here is to strike a balance between statistical performance and assay sensitivity. To demonstrate robust and reproducible behavior in a 96-well or higher density format, a signal-to-background ratio of at least 5 and a coefficient of variation of less than 10% are usually required. When tested in a 96-well format by fixing the ratio of EPAC2:8-NBD-cAMP at 1:1.2, the fluorescence intensity signal showed an excellent linearity as a function of EPAC2 concentration between 0 and 150 nM with a signal-to-background ratio of more than 11 at 50 nM of EPAC2 concentration.[44] Under such an assay condition, the assay is highly robust with a Z' score of more than 0.8. The Z' score, calculated based on the equation $1 - (3\sigma_{signal} + 3\sigma_{background})/|\mu_{signal} - \mu_{background}|$, is the most useful indication for assessing the statistical performance of an assay. An acceptable assay usually requires Z' values larger than 0.5. However, an assay with a high Z' score may not necessarily be equivalent to an optimal assay. Here is where the assay sensitivity comes into play. Typically, increases in substrate or ligand concentrations will lead to improved signal-to-noise ratio, thus better Z' scores. With this in mind, under an assay condition where the substrate or ligand concentration is set to be 10 times that of the K_m or K_d, competitive inhibitors with a IC_{50} more than 10% of the compound testing concentration will likely be below the detection limit. Ideally, HTS assays should be performed with substrate or ligand concentrations below their K_m or K_d, so that compounds with IC_{50} values identical to the testing concentration will show more than 50% inhibition. The ligand concentration used for 8-NBD-cAMP (60 nM) is significantly lower than the K_d value (270 nM), suggesting that our assay is also highly sensitive.

Because almost all the small compound libraries are dissolved in DMSO. The effect of DMSO on assay readout should be tested to work out a tolerable working DMSO concentration range. Under the assay conditions, the measured florescence signal at various concentrations of DMSO remain fairly constant with less than 5% change in the presence of 1% DMSO, indicating that our assay is well tolerated up to 1% of DMSO.

1.2.2.3 Dose-Dependent Responses to cAMP Analogs

If it is feasible, one final step in the development of the assay involves assay validation using a small collection of known pharmacologically active standards and their inactive analogs. This step will help identify screen compound concentration and hit selection criteria. Using a small collection of known cAMP analogs:

cAMP and 8-Cl-cAMP that can bind and activate EPAC as positive controls, as well as 2′-deoxy-cAMP and cXMP that bind EPAC weakly and are incapable of activating EPAC as negative controls, we found that cAMP and 8-Cl-cAMP led to dose-dependent inhibitions of fluorescence signals whereas 2′-deoxy-cAMP and cXMP had minimal effects. The degree of signal decrease was consistent with the relative affinity of these analogs to EPAC, such that 8-Cl-cAMP led to a bigger decrease in signal compared with cAMP at the same concentrations as 8-Cl-cAMP binds EPAC tighter than cAMP. Based on the finding that at 100 μM concentration, cAMP resulted in an approximately 80% decrease in fluorescence signal, we decided to set the compound screening concentration at a single dose of 100 μM. Compounds that cause more than 80% decrease in NBD fluorescence intensity were selected as initial positive hits. In theory, such screening criteria will allow us to identify compounds with similar binding affinity to that of the native ligand cAMP.

1.2.3 Pilot Screening

It is often useful to conduct a pilot screening using a small diversity-based library before the actual HTS effort. Such an endeavor will allow one to estimate the hit rate and to evaluate assay performance and reproducibility under the actual screening environments, which provides an opportunity to refine and adjust screen conditions such as sample concentration if necessary. From a pilot screening of the National Cancer Institute (NCI) Developmental Therapeutics Program (DTP) diversity set library, which contains 1990 carefully selected small molecules for their high chemical and pharmacological diversity from the entire collection of 140,000 compounds at NCI, 18 compounds were identified to decrease the fluorescence signal by more than 80% at a test concentration of 100 μM. This corresponds to a hit rate slightly lower than 1%. These 18 compounds were further tested individually using a secondary functional assay that monitored the ability of EPAC to catalyze the nucleotide exchange activity of Rap1. Three compounds, NCS45576, NCS119911, and NSC686365, were shown to be able to inhibit EPAC2 GEF activity to basal levels at 25 μM concentration in the presence of equal concentrations of cAMP.[44] Overall, the pilot screen data confirm that the 8-NBD-cAMP–based HTS assay is robust and sensitive for the identification of EPAC-specific antagonist.

1.3 HIGH-THROUGHPUT SCREENING

1.3.1 Reagents

1.3.1.1 Recombinant Proteins

Recombinant full-length EPAC1, EPAC2, and C-terminal truncated Rap1B(1–167) were purified as reported previously[45–47] and also as described in great detail in Chapter 6. PKA catalytic subunit, RIα and RIIβ regulatory subunits were individually expressed in *Escherichia coli* and purified to homogeneity as reported.[48] Type I and II PKA holoenzymes were reconstituted from purified PKA R and C subunits.[49]

1.3.1.2 Reagents

8-NBD-cAMP, cXMP, 8-Cl-cAMP, and 2′-deoxy-cAMP were purchased from BioLog Life Science Institute (Bremen, Germany). MANT-GDP and BODIPY FL GDP were obtained from Molecular Probes, Invitrogen (Carlsbad, CA, USA). All other reagents were purchased through Sigma-Aldrich (St. Louis, MO, USA).

1.3.1.3 Compound Library

A Maybridge HitFinder library (Maybridge, Thermo Fisher Scientific, MA, USA) with a preplated collection of 14,400 drug-like small chemical compounds in 96-well plates was used for the HTS. The compounds were supplied as dry films and dissolved in DMSO at 10 mM concentration right before the screening.

1.3.2 Primary HTS

Primary HTS was performed in duplicates in black 384-well low volume microplates from Corning Costar (Cambridge, MA, USA). Based on the pilot screening performed using NCI DTP diversity set library in 96 wells, we further fine-tuned the assay conditions for optimal screening in 384-well plates. In brief, a protein solution containing 65 nM EPAC2 and 75 nM 8-NBD-cAMP was prepared in 20 mM Tris buffer (pH 7.5) containing 150 mM NaCl, 1 mM EDTA, and 1 mM DDT. BiomekFX Laboratory Automation Workstation equipped with 96-multichannel pipetting head (Beckman Coulter, CA, USA) was used to dispense samples into 384-well plates (35 μL/well). Test compounds (1 μL/well) were added from 10 mM stock solutions in DMSO from 96-well mother plates. DMSO or cAMP at 10 mM was used as negative or positive controls, respectively. Fluorescence intensity signal from 8-NBD-cAMP was recorded at room temperature before and after the tested compounds were added using SpectaMaxM5 microplate reader (Molecular Devices, Silicon Valley, CA, USA) with excitation/emission wavelengths set at 470/540 nm.

A total of 99 compounds from the Maybridge HitFinder library with 14,400 compounds that decreased the fluorescence signals to the levels of the positive controls were selected as initial hits. This represents a hit-rate of approximately 0.7%. The primary screen was again conducted in duplicate at 28 μM compound concentration for all 99 initial hits. A total of 33 compounds decreased fluorescence signals from the 8-NBD-cAMP by more than 60% and were designated as primary hits. By comparison, 28 μM of cAMP decreased fluorescence signal from the 8-NBD-cAMP by 45%.

1.3.3 Secondary Confirmation Assay

To confirm the compounds active in the primary screen as reproducible hits, it is necessary to validate the initial hits using secondary assays. Because the number of compounds at the validate stage is significantly reduced as compared with that of the primary HTS stage, it is not necessary to have HTS-compatible secondary assays. Low to medium throughput assays can also work if the compound number tested is relatively small. However, to maximize the chance of ruling out artifacts, it is preferable to have secondary assays using a format or readout that is different from that of the primary HTS assay.

To validate the 33 primary hits, we further tested these compounds with a functional confirmation assay that directly measured the in vitro GEF activity of EPAC using a well-known MANT-GDP–based exchange assay.[38] Briefly, 0.2 μM of Rap1B(1–167) loaded with the fluorescent GDP analog (Mant-GDP), was incubated with EPAC in 50 mM Tris buffer (pH 7.5), containing 50 mM NaCl, 5 mM $MgCl_2$, 1 mM DTT, and a 100-fold molar excess of unlabelled GDP (20 μM) in the presence of 25 μM tested compound and 25 μM cAMP. Exchange of Mant-GDP by GDP was measured as a decrease in fluorescence intensity over time using a FluoroMax-3 spectrofluorometer with excitation/emission wavelengths set at 366/450 nm. Typically, decay in the fluorescence intensity was recorded over a time course of 6000 s with data points taken every 60 s. We identified seven compounds (ESJ-04 to ESI-10) that were capable of inhibiting EPAC2 GEF activity to basal levels at 25 μM concentration in the presence of equal concentrations of cAMP. A working concentration of 25 μM cAMP was used to provide optimal sensitivity as the apparent AC_{50} for cAMP was determined at 20 μM.[46]

1.3.4 Counterscreening Assays

It is also necessary to test the hits against related targets using counterscreening assays to further validate and characterize the specificity of the hits. To ensure that our identified EPAC antagonists do not affect the PKA family intracellular cAMP sensors, the effect of hits on kinase activities of type I and II PKA holoenzymes were measured spectrophotometrically in a 96-well plate with a coupled enzyme assay. In this assay, the formation of the ADP is coupled to the oxidation of NADH by the pyruvate kinase/lactate dehydrogenase reactions so the reaction rate can be determined by following the oxidation of NADH, reflected by a decrease in absorbance at 340 nm. Briefly, the kinase reaction mixture (100 μL) contained 50 mM Mops (pH 7.0), 10 mM $MgCl_2$, 1 mM ATP, 1 mM PEP, 0.1 mM NADH, 8 U of pyruvate kinase, 15 U of lactate dehydrogenase, a fixed amount of type I or type II PKA holoenzyme, and 0.1 mM cAMP with or without 25 μM of the test compound. Reactions were pre-equilibrated at room temperature and initiated by adding the Kemptide substrate (final concentration, 0.26 mM). PKA activities measured in the presence of 25 μM H89, a selective PKA inhibitor, were used as positive controls of PKA inhibition. Additional counterscreening assays against PDEs, CNGs, and HCNs could also be performed.

1.3.5 Advantages and Limitations

1.3.5.1 Advantages

One of the major advantages of our 8-NBD-cAMP–based HTS assay is its simplicity. Only recombinant EPAC2 protein is required. The assay can be performed under a mix-and-measure format without extra steps such as centrifugation, filtration, or extraction, which is highly desired and preferable for HTS. Moreover, the assay is highly sensitive and robust with outstanding signal-to-noise ratio and dynamic range. More importantly, as discussed in detail in Section 1.4.2, this assay is able to identify isoform-specific allosteric inhibitors.

1.3.5.2 Limitations

Because the change in fluorescence signal in the presence of EPAC1 is much more modest, 8-NBD-cAMP–based HTS assay only works well for EPAC2. As such, we cannot run parallel screenings using both isoforms, and it will not be feasible to identify EPAC1-specific agonists using this assay. In addition, this cAMP analog–based competition assay is also unlikely to be useful for screening uncompetitive and noncompetitive inhibitors.

To overcome these limitations, we have designed a second generation EPAC HTS assay using a fluorescent GDP analog, BODIPY FL GDP, which allows us to directly follow the GEF activity of EPAC proteins. Using this new HTS assay and in collaboration with colleagues at the NIH Chemical Genomics Center, we have performed parallel, quantitative HTS using purified full-length recombinant EPAC1 and EPAC2 proteins. From a library of more than 480,000 compounds, we have identified multiple isoform inhibitors specific for either EPAC1 or EPAC2 in addition to many pan EPAC inhibitors. We are currently validating and characterizing these hits.

1.4 CHARACTERIZATION OF EPAC-SPECIFIC INHIBITORS

1.4.1 Relative Potency

To determine the relative binding affinity of seven EPAC antagonists identified from our screen, we performed dose-dependent titrations to test the ability of these compounds to compete with the binding of 8-NBD-cAMP to EPAC2. When various concentrations of cAMP or EPAC2 antagonists were added to a reaction mixture with fixed concentrations of EPAC2 and 8-NBD-cAMP, a dose-dependent decrease in 8-NBD-cAMP fluorescence was observed (Figure 1.2a). Although

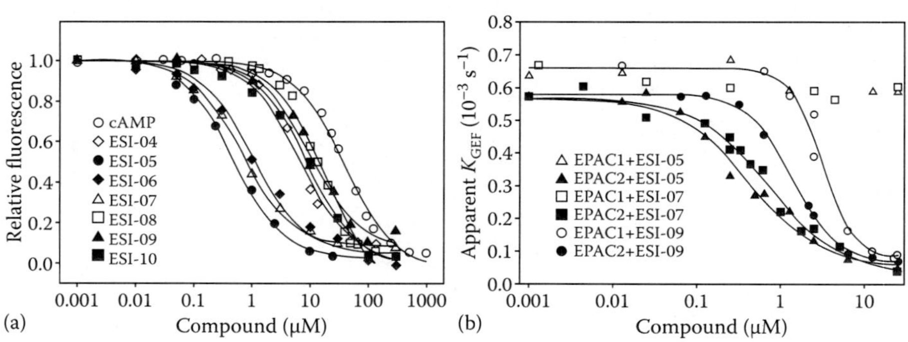

FIGURE 1.2 Relative potency of EPAC2-specific antagonists. (a) Dose-dependent competition of EPAC2-specific antagonists with 8-NBD-cAMP in binding to EPAC2. (b) Dose-dependent inhibition of EPAC1 (open symbols) or EPAC2 (closed symbols) GEF activity by EPAC2-specific antagonists in the presence of 25 μM of cAMP.

TABLE 1.1
Apparent IC_{50} Values of EPAC-Specific Antagonists in Competing with 8-NBD-cAMP Binding to EPAC2

Compound	Apparent IC_{50} (μM)	Relative Potency (RA)[a]
cAMP	39 ± 2.0	1.0
ESI-04	6.7 ± 0.7	5.8
ESI-05	0.48 ± 0.03	81
ESI-06	1.0 ± 0.2	39
ESI-07	0.67 ± 0.03	57
ESI-08	8.7 ± 1.1	4.5
ESI-09	10 ± 1.2	3.9
ESI-10	18 ± 2.0	2.2

[a] $RA = IC_{50,cAMP}/IC_{50,compound}$.

cAMP competed with 8-NBD-cAMP binding with an apparent IC_{50} of 39 μM, all selected EPAC2 antagonists showed an increased potency with apparent IC_{50} values ranging from 0.48 to 18 μM (Table 1.1). These data suggest that all seven identified EPAC-selective inhibitors interact with EPAC2 with a potency stronger than its native ligand, cAMP.

1.4.2 Isoform Specificity

Because these antagonists were identified using EPAC2 as a target, we tested if these compounds were also effective in suppressing cAMP-mediated EPAC1 GEF activity. All seven compounds, except ESI-05 and ESI-07, were also capable of suppressing EPAC1 GEF activity to basal levels at 25 μM concentration in the presence of an equal concentration of cAMP. For example, compound ESI-09 inhibited both EPAC1- and EPAC2-mediated Rap1-GDP exchange activity in a dose-dependent manner with an apparent IC_{50} of 3.2 or 1.4 μM,[50] respectively. By contrast, although compounds ESI-05 and ESI-07 potently blocked EPAC2-mediated Rap1-GDP exchange activity with apparent IC_{50} values of 0.4 or 0.7 μM, respectively, these compounds were completely ineffective in suppressing EPAC1-mediated GEF activity (Figure 1.2b).[51]

To determine the potential mechanism of this EPAC2 isoform-specific inhibiting action, we probed the mode of interaction between ESI-07 and EPAC2 protein by monitoring the rates of amide hydrogen exchange using a deuterium exchange mass spectrometry technique. Our study suggests that ESI-07 acts as an allosteric modulator by binding the interface between two CBDs of EPAC2 and locks the protein in its autoinhibitory conformation.[51] This model of action is consistent with the fact that EPAC1 only has one CBD and therefore lacks the putative allosteric site. A subsequent study confirmed that EPAC2 lacking the first CBD is insensitive to ESI-05 inhibition.[52]

1.4.3 Cellular Activity of EPAC-Specific Antagonists

To test if our EPAC antagonists were capable of modulating EPAC activation in living cells, we monitored the ability of these compounds in suppressing EPAC-mediated Rapl cellular activation. As shown in Figure 1.3a, when HEK293 cells that ectopically express full-length EPAC2 proteins were treated with an EPAC-selective cAMP analog 8-(4-chlorophenylthio)-2′-*O*-methyladenosine-3′,5′-cyclic monophosphate, acetoxymethyl ester (007-AM), an increase in the fraction of GTP-bound cellular Rapl was observed. Pretreatment of HEK293/EPAC2 cells with 10 μM of compounds ESI-05, ESI-07, and ESI-09 led to a significant reduction of 007-AM–induced Rapl activation. In contrast, when HEK293 cells that ectopically express full-length EPAC1 proteins were used, only compound ESI-09 was effective in blocking 007-AM–induced Rapl activation whereas compounds ESI-05 and ESI-07 were completely ineffective (Figure 1.3b). These results are consistent with the biochemical Rapl exchange data shown and further confirm that compounds ESI-05 and ESI-07 are EPAC2-specific antagonists whereas compounds ESI-09 is a pan-EPAC antagonist.

In addition to mediating cAMP-induced Rapl activation, EPAC proteins are also known to activate the Akt/PKB signaling pathways whereas PKA inhibits Akt/PKB activation.[53] To determine if EPAC antagonists are capable of blocking EPAC-mediated Akt activation, the phosphorylation status of T308 and S473 of Akt in AsPC-1 cells was followed using anti-phospho-Akt antibodies. As expected, although 007-AM led to an increase in Akt phosphorylation for both T308 and S473, ESI-09 inhibited 007-AM–stimulated Akt phosphorylation in a dose-dependent manner. Furthermore, consistent with observations based on EPAC1 knockdown

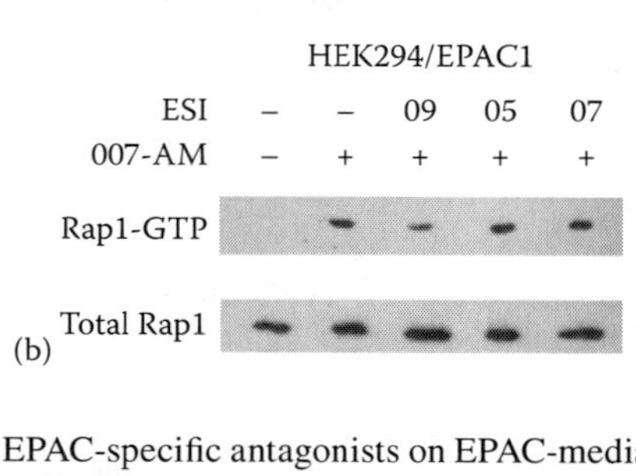

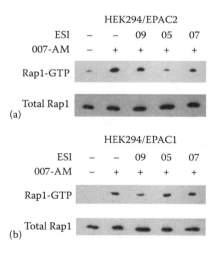

FIGURE 1.3 Effects of EPAC-specific antagonists on EPAC-mediated Rapl cellular activation. Serum-starved HEK293/EPAC2 cells (a) or HEK293/Epacl cells (b) with or without pretreatment of 10 μM Epac-specific antagonists were stimulated with 10 μM of Sp-8-CPT-2′-*O*-Me-cAMP. GTP-bound Rapl (RaplGTP), probed by a glutathione *S*-transferase pull-down assay and total cellular Rapl were detected by immunoblotting with Rapl-specific antibody.

experiments, ESI-09 treatment dose-dependently inhibited the migration and invasion ability of AsPC-1 and PANC-1 cells.[50]

1.4.4 Application of EPAC-Specific Antagonists In Vivo

To test if EPAC-specific antagonists can be applied in vivo to facilitate an understanding of the physiological functions and related disease mechanisms of EPAC, as well as the development of potential therapeutics, we performed preliminary pharmacokinetic analyses. ESI-09 showed an excellent bioavailability without overt toxicity. In light of our finding that deletion of EPAC1 in mice protects them from fatal rickettsioses, we decided to test if administration of ESI-09 in mice would also protect the animals from this disease. Our study showed that treatment of mice with a daily intraperitoneal injection of a 10 mg/kg dose of ESI-09 completely recapitulated the EPAC1 knockout phenotype. Mice treated with ESI-09 were resistant to rickettsial infection, showing a dramatic decrease in morbidity and mortality associated with fatal spotted fever rickettsiosis in a manner similar to EPAC1 KO mice. These studies demonstrated that EPAC1 is a potential target for the prevention and treatment of fatal rickettsioses. In addition, administration of ESI-09 in mouse models of chronic pain have also been shown to suppress hyperalgesia (see Chapter 7). Therefore, it seems that ESI-09 is a viable lead compound for developing therapeutics targeting EPAC signaling.

1.4.5 Lead Optimization

Although ESI-09 works very well in in vivo animal models, we would like to emphasize that hits from HTS rarely directly move into clinical applications. It often requires extensive compound modification and optimization to design drug candidates with high potency and specificity, as well as optimized drug-like properties. In fact, extensive medicinal chemistry efforts, which are beyond the scope of this chapter, have been performed to improve the pharmacological properties of identified EPAC-specific antagonists.[54]

1.5 CONCLUSION AND PERSPECTIVES

More than one decade of extensive research has firmly established the EPAC family proteins as important cAMP sensors with vital physiological functions.[55] A multitude of recent publications based on various EPAC knockout mouse models have also implicated the involvement of EPAC proteins in major diseases including cancer,[56] chronic pain,[21] diabetes,[57] heart failure,[22] and infectious disease.[20] The discovery of novel EPAC-specific modulators using HTS approaches as described in this chapter will significantly expand the pharmacological toolbox useful for fundamental research to advance scientific knowledge and for preclinical/clinical studies to explore disease biology and novel therapeutics targeting the cAMP/EPAC signaling axis. It is expected that novel and improved HTS assays, as well as new compound libraries based on both combinatorial small molecules and natural products, will be further developed. These new advances, coupled with medicinal chemistry and structural-based compound optimization, will lead to the identification of more potent and selective small molecule EPAC-specific modulators.

ACKNOWLEDGMENT

This work was supported by NIH grants GM066170 and GM106218.

REFERENCES

1. Rall, T.W., and Sutherland, E.W. Formation of a cyclic adenine ribonucleotide by tissue particles. *J Biol Chem* 232, 1065–1076 (1958).
2. Sutherland, E.W., and Rall, T.W. Fractionation and characterization of a cyclic adenine ribonucleotide formed by tissue particles. *J Biol Chem* 232, 1077–1091 (1958).
3. de Rooij, J. et al. EPAC is a Rap1 guanine-nucleotide-exchange factor directly activated by cyclic AMP. *Nature* 396, 474–477, doi:10.1038/24884 (1998).
4. Kawasaki, H. et al. A family of cAMP-binding proteins that directly activate Rap1. *Science* 282, 2275–2279 (1998).
5. Craven, K.B., and Zagotta, W.N. CNG and HCN channels: Two peas, one pod. *Annu Rev Physiol* 68, 375–401, doi:10.1146/annurev.physiol.68.040104.134728 (2006).
6. Walsh, D.A., Perkins, J.P., and Krebs, E.G. An adenosine 3′,5′-monophosphate-dependant protein kinase from rabbit skeletal muscle. *J Biol Chem* 243, 3763–3765 (1968).
7. Renstrom, E., Eliasson, L., and Rorsman, P. Protein kinase A-dependent and -independent stimulation of exocytosis by cAMP in mouse pancreatic B-cells. *J Physiol* 502 (Pt 1), 105–118 (1997).
8. Pedarzani, P., and Storm, J.F. Protein kinase A-independent modulation of ion channels in the brain by cyclic AMP. *Proc Natl Acad Sci U S A* 92, 11716–11720 (1995).
9. Wolfgang, W.J., Roberts, I.J., Quan, F., O'Kane, C., and Forte, M. Activation of protein kinase A-independent pathways by Gs alpha in Drosophila. *Proc Natl Acad Sci U S A* 93, 14542–14547 (1996).
10. Shintani, Y., and Marunaka, Y. Regulation of chloride channel trafficking by cyclic AMP via protein kinase A-independent pathway in A6 renal epithelial cells. *Biochem Biophys Res Commun* 223, 234–239, doi:10.1006/bbrc.1996.0877 (1996).
11. Murray, A.J., and Shewan, D.A. EPAC mediates cyclic AMP-dependent axon growth, guidance and regeneration. *Mol Cell Neurosci* 38, 578–588, doi:10.1016/j.mcn.2008.05.006 (2008).
12. Qiao, J., Mei, F.C., Popov, V.L., Vergara, L.A., and Cheng, X. Cell cycle-dependent subcellular localization of exchange factor directly activated by cAMP. *J Biol Chem* 277, 26581–26586, doi:10.1074/jbc.M203571200 (2002).
13. Niimura, M. et al. Critical role of the N-terminal cyclic AMP-binding domain of EPAC2 in its subcellular localization and function. *J Cell Physiol* 219, 652–658, doi:10.1002/jcp.21709 (2009).
14. Liu, C. et al. The interaction of EPAC1 and Ran promotes Rap1 activation at the nuclear envelope. *Mol Cell Biol* 30, 3956–3969, doi:10.1128/MCB.00242-10 (2010).
15. Gloerich, M. et al. Spatial regulation of cyclic AMP-EPAC1 signaling in cell adhesion by ERM proteins. *Mol Cell Biol* 30, 5421–5431, doi:10.1128/MCB.00463-10 (2010).
16. Gloerich, M. et al. The nucleoporin RanBP2 tethers the cAMP effector EPAC1 and inhibits its catalytic activity. *J Cell Biol* 193, 1009–1020, doi:10.1083/jcb.201011126 (2011).
17. Yan, J. et al. Enhanced leptin sensitivity, reduced adiposity, and improved glucose homeostasis in mice lacking exchange protein directly activated by cyclic AMP isoform 1. *Mol Cell Biol* 33, 918–926, doi:10.1128/MCB.01227-12 (2013).

18. Kai, A.K. et al. Exchange protein activated by cAMP 1 (EPAC1)-deficient mice develop beta-cell dysfunction and metabolic syndrome. *FASEB J* 27, 4122–4135, doi:10.1096/fj.13-230433 (2013).
19. Pereira, L. et al. EPAC2 mediates cardiac beta1-adrenergic-dependent sarcoplasmic reticulum Ca^{2+} leak and arrhythmia. *Circulation* 127, 913–922, doi:10.1161/circulationaha.12.148619 (2013).
20. Gong, B. et al. Exchange protein directly activated by cAMP plays a critical role in bacterial invasion during fatal rickettsioses. *Proc Natl Acad Sci U S A* 110, 19615–19620, doi:10.1073/pnas.1314400110 (2013).
21. Wang, H. et al. Balancing GRK2 and EPAC1 levels prevents and relieves chronic pain. *J Clin Invest* 123, 5023–5034, doi:10.1172/jci66241 (2013).
22. Okumura, S. et al. EPAC1-dependent phospholamban phosphorylation mediates the cardiac response to stresses. *J Clin Invest* 124, 2785–2801, doi:10.1172/jci64784 (2014).
23. Schwede, F., Maronde, E., Genieser, H., and Jastorff, B. Cyclic nucleotide analogs as biochemical tools and prospective drugs. *Pharmacol Ther* 87, 199–226 (2000).
24. Poppe, H. et al. Cyclic nucleotide analogs as probes of signaling pathways. *Nat Methods* 5, 277–278 (2008).
25. Enserink, J.M. et al. A novel EPAC-specific cAMP analogue demonstrates independent regulation of Rap1 and ERK. *Nat Cell Biol* 4, 901–906, doi:10.1038/ncb874 (2002).
26. Su, Y. et al. Regulatory subunit of protein kinase A: Structure of deletion mutant with cAMP binding domains. *Science* 269, 807–813 (1995).
27. Holz, G.G., Chepurny, O.G., and Schwede, F. EPAC-selective cAMP analogs: New tools with which to evaluate the signal transduction properties of cAMP-regulated guanine nucleotide exchange factors. *Cell Signal* 20, 10–20, doi:10.1016/j.cellsig.2007.07.009 (2008).
28. Christensen, A.E. et al. cAMP analog mapping of EPAC1 and cAMP kinase. Discriminating analogs demonstrate that EPAC and cAMP kinase act synergistically to promote PC-12 cell neurite extension. *J Biol Chem* 278, 35394–35402, doi:10.1074/jbc.M302179200 (2003).
29. Kang, G. et al. EPAC-selective cAMP analog 8-pCPT-2′-*O*-Me-cAMP as a stimulus for Ca^{2+}-induced Ca^{2+} release and exocytosis in pancreatic beta-cells. *J Biol Chem* 278, 8279–8285, doi:10.1074/jbc.M211682200 (2003).
30. Rangarajan, S. et al. Cyclic AMP induces integrin-mediated cell adhesion through EPAC and Rap1 upon stimulation of the beta 2-adrenergic receptor. *J Cell Biol* 160, 487–493, doi:10.1083/jcb.200209105 (2003).
31. Rehmann, H., Schwede, F., Døskeland, S.O., Wittinghofer, A., and Bos, J.L. Ligand-mediated activation of the cAMP-responsive guanine nucleotide exchange factor EPAC. *J Biol Chem* 278, 38548–38556, doi:10.1074/jbc.M306292200 (2003).
32. Enyeart, J.A., and Enyeart, J.J. Metabolites of an EPAC-selective cAMP analog induce cortisol synthesis by adrenocortical cells through a cAMP-independent pathway. *PLoS One* 4, e6088, doi:10.1371/journal.pone.0006088 (2009).
33. Enyeart, J.A., Liu, H., and Enyeart, J.J. cAMP analogs and their metabolites enhance TREK-1 mRNA and K+ current expression in adrenocortical cells. *Mol Pharmacol* 77, 469–482, doi:10.1124/mol.109.061861 (2010).
34. Enyeart, J.A., Liu, H., and Enyeart, J.J. 8-Phenylthio-adenines stimulate the expression of steroid hydroxylases, Cav3.2 Ca^{2+} channels, and cortisol synthesis by a cAMP-independent mechanism. *Am J Physiol Endocrinol Metab* 301, E941–E954, doi:10.1152/ajpendo.00282.2011 (2011).
35. Sand, C. et al. 8-pCPT-conjugated cyclic AMP analogs exert thromboxane receptor antagonistic properties. *Thromb Haemost* 103, 662–678, doi:10.1160/TH09-06-0341 (2010).

36. Brown, L.M., Rogers, K.E., McCammon, J.A., and Insel, P.A. Identification and validation of modulators of exchange protein activated by cAMP (EPAC) activity: Structure–function implications for EPAC activation and inhibition. *J Biol Chem* 289, 8217–8230, doi:10.1074/jbc.M114.548636 (2014).
37. Jiang, L.I. et al. Use of a cAMP BRET sensor to characterize a novel regulation of cAMP by the sphingosine 1-phosphate/G13 pathway. *J Biol Chem* 282, 10576–10584, doi:10.1074/jbc.M609695200 (2007).
38. van den Berghe, N., Cool, R.H., Horn, G., and Wittinghofer, A. Biochemical characterization of C3G: An exchange factor that discriminates between Rap1 and Rap2 and is not inhibited by Rap1A(S17N). *Oncogene* 15, 845–850, doi:10.1038/sj.onc.1201407 (1997).
39. Cremo, C.R. Fluorescent nucleotides: Synthesis and characterization. *Methods Enzymol* 360, 128–177 (2003).
40. Moll, D. et al. Biochemical characterization and cellular imaging of a novel, membrane permeable fluorescent cAMP analog. *BMC Biochem* 9, 18, doi:10.1186/1471-2091-9-18 (2008).
41. Alfonso, A., Estevez, M., Louzao, M.C., Vieytes, M.R., and Botana, L.M. Determination of phosphodiesterase activity in rat mast cells using the fluorescent cAMP analogue anthraniloyl cAMP. *Cell Signal* 7, 513–518 (1995).
42. Biskup, C. et al. Relating ligand binding to activation gating in CNGA2 channels. *Nature* 446, 440–443, doi:10.1038/nature05596 (2007).
43. Kraemer, A. et al. Dynamic interaction of cAMP with the Rap guanine-nucleotide exchange factor EPAC1. *J Mol Biol* 306, 1167–1177, doi:10.1006/jmbi.2001.4444 (2001).
44. Tsalkova, T., Mei, F.C., and Cheng, X. A fluorescence-based high-throughput assay for the discovery of exchange protein directly activated by cyclic AMP (EPAC) antagonists. *PLoS One* 7, e30441, doi:10.1371/journal.pone.0030441 (2012).
45. Li, S. et al. Mechanism of intracellular cAMP sensor EPAC2 activation: cAMP-induced conformational changes identified by amide hydrogen/deuterium exchange mass spectrometry (DXMS). *J Biol Chem* 286, 17889–17897, doi:10.1074/jbc.M111.224535 (2011).
46. Tsalkova, T., Blumenthal, D.K., Mei, F.C., White, M.A., and Cheng, X. Mechanism of EPAC activation: Structural and functional analyses of EPAC2 hinge mutants with constitutive and reduced activities. *J Biol Chem* 284, 23644–23651, doi:10.1074/jbc.M109.024950 (2009).
47. Mei, F.C., and Cheng, X. Interplay between exchange protein directly activated by cAMP (EPAC) and microtubule cytoskeleton. *Mol Biosyst* 1, 325–331, doi:10.1039/b511267b (2005).
48. Cheng, X., Phelps, C., and Taylor, S. Differential binding of cAMP-dependent protein kinase regulatory subunit isoforms I alpha and II beta to the catalytic subunit. *J Biol Chem* 276, 4102–4108, doi:10.1074/jbc.M006447200 (2001).
49. Yu, S., Mei, F., Lee, J., and Cheng, X. Probing cAMP-dependent protein kinase holoenzyme complexes I alpha and II beta by FT-IR and chemical protein footprinting. *Biochemistry* 43, 1908–1920, doi:10.1021/bi0354435 (2004).
50. Almahariq, M. et al. A novel EPAC-specific inhibitor suppresses pancreatic cancer cell migration and invasion. *Mol Pharmacol* 83, 122–128, doi:10.1124/mol.112.080689 (2013).
51. Tsalkova, T. et al. Isoform-specific antagonists of exchange proteins directly activated by cAMP. *Proc Natl Acad Sci U S A* 109, 18613–18618, doi:10.1073/pnas.1210209109 (2012).
52. Rehmann, H. EPAC-inhibitors: Facts and artefacts. *Sci Rep* 3, 3032, doi:10.1038/srep03032 (2013).

53. Mei, F. et al. Differential signaling of cyclic AMP—Opposing effects of exchange protein directly activated by cyclic AMP and cAMP-dependent protein kinase on protein kinase B activation. *J Biol Chem* 277, 11497–11504, doi:10.1074/jbc.M110856200 (2002).
54. Chen, H. et al. Recent advances in the discovery of small molecules targeting exchange proteins directly activated by cAMP (EPAC). *J Med Chem* 57, 3651–3665, doi:10.1021/jm401425e (2014).
55. Schmidt, M., Dekker, F.J., and Maarsingh, H. Exchange protein directly activated by cAMP (EPAC): A multidomain cAMP mediator in the regulation of diverse biological functions. *Pharmacol Rev* 65, 670–709, doi:10.1124/pr.110.003707 (2013).
56. Onodera, Y., Nam, J.M., and Bissell, M.J. Increased sugar uptake promotes oncogenesis via EPAC/RAP1 and O-GlcNAc pathways. *J Clin Invest* 124, 367–384, doi:10.1172/jci63146 (2014).
57. Song, W.J., Mondal, P., Li, Y., Lee, S.E., and Hussain, M.A. Pancreatic beta-cell response to increased metabolic demand and to pharmacologic secretagogues requires EPAC2A. *Diabetes*, 62, 2796–2807, doi:10.2337/db12-1394 (2013).

2 Cyclic Nucleotide Analogs as Pharmacological Tools for Studying Signaling Pathways

Elke Butt

CONTENTS

2.1 INTRODUCTION

Since the discovery of adenosine-3′,5′-cyclic monophosphate (cAMP)[1] and guanosine-3′,5′-cyclic monophosphate (cGMP)[2] nearly 60 years ago, cyclic nucleotides have been envisioned as one of the most universal and versatile second messengers and have been detected from *elephants down to E. coli*. In the following years, the corresponding cAMP-dependent protein kinases (PKA I and II; for review, see Kim et al.[3]) and the cGMP-dependent protein kinases (PKG Iα, Iβ, and II; for review, see Francis et al.[4]) were identified and characterized.

cAMP and cGMP are both present in many cell types. cGMP levels have been determined to be approximately 10^{-8} M in most tissues whereas intracellular cAMP concentrations are 10-fold to 100-fold higher. cAMP and cGMP are synthesized by adenylyl[5] and guanylyl cyclases, respectively.[6,7] Their degradation is controlled by cyclic nucleotide phosphodiesterases (PDEs) with 11 known PDE isoforms to date.[8,9] In addition to cyclic nucleotide hydrolysis, several PDEs are under the direct control of cGMP, for example, the cGMP-stimulated PDE2 and the cGMP-inhibited PDE3. Both isozymes preferentially hydrolyze cAMP, albeit at different activation/inhibition constants and are exemplary for cyclic nucleotide cross talk.[10]

Other direct intracellular targets for cAMP and cGMP are cyclic nucleotide–gated (CNG) cation channels particularly involved in olfactory and visual systems[11,12] and hyperpolarization-activated, cyclic nucleotide–modulated (HCN) channels mainly expressed in the heart and the nervous system.[13,14] More recently, cAMP has been reported to directly activate guanine nucleotide exchange factors for small G proteins (exchange proteins directly activated by cAMP, Epac 1, and 2).[15,16] Overall, all players are expressed in multiple tissues and exert their biological functions either alone or in concert with other cyclic nucleotide–binding (CNB) partners. Therefore, when using inhibitors or activators for these proteins, knowledge about the concentrations and binding properties is important for the proper interpretation of results. For example, exposure of cells to NO increases the level of cGMP and activates the respective kinase, but can also inhibit PDE3, resulting in cAMP elevation and increased signaling through cAMP/PKA or EPAC pathways as observed in platelets.[17]

In this chapter, we summarize published information and new data on activation/inhibition constants, hydrolysis rates, cell permeability and concentrations of cyclic nucleotides, and propose guidelines to help interpret observations based on PKG and PKA signal transduction in cell models using cyclic nucleotide analogs.

2.2 PROTEIN KINASE A

Biochemical studies have identified two distinct cAMP-dependent protein kinase isoforms that are present in most cells, PKA I and PKA II. The tetrameric PKA holoenzymes are composed of two catalytic isoforms (Cα, -β, or -γ) and two regulatory subunits (RIα, -β or RIIα, -β). Each R subunit has two cAMP-binding domains (site A and site B). Upon binding of two cAMP molecules on both R subunits, the inactive tetramer is dissociated into one dimer of R subunits and two active C subunits, which then phosphorylate various cytoplasmic and nuclear target proteins.[3] The binding sites differ in rates of association and exchange of cAMP. In general, the amino terminal site (A) exhibits rapid exchange of bound cAMP and partially excludes 8-amino-substituted cAMP analogs in both RI and RII, and most 8-substituted analogs in RII. Site B exhibits slower exchange of cAMP and partially excludes cAMP analogs with bulky substituent at the N^6 position of the adenine ring.[18,19]

2.2.1 Protein Kinase A Activators

To date, the most PKA-selective and PDE-resistant cAMP analogs that preferentially activate PKA over Epac are N^6-benzoyl-cAMP (6-Bnz-cAMP) and N^6-monobutyryl-cAMP (6-MB-cAMP).[20] By appropriately pairing of cAMP analogs, it is possible to preferentially activate either PKA I or PKA II; allowing a prognosis for the type of PKA isozyme mainly responsible for the biological effect observed. For PKA I activation, a combination of 8-AHA-cAMP/8-PIP-cAMP[21] or the couple 6-Bnz-8-PIP-cAMP/2-Cl-8-AHA-cAMP with a 1:100 ratio is recommended.[22] For PKA II, the analog pair 6-MBC-cAMP/Sp-5,6-DCl-cBIMPS with a 1:60 ratio is suggested.

The frequently used derivative Sp-5,6-DCl-cBIMPS was described as a potent activator of PKA in vivo with resistance to PDE-mediated hydrolysis.[23] Unfortunately, PDE1B, PDE2, PDE4, and PDE6 are inhibited by Sp-5,6-DCl-cBIMPS in *in vitro* assays with K_i values that are of the same order of magnitude as the K_m for their natural substrates.[24] Furthermore, Sp-5,6-DCl-cBIMPS activates Epac with a higher affinity than cAMP (14 μM versus 45 μM; Table 2.1).[24]

8-Br-cAMP and 8-pCPT-cAMP have traditionally been considered as selective activators of PKA I and II; however, both compounds are also highly effective activators of Epac (Table 2.1).[20] In addition, 8-pCPT-cAMP activates PKA and PKG with similar activation constants (K_a = 0.05 μM versus K_a = 0.11 μM, respectively) whereas 8-Br-cAMP is hydrolyzed by all PDEs even by cGMP-selective PDEs (Table 2.2).

2.2.2 Protein Kinase A Inhibitors (Cyclic Nucleotide–Based)

Rp-modified cAMPS analogs are known to inhibit PKA.[25] Although Rp-cAMPS prefers PKA II,[26] Rp-8-Br-cAMPS efficiently inhibits PKA I in biological systems such as fibroblasts, leukemia cells, hepatocytes, and platelets.[19] A combination of both Rp analogs can be used for global PKA inhibition. These derivatives are not hydrolyzed by any of the PDEs tested. However, Rp-8-Br-cAMPS is able to inhibit PDEs in *in vitro* assays albeit the K_i for Rp-8-Br-cAMPS for all PDEs tested is at least one order of magnitude higher than the K_m for cAMP (Table 2.2). Nevertheless, many investigators use high concentrations of this analog in the culture medium so caution is advised under these circumstances.

2.3 PROTEIN KINASE G

All three cGMP-dependent protein kinases act as homodimers and bind two cGMP molecules per monomer. The soluble forms PKG Ia and Iβ are products of alternative splicing and differ only in their N-terminal region. Although PKG II is myristoylated and membrane bound,[27] the PKG I isoforms are thought to be soluble. However, several studies support a compartmentalization of cGMP signaling.[28–30]

Activation involves allosteric cGMP binding to both cGMP binding sites and is associated with a conformational change involving molecular elongation. Despite identical regulatory cGMP binding sites, the activation constants for PKG Iα (K_a = 0.1 μM) and PKG Iβ (K_a = 0.9 μM) differ 10-fold and allow differentiation by cyclic nucleotide analogs.[24]

2.3.1 Protein Kinase G Activators

The analogs most commonly used as PKG activators are 8-Br-cGMP, 8-pCPT-cGMP, and 8-Br-PET-cGMP. As shown in Table 2.1, PKG II is best activated by 8-pCPT-cGMP

TABLE 2.1

Activation (K_a), Inhibition (K_i/IC_{50}), and Binding (K_d) Constants of CNB Domain Containing Proteins by cAMP and cGMP Analogs

Activators	PKG Iα K_a (μM)	PKG Iβ K_a (μM)	PKG II K_a (μM)	PKA I K_a (μM)	PKA II K_a (μM)	Epac 1 K_d (μM)	Epac 1 K_a (μM)	Epac 2 K_a (μM)	CNGs K_a (μM)	HCN2 EC_{50} (μM)	Lipophilicity log K_w[31]
cAMP	39[a]	18[a]	12[a]	0.09	0.08	2.9[i]	45[b]	20[n]	1.8[c] (olf) 3.6[k] (olf) 1500[e] (rod)	0.5[d]	1.09
8-Br-cAMP	5.8[a]	10[a]	6[a]	0.08	0.1	0.36[i]	7[b]		1760[e] (rod)		1.35
Sp-5,6-DCl-cBIMPS	10[a]	27[a]	7 > 100 inh.	1.7	0.1	4.3[i]	14[i]				2.99
6-Bnz-cAMP	14[i]	70[i]	5.5[i]	0.08[s]	0.2[s]	2.23[i]	#				1.90
6-MB-cAMP				0.03[s]	0.04[s]		0.77[r]				1.64
8-pCPT-2′-O-Me-cAMP	1000[i]	>1000[i]	>1000[i]	16	15	0.63[i]	1.8[b]		22[k] (olf)		2.94
Sp-8-pCPT-2′-O-Me-cAMPS	>1000[i]	>1000[i]	>1000[i]	45	77	>250[i]	7.3[i]				3.05
cGMP	0.1[a]	0.9[a]	0.07[a]	11	42	40[i]	#[b]		2.3[c] (olf) 40[f] (rod)	6[d]	0.77
8-APT-cGMP	0.01	3.1	0.06		20						2.38
8-Br-cGMP	0.01[a]	1.0[a]	0.025[a]	31	29	25[i]	#[b]		6.7[f] (rod)		1.17
8-pCPT-cGMP	0.04[a]	0.9[a]	0.004[a]	5	7	0.9[i]	#[b]		0.07[k] (olf) 0.5[f] (rod)		2.52
8-Br-PET-cGMP	0.013[g]	0.009[g]	0.02[i]	22	30	12[i]	#[b]		IC_{50} 64[f] (rod)		2.83
Rp-cAMPS	53[a]	14[a]	20[a]		3.7		#[b]				1.21

(Continued)

TABLE 2.1 (CONTINUED)

Activation (K_a), Inhibition (K_i/IC_{50}), and Binding (K_d) Constants of CNB Domain Containing Proteins by cAMP and cGMP Analogs

Inhibitors	PKG Iα K_i (μM)	PKG Iβ K_i (μM)	PKG II K_i (μM)	PKA I K_i (μM)	PKA II K_i (μM)	Epac 1 K_d (μM)	Epac 1 IC_{50} (μM)	Epac 2 IC_{50} (μM)	CNGs (μM)	HCN2 EC_{50} (μM)	Lipophilicity log K_w[31]
Rp-8-Br-cAMPS	35[i]	7[i]	20[i]		3.8	195	#[b]				1.47
Rp-cGMPS	20[h]	15	0.5		20[h]		#[b]				0.89
Rp-8-Br-cGMPS	3.0	15			20				EC_{50} 174[f (rod)]		1.29
Rp-8-pCPT-cGMPS	0.5[a]	0.45[i]	0.7[i]		8.3[a]	49	#[b]				2.6
Rp-8-Br-PET-cGMPS	0.035[h]	0.03[h]	0.9		11[h]	>100	#[b]		IC_{50} 25[f (rod)]		2.83
(R)-CE3F4 (Epac 1 inhibitor)							36[o]	2.3[o]			n.d.
ESI-05 (Epac 2 inhibitor)				No effect[m]			No effect[m]	0.43[m]			n.d.

Source: [a]Pohler, D. et al., *FEBS Letters* 374, 419–425, 1995. [b]Rehmann, H. et al., *Nature Structural Biology* 10, 26–32, 2003. [c]Ludwig, J. et al., *FEBS Letters* 270, 24–29, 1990. [d]Ludwig, A. et al., *Nature* 393, 587–591, 1998. [e]Tanaka, J.C. et al., *Biochemistry* 28, 2776–2784, 1989. [f]Wei, J.Y. et al., *Journal of Molecular Neuroscience* 10, 53–64, 1998. [g]Sekhar, K.R. et al., *Molecular Pharmacology* 42, 103–108, 1992. [h]Butt, E. et al., *British Journal of Pharmacology* 116, 3110–3116, 1995. [i]Poppe, H. et al., *Nature Methods* 5, 277–278, 2008. [j]Strassmaier, T., and Karpen, J.W., *Journal of Medicinal Chemistry* 50, 4186–4194, 2007. [k]Tsalkova, T. et al., *Proceedings of the National Academy of Sciences of the United States of America* 109, 18613–18618, 2012. [l]Rehmann, H. et al., *Nature* 439, 625–628, 2006. [m]Courilleau, D. et al., *Biochemical and Biophysical Research Communications* 440, 443–448, 2013. [n]Christensen, A.E. et al., *Journal of Biological Chemistry* 278, 35394–35402, 2003. [o]Ogreid, D. et al., *European Journal of Biochemistry/FEBS* 150, 219–227, 1985.

Notes: #, at a cyclic nucleotide concentration of 500 μM, the induced activity remains less than 10% of the K_{max} of cAMP; CNG, cyclic nucleotide–gated channel; HCN 2, hyperpolarization-activated cyclic nucleotide–gated channel 2; n.d., not detected; olf, olfactory channel; rod, rod photoreceptor channel. Nonreferenced data are all recorded and verified in the participating laboratories of the Cyclic Nucleotide Project (http://www.cyclic-nucleotides.org).

TABLE 2.2
Hydrolysis Data of cAMP and cGMP Analogs by Phosphodiesterases

	PDE1A			PDE1B			PDE1C			PDE2			PDE4			PDE5			PDE6			PDE10		
Analog	K_m	V_{max}	K_i	K_m	V_{max}	K_i	K_m	V_{max}	K_i	K_m	V_{max}	K_i	K_m	V_{max}	K_i	K_m	V_{max}	K_i	K_m	V_{max}	K_i	K_m	V_{max}	K_i
cAMP	93	41		33	1.5		3.2	16		112	215		5.5	63		201	20		823	640		0.24	0.7	
											Allosteric													
8-Br-cAMP	33	3.6		63	0.75		18	10		39	6.0		54	2.9		23	6.2		68.8	100		4.8	0.8	
Sp-5,6-DCl-cBiMPS			90			2.3			56			18			19			15			25			4.8
		Competitive			Linear mixed			Competitive			Competitive			Competitive			Linear mixed			Competitive			Linear mixed	
6-Bnz-cAMP		No effect				10			323			240			49			69			65	45	0.4	
					Linear mixed			Noncompetitive			Competitive			Linear mixed			Linear mixed			Competitive				
8-pCPT-2′-O-Me-			51			8.6			44			15			895	3.1	6.4				3.5	5.8	4.5	
cAMP		Competitive			Competitive			Competitive			Competitive			Competitive						Competitive				
Sp-8-pCPT-2′-O-			0.4			2.0			17						0.8			0.4			1.0			2.4
Me-cAMPS		Linear mixed			Linear mixed			Competitive			Not detected			Linear mixed			Linear mixed			Linear mixed			Linear mixed	
Rp-8-Br-cAMPS		No effect				5.6			38			23			29			95			106			28
					Linear mixed			Competitive			Competitive			Linear mixed			Noncompetitive			Competitive			Competitive	
cGMP	8.2	20		5.3	2.63		4.6	14		31	176				25	2.0	6.9		10	410		1.1	1.5	
														Competitive										
8-Br-cGMP			47			4.5			63			91			30			79			33	24	0.5	
		Competitive			Competitive			Competitive			Competitive			Linear mixed			Linear mixed			Competitive				
8-pCPT-cGMP		No effect				8.6			47			40	53.4	2		28	0.74				41	6.0	3.6	
		Partial activity			Competitive			Competitive			Competitive									Competitive				
8-Br-Pet-cGMP			2.4			2.1	73	3.5		98	175				8.8			5.3			11			34
		Linear mixed			Linear mixed									Linear mixed			Linear mixed			Linear mixed			Competitive	
Rp-8-pCPT-cGMPS			42			2.1			33		No effect				137			50			26			38
		Noncompetitive			Linear mixed			Competitive			Partial activity			Competitive			Competitive			Competitive			Competitive	
Rp-8-Br-Pet-cGMPS		No effect				2.5			55			0.8			8.1			4.1			14			5.0
					Linear mixed			Competitive			Linear mixed			Competitive			Linear mixed			Linear mixed			Linear mixed	

Source: Poppe, H. et al., *Nature Methods* 5, 277–278, 2008.
Note: K_m (μM); V_{max} (μmol/min/mg); K_i (μM).

whereas 8-Br-PET-cGMP has the highest affinity for PKG Iα and PKG Iβ. 8-APT-cGMP is suited to discriminate between the isoforms of PKG I because it activates PKG Iα with a K_a of 0.01 μM and PKG Iβ at 3.1 μM. 8-Br-cGMP is a good activator of PKG Iα and PKG II; however, this derivative can be hydrolyzed by PDE5 and PDE10.[24]

8-Br-PET-cGMP is resistant to degradation by all PDEs tested but inhibits PDE2, PDE4, and PDE6 in in vitro assays with a K_i in the same order of magnitude as the K_m for their natural substrate. 8-pCPT-cGMP can be hydrolyzed by PDE5 and PDE10, and surprisingly, to some extent, also by the *cAMP-specific* PDE4. 8-pCPT-cGMP can also bind to Epac 1 with a K_d of 0.9 μM (K_d for cAMP: 2.9 μM) but is incapable of activating the protein (Table 2.1). Therefore, this compound might interfere with Epac-mediated effects by competing with cAMP for the binding site.

2.3.2 Protein Kinase G Inhibitors (Cyclic Nucleotide–Based)

Rp isomers of guanosine-3′,5′-cyclic monophosphorothioate usually act as inhibitors of cyclic nucleotide protein kinase activation. These compounds bind to PKG, but apparently do not evoke the conformational change required for activation.[44,45] Rp-8-pCPT-cGMPS and Rp-8-Br-PET-cGMPS were found to be the most potent inhibitors for both PKGs (Table 2.1) with Rp-8-Br-PET-cGMPS favoring PKG I over PKG II (K_i = 0.03 μM versus K_i = 0.9 μM, respectively). In rat tail arteries, human platelets,[39] and intestinal mucosa,[46] these Rp derivatives have been tested successfully to selectively inhibit PKG-mediated effects. However, although not hydrolyzed by PDE, both Rp analogs are inhibitors of all tested PDEs (Table 2.2). With PDE2, PDE5, PDE6, and PDE10, a linear mixed type of inhibition is observed, suggesting a second binding site in addition to the catalytic site.

In contrast to protein kinases, Rp-8-Br-cGMPS is an activator on CNG channels (EC_{50} = 174 μM) whereas Rp-8-Br-PET-cGMPS inhibits rod channels with an IC_{50} of 25 μM.[37] At higher concentrations (>300 μM), a noncompetitive antagonistic effect is observed.

2.4 EPAC

In 1998, Epac 1 and Epac 2 were identified independently on a database search aimed to unravel the PKA-independent activation of small GTPases Rap1 and Rap2[47] and by a differential display screen for novel CNB domain-bearing proteins.[48] Whereas Epac 1 has only one CNB domain, Epac 2 harbours two CNBs that share significant sequence identity with the slow and fast cAMP-binding sites of PKA.[47] Although binding of cAMP to the high-affinity CNB domains in Epac 1 and Epac 2 leads to the opening of the catalytic site, cAMP binding to the lower affinity N-terminal CNB site in Epac 2 is not required, neither for maintaining the autoinhibitory state nor for activation.[7,49] Most likely, binding of cAMP at this site controls the intracellular localization of the protein.[50]

The physiological potency of cAMP for Epac and PKA is quite similar (≈2.8 μM)[51] and discriminating between PKA- and Epac-dependent signaling pathways in cells was initially hampered by the lack of any specific agent that selectively activates each of these two proteins. In 2002, a cAMP analog [8-pCPT-2′-O-Me-cAMP (8-(4-chloro-phenylthio)-2′-*O*-methyladenosine-3′,5′-cAMP), also named 007] that exhibited high affinity for Epac 1 (K_a = 1.8 μM) over PKA (K_a = 15 μM) was developed and characterized.[20,49,52]

Introducing an acetoxymethyl (AM) ester group masks the negatively charged phosphate group, generating a prodrug that is more efficiently delivered into cells than its parent compound.[53] Intracellularly, the AM group is hydrolyzed by esterases to release the parent compound 8-pCPT-2′-O-Me-cAMP. Thus far, there is no data available to indicate whether the Epac agonist could discriminate between Epac 1 and Epac 2. In addition, it should be taken into account that 8-pCPT-2′-O-Me-cAMP or its nonhydrolyzable form, Sp-8-pCPT-2′-O-Me-cAMPS, may activate cAMP/PKA or cGMP/PKG pathways through the inhibition of PDE1, PDE2, and PDE6,[24] and that metabolites of 8-pCPT-2′-O-Me-cAMP may have off-target effects on gene expression and cell function.[54,55]

To date, no cyclic nucleotide analog for competitive Epac inhibition has been developed successfully. The identification and characterization of several novel noncyclic nucleotide EPAC antagonists dubbed ESI-09, ESI-05, and HJCO-197, which are supposed to specifically block intracellular EPAC-mediated Rapl activation, was recently reported.[56] However, a new study indicates that ESI-09 does not act as a selective Epac inhibitor but rather shows nonspecific protein denaturing effects, albeit at nonphysiologically high concentrations.[57] In this study, a second, noncyclic nucleotide reported Epac inhibitor (ESI-05)[41] was validated to be positive for Epac 2–specific inhibition (IC_{50} = 1 μM). The inhibitor exerts its isoform specificity by binding to a unique allosteric site at the interface of the two cAMP binding domains in Epac 2, which is not present in Epac 1. A tetrahydroquinoline analog, (R)-CE3F4, has been characterized as a specific Epac I inhibitor with low micromolar potency (0.43 μM) and a 10-fold preference for Epac 1 versus Epac 2.[43,58]

2.5 CNG CATION CHANNELS

The family of cyclic nucleotide–modulated channels is composed of the CNG channels and the HCN channels.

2.5.1 HCN

The four mammalian HCN channels comprise a small subfamily of HCN channels (HCN1–4) that are mainly expressed in the heart and nervous system.[13,14] The channels are permeable to Na^+ and K^+, they are primarily activated upon membrane depolarization, as well as by binding intracellular cAMP. Activation kinetics revealed the strongest binding of cAMP to HCN4, the pacemaker channel (EC_{50} cAMP = 0.5 μM), followed by HCN2 (EC_{50} cAMP = 1 μM, EC_{50} cGMP = 6 μM) whereas HCN1 and HCN3 channel properties are almost unaffected by cAMP.[59] Within the HCN channel family, HCN2 and HCN4 also display profound regulation by cCMP activation (EC_{50} cCMP ≈ 30 μM) with an efficacy of around 60% compared with cAMP activation.[60] High levels of cCMP are found in several mammalian cell lines (e.g., HEK 293) and astrocytes.[61]

Interestingly, PKG II phosphorylates HCN2 at the C-terminal end of the cyclic nucleotide binding site, thus shifting the voltage dependence to more negative potentials and hence counteracting the stimulatory effects of cGMP on gating.[62] Therefore, it should be borne in mind that at low cGMP concentrations, the inhibitory effects of cGMP via PKG II is predominant whereas, at higher concentrations, the direct channel stimulation of cGMP is outweighed, indicating a precise modulation of HCN2 by cGMP.

2.5.2 CNG

Na^+ and Ca^{2+} channels directly regulated by cyclic nucleotides were first identified in rod and cone photoreceptors and in olfactory sensory neurons.[11] Rod photoreceptor channels open in response to cGMP binding. PDE 6 hydrolyzes cGMP, leading to the closure of the channel. In olfactory sensory neurons, binding of an odorant results in the synthesis of cAMP via G_s-coupled receptors and opening of the channel, which leads to membrane depolarization and release of neurotransmitters.

Thus far, seven vertebrate channel subunits have been described: rod photoreceptors (CNGA1 and CNGB1), cone photoreceptor (CNGA3 and CNGB3), and olfactory epithelia (CNGA2, CNGA4, and CNGB1b). Although photoreceptor channels strongly discriminate between cGMP and cAMP, the olfactory channel is almost equally sensitive to both cyclic nucleotides (Table 2.3).[50]

For the rod and cone channels, several studies with cGMP analogs have been performed. The data are summarized in Table 2.1.[37,40,63] For these channels, it should be noticed that the PKG activator 8-Br-PET-cGMP behaves as a very weak partial agonist at high concentrations and as an antagonist (IC_{50} = 64 μM) at lower concentrations when coapplied with cGMP.[63]

TABLE 2.3
Activation (K_a) Constants of CNG Channels by cAMP and cGMP

Channel	K_a cGMP (μM)	K_a cAMP (μM)	References
CNGA1	60–80	High millimolar range (pA, 1% of I_{cGMP})	64–66
CNGA2	2	40–80	67–69
CNGA3	10–20	1700–2000 (pA, <20% of I_{cGMP})	70–72
CNGA1/CNGB1a (native rod channel)	40–80	2100 (pA)	66,73
CNGA3/CNGB3 (native cone channel)	20	700 (pA, 60% of I_{cGMP})	70
CNGA2/A4/B1b (native olfactory channel)	2	5	68,69

Notes: CNGB1a/b (CNG4), CNGA4 (CNG5), and CNGB3 (CNG6) do not form homomeric channels and can only be studied when coexpressed with A1 to A3 subunits. Homomeric A1 to A3 can be expressed in cell lines as homomers but have not been detected in vivo thus far. The only established native channels *in vivo* are the heteromeric channels of rod/cone photoreceptors and olfactory neurons as listed in the table. K_a at positive values (range, +40 to +80 mV); pA, partial agonist.

2.6 PHOSPHODIESTERASES

PDEs are enzymes that regulate the degradation of the second messengers cAMP and cGMP by hydrolyzing the phosphodiester bond and converting the cyclic nucleotide to its 5′-monophosphate. It has been known for several years that substitution of a sulfur atom for one of the exocyclic oxygens of the cyclic phosphate and thus generating Sp and Rp derivatives will hinder the degradation of the cyclic nucleotide analogs.[74] This inhibition of PDEs might cause an increase of basal cAMP or cGMP levels followed by an activation of cAMP or cGMP receptor proteins.

In an *in vitro* study, we tested 8 representative isoforms out of 11 PDE families (Table 2.2).[24] The K_m, V_{max}, and K_i values obtained for the analogs tested in this study demonstrate that some of these compounds, preferentially 8-Br-cAMP and 8-pCPT-cGMP, could be hydrolyzed by several PDE isoforms, especially PDE5 and PDE10. Moreover, due to their bulky substituents, many more of these analogs are effective competitive inhibitors that can cause the elevation of cAMP or cGMP levels, and therefore either attenuate or further enhance their main physiological effects in cell culture or tissue.

This dilemma becomes evident for 8-pCPT-2′-O-Me-cAMP, a specific activator of Epac. 8-pCPT-2′-O-Me-cAMP is hydrolyzed by PDE5 and PDE10 and thus, in certain tissues or cell lines, the lifetime of the analog may be drastically reduced. To improve hydrolysis resistance, Sp-8-pCPT-2′-O-Me-cAMPS was generated, and indeed, this compound is not degraded by PDEs. However, Sp-8-pCPT-2′-O-Me-cAMPS inhibits all 8 tested PDEs with high efficiency (Table 2.3) and therefore, in addition to its primary action specific to Epac, it might activate PKA or PKG signaling indirectly by inhibiting PDEs.

2.7 LIPOPHILICITY AND CELL PERMEABILITY

Cell membranes are bilayer structures principally formed by phospholipids with hydrophilic ionic phosphate head groups attached to two long hydrophobic alkyl tails. Passive diffusion is the major route for cell permeation of most drugs and is predominantly governed by lipophilicity, polarity, charge, and size of the respective molecule. Due to its negatively charged cyclic phosphate, cAMP and cGMP itself are too hydrophilic to passage through the cell membrane whereas most cyclic nucleotide analogs with hydrophobic substituents have diminished water solubility and are able to enter the cell. Cell permeability of cyclic nucleotide analogs is variable and depends on lipophilicity, namely, its octanol/water partition coefficient. More recently, a gradient HPLC-based method for the determination of nucleoside and nucleotide lipophilicity was developed.[31] In general, the intracellular cyclic nucleotide concentrations correlate with lipophilicity and permeability (Table 2.4).

Compounds with low lipophilicity ($\log K_w < 1$; cAMP, cGMP) are too hydrophilic for any diffusion into the cell. At a $\log K_w$ of 1 to 2, approximately 10% of the extracellularly applied cyclic nucleotides (8-Br-cGMP, Rp-cAMPS) are detected inside cells. At higher lipophilicity ($\log K_w \approx 2.5$), passive diffusion increases and intracellular cyclic nucleotide concentrations increase by up to 20% (8-pCPT-cGMP) and even 30% (8-Br-PET-cGMP; $\log K_w$ 2.8). Extrapolation predicts 45% diffusion for compounds with $\log K_w \approx 3$.[75]

TABLE 2.4
Lipophilicity and Cell Membrane Permeability of Cyclic Nucleotide Analogs

Analog	Lipophilicity log K_w[31]	Permeability (%)
cGMP	0.77	0
cAMP	1.09	0
8-Br-cGMP	1.17	12.1
Rp-cAMPS	1.21	12.2
Rp-cGMPS	1.32	
8-Br-cAMP	1.35	8.0[34]
Sp-cGMPS	1.78	
6-MB-cAMP	1.64	
6-Bnz-cAMP	1.90	
8-pCPT-cGMP	2.52	19.6
8-pCPT-cAMP	2.65	22.0[b]
8-Br-PET-cGMP	2.83	30.9
Rp-8-Br-PET-cGMPS	2.83	
8-pCPT-2′-O-Me-cAMP	2.94	
Sp-5,6-DCl-cBIMPS	2.99	

Source: Werner, K. et al., *Naunyn-Schmiedeberg's Archives of Pharmacology* 384, 169–176, 2011.

Note: Lipophilicity is expressed as a partition coefficient (log K_w) of the analog based on gradient elution chromatography between methanol and water. Permeability represents the percentage of intracellularly measured cyclic nucleotide content as a function of external applied concentration. Combined cyclic nucleotides show comparable permeability scores.

In general, starting with a cyclic nucleotide concentration in the range of 10 to 100 μM and a 10 to 20 min preincubation time is recommended for most tissue and cell culture conditions. Immediate reaction of an analog is most likely due to unspecific binding to surface receptors.[76]

For highly lipophilic compounds (log $K_w > 2.8$) the use of serum-free medium during preincubation is advised. This will reduce lipid vesicle complex formation of the cyclic nucleotides in the medium. To increase the lipophilicity of cAMP or cGMP analogs in the low log K_w range, *caged* compounds have been developed that release the biologically active compound after the UV flash.[77] Caged 8-Br-cGMP was successfully used in studies with CNG channels.[78] A second approach introduces an AM ester at the cyclic phosphate ring. This masks the negatively charged phosphate and increases lipophilicity.[79] Inside the cell, the AM ester is hydrolyzed by esterases into acetic acid, formaldehyde, and cyclic nucleotide. To date, several AM derivatives have been synthesized and used in intact cells, including the new Epac-selective activator, 8-pCPT-2′-O-Me-cAMP-AM.[53,56]

2.8 CONCLUSION

Cross-activation in cyclic nucleotide signaling pathways is a major technical problem when investigating CNB proteins. Therefore, controlling the presence and activity of these proteins in the cell of interest should be the first step before using cyclic nucleotide analogs to address any functional question. In most cases, the presence of a protein can be evaluated by specific antibodies. For example, in mouse cardiac myocytes, the PKG Iα level has been calculated at 0.11 μM (unpublished results), whereas in human platelets, PKA II and PKG Iβ concentrations are in the range of 3 and 7 μM, respectively.[80]

Activation of PKG and PKA can be assessed by analyzing the phosphorylation sites in the protein VASP. Although PKG phosphorylates predominantly Ser-239, Ser-157 is more rapidly phosphorylated by PKA.[81] In addition to VASP phosphorylation, PDE5 phosphorylation at Ser-92 is an effective monitor of intracellular PKG activation.[24,82–84] In summary, many cAMP and cGMP analogs have multiple targets. However, the intelligent choice of a selective cyclic nucleotide analog in combination with elaborated control experiments is the best way to obtain interpretable results in a given cell system.

ACKNOWLEDGMENTS

Many thanks to Dr. Frank Schwede for cross-reading the manuscript and Dr. Martin Biel for supplying the CNG channel data in Table 2.3.

REFERENCES

1. Sutherland, E.W., and Rall, T.W. Fractionation and characterization of a cyclic adenine ribonucleotide formed by tissue particles. *Journal of Biological Chemistry* 232, 1077–1091 (1958).
2. Ashman, D.F., Lipton, R., Melicow, M.M., and Price, T.D. Isolation of adenosine 3′,5′-monophosphate and guanosine 3′,5′-monophosphate from rat urine. *Biochemical and Biophysical Research Communications* 11, 330–334 (1963).
3. Kim, C., Vigil, D., Anand, G., and Taylor, S.S. Structure and dynamics of PKA signaling proteins. *European Journal of Cell Biology* 85, 651–654 (2006).
4. Francis, S.H., Busch, J.L., Corbin, J.D., and Sibley, D. cGMP-dependent protein kinases and cGMP phosphodiesterases in nitric oxide and cGMP action. *Pharmacological Reviews* 62, 525–563 (2010).
5. Cooper, D.M. Regulation and organization of adenylyl cyclases and cAMP. *Biochemical Journal* 375, 517–529 (2003).
6. Mullershausen, F., Koesling, D., and Friebe, A. NO-sensitive guanylyl cyclase and NO-induced feedback inhibition in cGMP signaling. *Frontiers in Bioscience* 10, 1269–1278 (2005).
7. Potter, L. Guanylyl cyclase structure, function and regulation. *Cell Signal* 23, 1921–1926 (2011).
8. Bender, A.T., and Beavo, J.A. Cyclic nucleotide phosphodiesterases: Molecular regulation to clinical use. *Pharmacological Reviews* 58, 488–520 (2006).
9. Maurice, D.H., Ke, H., Ahmad, F., Wang, Y., Chung, J., and Manganiello, V.C. Advances in targeting cyclic nucleotide phosphodiesterases. *Nature Reviews* 13, 290–314 (2014).
10. Haslam, R.J., Dickinson, N.T., and Jang, E.K. Cyclic nucleotides and phosphodiesterases in platelets. *Thrombosis and Haemostasis* 82, 412–423 (1999).

11. Biel, M., and Michalakis, S. Cyclic nucleotide-gated channels. *Handbook of Experimental Pharmacology*, 111–136 (2009).
12. Podda, M.V., and Grassi, C. New perspectives in cyclic nucleotide-mediated functions in the CNS: The emerging role of cyclic nucleotide-gated (CNG) channels. *Pflügers Archiv* (2013).
13. Biel, M., Wahl-Schott, C., Michalakis, S., and Zong, X. Hyperpolarization-activated cation channels: From genes to function. *Physiological Reviews* 89, 847–885 (2009).
14. Benarroch, E.E. HCN channels: Function and clinical implications. *Neurology* 80, 304–310 (2013).
15. Gloerich, M., and Bos, J.L. Epac: Defining a new mechanism for cAMP action. *Annual Review of Pharmacology and Toxicology* 50, 355–375 (2010).
16. Schmidt, M., Dekker, F.J., and Maarsingh, H. Exchange protein directly activated by cAMP (epac): A multidomain cAMP mediator in the regulation of diverse biological functions. *Pharmacological Reviews* 65, 670–709 (2013).
17. Maurice, D.H., and Haslam, R.J. Molecular basis of the synergistic inhibition of platelet function by nitrovasodilators and activators of adenylate cyclase: Inhibition of cyclic AMP breakdown by cyclic GMP. *Molecular Pharmacology* 37, 671–681 (1990).
18. Dostmann, W.R. et al. Probing the cyclic nucleotide binding sites of cAMP-dependent protein kinases I and II with analogs of adenosine 3′,5′-cyclic phosphorothioates. *Journal of Biological Chemistry* 265, 10484–10491 (1990).
19. Schwede, F. et al. 8-Substituted cAMP analogues reveal marked differences in adaptability, hydrogen bonding, and charge accommodation between homologous binding sites (AI/AII and BI/BII) in cAMP kinase I and II. *Biochemistry* 39, 8803–8812 (2000).
20. Christensen, A.E. et al. cAMP analog mapping of Epac1 and cAMP kinase. Discriminating analogs demonstrate that Epac and cAMP kinase act synergistically to promote PC-12 cell neurite extension. *Journal of Biological Chemistry* 278, 35394–35402 (2003).
21. Ogreid, D. et al. Activation of protein kinase isozymes by cyclic nucleotide analogs used singly or in combination. Principles for optimizing the isozyme specificity of analog combinations. *European Journal of Biochemistry/FEBS* 150, 219–227 (1985).
22. Gausdal, G. et al. Cyclic AMP can promote APL progression and protect myeloid leukemia cells against anthracycline-induced apoptosis. *Cell Death & Disease* 4, e516.
23. Sandberg, M. et al. Characterization of Sp-5,6-dichloro-1-beta-D-ribofuranosylbenzimidazole-3′,5′-monophosphorothioate (Sp-5,6-DCl-cBiMPS) as a potent and specific activator of cyclic-AMP-dependent protein kinase in cell extracts and intact cells. *Biochemical Journal* 279 (Pt 2), 521–527 (1991).
24. Poppe, H. et al. Cyclic nucleotide analogs as probes of signaling pathways. *Nature Methods* 5, 277–278 (2008).
25. Dostmann, W.R., and Taylor, S.S. Identifying the molecular switches that determine whether (Rp)-cAMPS functions as an antagonist or an agonist in the activation of cAMP-dependent protein kinase I. *Biochemistry* 30, 8710–8716 (1991).
26. Jensen, B.O., Selheim, F., Doskeland, S.O., Gear, A.R., and Holmsen, H. Protein kinase A mediates inhibition of the thrombin-induced platelet shape change by nitric oxide. *Blood* 104, 2775–2782 (2004).
27. Vaandrager, A.B., Ehlert, E.M., Jarchau, T., Lohmann, S.M., and de Jonge, H.R. N-terminal myristoylation is required for membrane localization of cGMP-dependent protein kinase type II. *Journal of Biological Chemistry* 271, 7025–7029 (1996).
28. Piggott, L.A. et al. Natriuretic peptides and nitric oxide stimulate cGMP synthesis in different cellular compartments. *Journal of General Physiology* 128, 3–14 (2006).
29. Wilson, L.S., Elbatarny, H.S., Crawley, S.W., Bennett, B.M., and Maurice, D.H. Compartmentation and compartment-specific regulation of PDE5 by protein kinase G allows selective cGMP-mediated regulation of platelet functions. *Proceedings of the National Academy of Sciences of the United States of America* 105, 13650–13655 (2008).

30. Castro, L.R., Schittl, J., and Fischmeister, R. Feedback control through cGMP-dependent protein kinase contributes to differential regulation and compartmentation of cGMP in rat cardiac myocytes. *Circulation Research* 107, 1232–1240 (2010).
31. Krass, J.D., Jastorff, B., and Genieser, H.G. Determination of lipophilicity by gradient elution high-performance liquid chromatography. *Analytical Chemistry* 69, 2575–2581 (1997).
32. Pohler, D. et al. Expression, purification, and characterization of the cGMP-dependent protein kinases I beta and II using the baculovirus system. *FEBS Letters* 374, 419–425 (1995).
33. Rehmann, H., Schwede, F., Doskeland, S.O., Wittinghofer, A., and Bos, J.L. Ligand-mediated activation of the cAMP-responsive guanine nucleotide exchange factor Epac. *Journal of Biological Chemistry* 278, 38548–38556 (2003).
34. Ludwig, J., Margalit, T., Eismann, E., Lancet, D., and Kaupp, U.B. Primary structure of cAMP-gated channel from bovine olfactory epithelium. *FEBS Letters* 270, 24–29 (1990).
35. Ludwig, A., Zong, X., Jeglitsch, M., Hofmann, F., and Biel, M. A family of hyperpolarization-activated mammalian cation channels. *Nature* 393, 587–591 (1998).
36. Tanaka, J.C., Eccleston, J.F., and Furman, R.E. Photoreceptor channel activation by nucleotide derivatives. *Biochemistry* 28, 2776–2784 (1989).
37. Wei, J.Y., Cohen, E.D., Genieser, H.G., and Barnstable, C.J. Substituted cGMP analogs can act as selective agonists of the rod photoreceptor cGMP-gated cation channel. *Journal of Molecular Neuroscience* 10, 53–64 (1998).
38. Sekhar, K.R. et al. Relaxation of pig coronary arteries by new and potent cGMP analogs that selectively activate type I alpha, compared with type I beta, cGMP-dependent protein kinase. *Molecular Pharmacology* 42, 103–108 (1992).
39. Butt, E., Pohler, D., Genieser, H.G., Huggins, J.P., and Bucher, B. Inhibition of cyclic GMP-dependent protein kinase-mediated effects by (Rp)-8-bromo-PET-cyclic GMPS. *British Journal of Pharmacology* 116, 3110–3116 (1995).
40. Strassmaier, T., and Karpen, J.W. Novel N7- and N1-substituted cGMP derivatives are potent activators of cyclic nucleotide-gated channels. *Journal of Medicinal Chemistry* 50, 4186–4194 (2007).
41. Tsalkova, T. et al. Isoform-specific antagonists of exchange proteins directly activated by cAMP. *Proceedings of the National Academy of Sciences of the United States of America* 109, 18613–18618 (2012).
42. Rehmann, H., Das, J., Knipscheer, P., Wittinghofer, A., and Bos, J.L. Structure of the cyclic-AMP-responsive exchange factor Epac2 in its auto-inhibited state. *Nature* 439, 625–628 (2006).
43. Courilleau, D., Bouyssou, P., Fischmeister, R., Lezoualc'h, F., and Blondeau, J.P. The (R)-enantiomer of CE3F4 is a preferential inhibitor of human exchange protein directly activated by cyclic AMP isoform 1 (Epac1). *Biochemical and Biophysical Research Communications* 440, 443–448 (2013).
44. Dostmann, W.R. (RP)-cAMPS inhibits the cAMP-dependent protein kinase by blocking the cAMP-induced conformational transition. *FEBS Letters* 375, 231–234 (1995).
45. Zhao, J. et al. Progressive cyclic nucleotide-induced conformational changes in the cGMP-dependent protein kinase studied by small angle X-ray scattering in solution. *Journal of Biological Chemistry* 272, 31929–31936 (1997).
46. Vaandrager, A.B. et al. Endogenous type II cGMP-dependent protein kinase exists as a dimer in membranes and can be functionally distinguished from the type I isoforms. *Journal of Biological Chemistry* 272, 11816–11823 (1997).
47. de Rooij, J. et al. Epac is a Rap1 guanine-nucleotide-exchange factor directly activated by cyclic AMP. *Nature* 396, 474–477 (1998).
48. Kawasaki, H. et al. A family of cAMP-binding proteins that directly activate Rap1. *Science* 282, 2275–2279 (1998).

49. Rehmann, H. et al. Structure and regulation of the cAMP-binding domains of Epac2. *Nature Structural Biology* 10, 26–32 (2003).
50. Niimura, M. et al. Critical role of the N-terminal cyclic AMP-binding domain of Epac2 in its subcellular localization and function. *Journal of Cellular Physiology* 219, 652–658 (2009).
51. Dao, K.K. et al. Epac1 and cAMP-dependent protein kinase holoenzyme have similar cAMP affinity, but their cAMP domains have distinct structural features and cyclic nucleotide recognition. *Journal of Biological Chemistry* 281, 21500–21511 (2006).
52. Enserink, J.M. et al. A novel Epac-specific cAMP analogue demonstrates independent regulation of Rap1 and ERK. *Nature Cell Biology* 4, 901–906 (2002).
53. Vliem, M.J. et al. 8-pCPT-2′-O-Me-cAMP-AM: An improved Epac-selective cAMP analogue. *Chembiochem* 9, 2052–2054 (2008).
54. Enyeart, J.A., and Enyeart, J.J. Metabolites of an Epac-selective cAMP analog induce cortisol synthesis by adrenocortical cells through a cAMP-independent pathway. *PLoS One* 4, e6088 (2009).
55. Herfindal, L. et al. Off-target effect of the Epac agonist 8-pCPT-2′-O-Me-cAMP on P2Y12 receptors in blood platelets. *Biochemical and Biophysical Research Communications* 437, 603–608 (2013).
56. Almahariq, M. et al. A novel EPAC-specific inhibitor suppresses pancreatic cancer cell migration and invasion. *Molecular Pharmacology* 83, 122–128 (2013).
57. Rehmann, H. Epac-inhibitors: Facts and artefacts. *Scientific Reports* 3, 3032 (2013).
58. Courilleau, D. et al. Identification of a tetrahydroquinoline analog as a pharmacological inhibitor of the cAMP-binding protein Epac. *Journal of Biological Chemistry* 287, 44192–44202 (2012).
59. Stieber, J., Stockl, G., Herrmann, S., Hassfurth, B., and Hofmann, F. Functional expression of the human HCN3 channel. *Journal of Biological Chemistry* 280, 34635–34643 (2005).
60. Zong, X. et al. Regulation of hyperpolarization-activated cyclic nucleotide-gated (HCN) channel activity by cCMP. *Journal of Biological Chemistry* 287, 26506–26512 (2012).
61. Hartwig, C. et al. cAMP, cGMP, cCMP and cUMP concentrations across the tree of life: High cCMP and cUMP levels in astrocytes. *Neuroscience Letters* 579, 183–187 (2014).
62. Hammelmann, V., Zong, X., Hofmann, F., Michalakis, S., and Biel, M. The cGMP-dependent protein kinase II Is an inhibitory modulator of the hyperpolarization-activated HCN2 channel. *PLoS One* 6, e17078 (2011).
63. Wei, J.Y., Cohen, E.D., Yan, Y.Y., Genieser, H.G., and Barnstable, C.J. Identification of competitive antagonists of the rod photoreceptor cGMP-gated cation channel: Beta-phenyl-1,N2-etheno-substituted cGMP analogues as probes of the cGMP-binding site. *Biochemistry* 35, 16815–16823 (1996).
64. Kaupp, U.B. et al. Primary structure and functional expression from complementary DNA of the rod photoreceptor cyclic GMP-gated channel. *Nature* 342, 762–766 (1989).
65. Gordon, S.E., and Zagotta, W.N. A histidine residue associated with the gate of the cyclic nucleotide-activated channels in rod photoreceptors. *Neuron* 14, 177–183 (1995).
66. Chen, T.Y. et al. A new subunit of the cyclic nucleotide-gated cation channel in retinal rods. *Nature* 362, 764–767 (1993).
67. Dhallan, R.S., Yau, K.W., Schrader, K.A., and Reed, R.R. Primary structure and functional expression of a cyclic nucleotide-activated channel from olfactory neurons. *Nature* 347, 184–187 (1990).
68. Sautter, A., Zong, X., Hofmann, F., and Biel, M. An isoform of the rod photoreceptor cyclic nucleotide-gated channel beta subunit expressed in olfactory neurons. *Proceedings of the National Academy of Sciences of the United States of America* 95, 4696–4701 (1998).

69. Bonigk, W. et al. The native rat olfactory cyclic nucleotide-gated channel is composed of three distinct subunits. *Journal of Neuroscience* 19, 5332–5347 (1999).
70. Gerstner, A., Zong, X., Hofmann, F., and Biel, M. Molecular cloning and functional characterization of a new modulatory cyclic nucleotide-gated channel subunit from mouse retina. *Journal of Neuroscience: The Official Journal of the Society for Neuroscience* 20, 1324–1332 (2000).
71. Weyand, I. et al. Cloning and functional expression of a cyclic-nucleotide-gated channel from mammalian sperm. *Nature* 368, 859–863 (1994).
72. Biel, M. et al. Another member of the cyclic nucleotide-gated channel family, expressed in testis, kidney, and heart. *Proceedings of the National Academy of Sciences of the United States of America* 91, 3505–3509 (1994).
73. Korschen, H.G. et al. A 240 kDa protein represents the complete beta subunit of the cyclic nucleotide-gated channel from rod photoreceptor. *Neuron* 15, 627–636 (1995).
74. Van Haastert, P.J. et al. Substrate specificity of cyclic nucleotide phosphodiesterase from beef heart and from Dictyostelium discoideum. *European Journal of Biochemistry/ FEBS* 131, 659–666 (1983).
75. Werner, K., Schwede, F., Genieser, H.G., Geiger, J., and Butt, E. Quantification of cAMP and cGMP analogs in intact cells: Pitfalls in enzyme immunoassays for cyclic nucleotides. *Naunyn-Schmiedeberg's Archives of Pharmacology* 384, 169–176 (2011).
76. Do, T., Sun, Q., Beuve, A., and Kuzhikandathil, E.V. Extracellular cAMP inhibits D1 dopamine receptor expression in CAD catecholaminergic cells via A2a adenosine receptors. *Journal of Neurochemistry* 101, 619–631 (2007).
77. Adams, S.R., and Tsien, R.Y. Controlling cell chemistry with caged compounds. *Annual review of physiology* 55, 755–784 (1993).
78. Hagen, V. et al. [7-(Dialkylamino)coumarin-4-yl]methyl-caged compounds as ultrafast and effective long-wavelength phototriggers of 8-bromo-substituted cyclic nucleotides. *Chembiochem* 4, 434–442 (2003).
79. Schultz, C. et al. Acetoxymethyl esters of phosphates, enhancement of the permeability and potency of cAMP. *Journal of Biological Chemistry* 268, 6316–6322 (1993).
80. Eigenthaler, M., Nolte, C., Halbrugge, M., and Walter, U. Concentration and regulation of cyclic nucleotides, cyclic-nucleotide-dependent protein kinases and one of their major substrates in human platelets. Estimating the rate of cAMP-regulated and cGMP-regulated protein phosphorylation in intact cells. *European Journal of Biochemistry/ FEBS* 205, 471–481 (1992).
81. Butt, E. et al. cAMP- and cGMP-dependent protein kinase phosphorylation sites of the focal adhesion vasodilator-stimulated phosphoprotein (VASP) in vitro and in intact human platelets. *Journal of Biological Chemistry* 269, 14509–14517 (1994).
82. Mo, E., Amin, H., Bianco, I.H., and Garthwaite, J. Kinetics of a cellular nitric oxide/ cGMP/phosphodiesterase-5 pathway. *Journal of Biological Chemistry* 279, 26149–26158, doi:10.1074/jbc.M400916200 (2004).
83. Shimizu-Albergine, M. et al. Individual cerebellar Purkinje cells express different cGMP phosphodiesterases (PDEs): In vivo phosphorylation of cGMP-specific PDE (PDE5) as an indicator of cGMP-dependent protein kinase (PKG) activation. *Journal of Neuroscience: The Official Journal of the Society for Neuroscience* 23, 6452–6459 (2003).
84. Begonja, A.J. et al. Differential roles of cAMP and cGMP in megakaryocyte maturation and platelet biogenesis. *Experimental Hematology* 41, 91–101.e104, doi:10.1016/j.exphem.2012.09.001 (2013).

3 High-Throughput FRET Assays for Fast Time-Dependent Detection of Cyclic AMP in Pancreatic β Cells

George G. Holz, Colin A. Leech,
Michael W. Roe, and Oleg G. Chepurny

CONTENTS

3.1 INTRODUCTION

Pancreatic β cells located in the islets of Langerhans secrete insulin in response to the increase of blood glucose concentration that occurs after a meal.[1] Insulin secreted in this manner acts as a circulating hormone to suppress hepatic glucose production while also promoting glucose uptake into striated muscle and fat.[2] When this process of systemic glucose homeostasis is disrupted, chronic hyperglycemia ensues, as is the case for patients diagnosed with type 2 diabetes mellitus (T2DM).[3] Current drug discovery efforts for the treatment of T2DM seek to identify new agents that will lower the levels of blood glucose by stimulating insulin secretion.[4] Here, we summarize efforts to achieve this goal, with emphasis on the identification of adenosine-3′,5′-cyclic monophosphate (cAMP) elevating agents that directly stimulate insulin release from β cells.[5] We also describe a new fluorescence resonance energy transfer (FRET) assay with which to implement a high-throughput screen of β cell cAMP-elevating agents. This assay uses cell lines transfected with genetically encoded cAMP biosensors that allow the detection of cAMP in a fast and time-dependent manner.

3.2 cAMP SIGNALING IN PANCREATIC β CELLS

Glucose is the primary stimulus for insulin secretion from pancreatic β cells,[6] and this action of glucose is potentiated by cAMP-elevating agents that activate G protein–coupled receptors (GPCRs) expressed on β cells (Figure 3.1).[7] Two such cAMP-elevating agents are exenatide (Byetta) and liraglutide (Victoza), both of which are prescribed

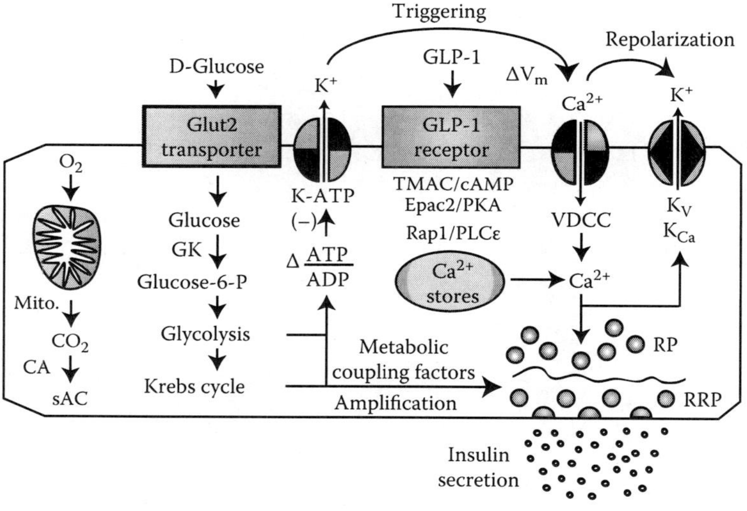

FIGURE 3.1 Glucose metabolism in β cells stimulates depolarization (ΔV_m) and entry of Ca^{2+} through voltage-dependent Ca^{2+} channels (VDCCs). Ca^{2+} triggers the exocytosis of insulin from secretory granules located in a readily releasable pool (RRP) and a reserve pool (RP). A cortical actin barrier (wavy line) separates the RRP and RP, and it must undergo *remodeling* for glucose to stimulate exocytosis of the RP. GLP-1 activates its GPCR to potentiate glucose-stimulated insulin secretion in a PKA- and Epac2-mediated manner. Downstream effectors of Epac2 include Rapl GTPase and a Rapl-regulated phospholipase C-epsilon (PLCε). Membrane repolarization occurs in response to the activation of voltage-dependent K^+ channels (K_V) and Ca^{2+}-activated K^+ channels (K_{Ca}). Mitochondrial (Mito.) respiration generates CO_2 that is converted by CA to generate HCO_3^- that activates sAC.

as blood glucose–lowering agents for the treatment of T2DM.[8] Exenatide and liraglutide are insulin secretagogues with blood glucose–lowering properties because they activate a GPCR that normally binds the cAMP-elevating hormone glucagon-like peptide-1 (GLP-1) in β cells.[9] GLP-1 is an insulinotropic peptide released from intestinal L-cells in response to the ingestion of a meal, and it complements the action of intestinally absorbed glucose to stimulate pancreatic insulin secretion.[10]

Binding of GLP-1 to the GLP-1 receptor (GLP-1R) activates heterotrimeric G_s proteins, thereby stimulating the activities of transmembrane adenylyl cyclases (TMACs) in β cells (Figure 3.1).[11] TMACs catalyze the synthesis of cAMP, a cytosolic second messenger that activates two distinct classes of cAMP-binding proteins: (1) protein kinase A (PKA, a serine/threonine protein kinase), and (2) cAMP-regulated guanine nucleotide exchange factors designated as Epac1 and Epac2.[12–15] Evidence exists that PKA and Epac2 mediate the actions of GLP-1R agonists to increase the efficacy and potency of glucose as a stimulus for insulin secretion.[12–15] Furthermore, PKA and Epac2 seem to mediate an action of GLP-1R agonists to repair defective stimulus-secretion coupling in the β cells of patients with T2DM.[12–15] Thus, GLP-1R agonists are not simply β cell glucose sensitizers, but they also induce β cell *glucose competence* under conditions of metabolic stress.[16–21] Potentially, the PKA and Epac2 signaling mechanisms in β cells can be targeted to identify new pharmacological agents for use in the treatment of T2DM.

Because β cells express a surprisingly diverse assortment of GPCRs,[22] there is good reason to expect that new classes of insulin secretagogues will be identifiable in high-throughput screens that are based on the ability of a candidate compound to stimulate cAMP production in a GPCR-mediated manner. Likely GPCRs that can be targeted in β cells are the cAMP-elevating GPCRs for fatty acid amides (GPR119), bile acids (TGR5), L-arginine (GPRC6A), pituitary adenylyl cyclase–activating polypeptide (PACAP-R), glucose-dependent insulinotropic peptide (GIP-R), and glucagon (Gluc-R).[4,7,22] Because β cells also express GPCRs that inhibit cAMP production,[23] it might be possible to identify antagonists of GPCRs that indirectly increase levels of cAMP to stimulate insulin secretion. Once identified, the properties of all such cAMP-elevating agents can be evaluated in secondary screens of glucose-stimulated insulin secretion (GSIS) using isolated islets,[24–26] or in single-cell assays of β cell ion channel function, Ca^{2+} handling, and secretory granule exocytosis.[27–37]

3.3 SELECTION OF β CELL LINES

High-throughput screening for detection of cAMP in primary cultures of pancreatic β cells is limited by the fact that it is difficult to obtain sufficient numbers of cells for this purpose. However, β cell lines are available, and they provide model systems with which to perform such assays.[38] Three commonly used cell lines are rat INS-1 cells,[39] mouse MIN6 cells,[40] and hamster HIT-T15 cells,[41] all of which are cancer cell lines derived from insulinomas. Additional cell lines are BRIN-D11 cells derived from electrofusion of RINm5f insulinoma cells and primary rat β cells,[42] as well as human insulin–secreting cell lines of β cell origin.[43,44] It is also possible to generate insulin-secreting cells from mouse and human embryonic stem cells,[45–47] although these cell lines have limited insulin secretory capacity. Because insulinoma-derived insulin-secreting cell lines are transformed and immortalized, they have limitations owing to the fact that they tend to dedifferentiate with increasing passage number.[48,49] Furthermore, some insulin-secreting cell lines do not secrete insulin exclusively, but also secrete islet peptides not normally found in β cells.[50,51] Optimization of an insulin-secreting cell line can be achieved by clonal selection to generate subclones such as INS-1E, INS-1 832/13, and MIN6-K that closely resemble β cells.[52–54] These subclones secrete insulin in response to physiologically relevant concentrations of glucose that span a concentration range of 2.8 to 16.7 mM. Furthermore, these subclones express target GPCRs of interest, in addition to standard genetic markers for β cell glucose sensing such as the Glut2 facilitative glucose transporter and the type IV hexokinase (glucokinase, GK; Figure 3.1). Independent confirmation of a positive *hit* in high-throughput assays using β cell lines is obtainable using human or rodent β cells that are virally transduced with cAMP biosensors so that live-cell imaging can be performed in real time under conditions that allow fluctuations of cAMP concentration to be measured in response to a GPCR agonist.[25]

An additional factor to consider when selecting a β cell line for high-throughput screening is that these cell lines express multiple isoforms of TMACs.[55,56] Ideally, the β cell line that is chosen will replicate the pattern of TMAC expression found in primary β cells. In this regard, rodent β cells express a Ca^{2+}/calmodulin-regulated type 8 adenylyl cyclase (AC-8) that is especially important to the cAMP-dependent stimulation of insulin secretion by GPCR agonists such as GLP-1.[55,57–59] AC-8 acts as a molecular coincidence detector because its activity is stimulated not only by GPCR agonists but also by depolarization-induced Ca^{2+} influx that occurs in response to β cell glucose metabolism.[60–63] Thus, a requirement for glucose metabolism will likely exist when evaluating the efficacy and potency of GPCR agonists to stimulate cAMP production in β cell lines. This requirement can be met using subclones of INS-1 and MIN6 cells that retain their glucose responsiveness.[52–54]

β Cells also express a soluble adenylyl cyclase (sAC) that is activated by the glucose metabolism independently of GPCRs (Figure 3.1).[64,65] The existence of this alternative source of cAMP is understandable because the glucose metabolism in β cells stimulates mitochondrial CO_2 production that upregulates sAC activity. CO_2 is converted to carbonic acid by carbonic anhydrase (CA), and at physiological pH, carbonic acid dissociates to yield bicarbonate ion $\left(HCO_3^-\right)$ that binds to and directly stimulates sAC.[66,67] The CA inhibitor acetazolamide reduces GSIS from isolated islets,[68,69] and a knockout (KO) of sAC gene expression in mice leads to reduced GSIS that is accompanied by reduced glucose tolerance and elevated levels of blood glucose in vivo.[65] Furthermore, the selective sAC inhibitors KH7 and catechol estrogen inhibit GSIS from INS-1E β cells.[64,65] Therefore, the potential role of sAC activity should be taken into consideration when evaluating how investigational compounds stimulate cAMP production in β cell lines.

3.4 BIOSENSORS BASED ON PKA HOLOENZYME DISSOCIATION

The first FRET-based cAMP biosensor was FlCRhR, a reporter developed by the R.Y. Tsien laboratory (Table 3.1).[70] Its design was based on the PKA holoenzyme and

TABLE 3.1
Genetically Encoded Biosensors

	EC_{50} cAMP	References
PKA-Based Biosensors		
FlCRhR (RIα/Cα)	90 nM	Adams et al.[70]
RIIβ EBFP/CαS65T	0.5–0.9 μM	Zaccolo et al.[71]
AKAR3 (ECFP/Venus)	n.a.	Allen and Zhang[72]
AKAR4 (Cerulean/Venus)	n.a.	Depry et al.[73]
Lyn-AKAR4 (Lipid rafts)	n.a.	Depry et al.[73]
AKAR4-Kras (Nonlipid rafts)	n.a.	Depry et al.[73]
PKA-camps (EYFP-RIIβ-ECFP)	1.9 μM	Nikolaev et al.[74]
delta-RIIβ-CFP-CAAX/Cα-YFP	n.r.	Dyachok et al.[75,76]
Epac-Based Biosensors		
Epac1-camps (EYFP-CNBD-ECFP)	2.4 μM	Nikolaev et al.[74]
ICUE1 (ECFP-FL-Epac1-Citrine)	10–50 μM	DiPilato et al.[77]
CFP-FL-Epac1-YFP	50 μM	Ponsioen et al.[78]
CFP-Epac1-ΔDEP-CD-YFP	14 μM	Ponsioen et al.[78]
mCerulean-Epac1-ΔDEP-CD-mCitrine (CEPAC, reduced ion sensitivity)	24 μM	Salonikidis et al.[79]
mTurq.delta-Epac1-[CD, ΔDEP]-cp173-Venus-Venus (expanded dynamic range)	n.r.	Klarenbeek et al.[80]
Epac2-camps (EYFP-CNBD-B-ECFP)	0.9 μM	Nikolaev et al.[74]
Cerulean-FL-Epac2-Venus	n.r.	Herbst et al.[81]
ECFP-FL-Epac2-EYFP	n.r.	Zhang et al.[82]
Citrene-Epac2-camps-CeruleanFP (reduced pH sensitivity)	545 nM	Everett and Cooper[83]
Citrene-Epac2-camps-CeruleanFP-AC8 (fusion to adenylyl cyclase 8)	356 nM	Everett and Cooper[83]

Note: FL, full-length; n.a., not applicable; n.r., not reported. All EC_{50} values are approximate.

incorporates two catalytic subunits (Cα) conjugated to fluorescein, and two regulatory subunits (RIα) conjugated to rhodamine. FlCRhR is not a genetically encoded cAMP biosensor, so it must be introduced into cells as a protein, typically by microinjection. However, understanding the properties of FlCRhR allows an introduction to the genetically encoded cAMP biosensors described below. In FlCRhR, the fluorescein moiety of the Cα subunit (FlC) serves as the donor chromophore for FRET, whereas the rhodamine moiety of the RIα subunit (RhR) serves as the acceptor chromophore. Using 488 nm excitation light, the fluorescein chromophore of FlCRhR is excited while also monitoring the fluorescein and rhodamine emission light intensities at 520 and 580 nm, respectively. In the absence of cAMP, FlCRhR exists as a tetramer ($C\alpha_2RI\alpha_2$) so that FRET will occur between nearby FlC and RhR subunits. However, binding of cAMP to FlCRhR induces its dissociation, and the resultant decrease of FRET is measurable as an increase of the 520/580 nm emission ratio. Using FlCRhR in combination with the Ca^{2+} indicator fura-2, DeBernardi and Brooker were the first to demonstrate that digital imaging techniques could be used to investigate the Ca^{2+}-dependent control of cAMP biosynthesis and degradation in living cells.[84] Subsequently, Zaccolo et al. designed a genetically encoded cAMP biosensor for use in assays of FRET (Table 3.1).[71] By transfecting cells with cDNAs that directed the expression of fluorescent PKA regulatory (RIIβ) and catalytic (Cα) subunits, Zaccolo et al. achieved the reconstitution of a tetrameric cAMP biosensors in situ.[71] In this biosensor, the RIIβ subunit is a fusion protein incorporating a blue fluorescent protein chromophore (RIIβ-EBFP) serving as the FRET donor, whereas the Cα subunit is fused to a green fluorescent protein chromophore (CαS65T) serving as the FRET acceptor.[71]

With the advent of genetically encoded biosensors, it soon became apparent that compartmentalized cAMP signaling could be monitored using biosensors that incorporate select targeting sequences. For studies of β cells, Dyachok et al. used a bimolecular PKA-based biosensor that does not monitor FRET, but that instead monitors reversible cAMP-dependent translocation of the Cα catalytic subunit from the plasma membrane (PM) to the cytosol (Table 3.1).[75] In this biosensor, the RIIβ subunit is truncated to remove the regulatory subunit dimerization domain and the binding domain for A-kinase anchoring proteins. This truncated ΔRIIβ is fused at its C-terminus to cyan fluorescent protein (CFP), which is itself C-terminally extended to include PM-targeting polybasic and CAAX motifs. The resultant ΔRIIβ-CFP-CAAX fusion protein localizes to the PM where it recruits its molecular partner, a Cα catalytic subunit fused at its C-terminus with yellow fluorescent protein (YFP).

Using total internal reflection microscopy (TIRF), it is possible to image this PKA-based biosensor at the inner surface of the PM, and to also monitor cAMP-induced dissociation of Cα-YFP away from the PM. When combined with the Ca^{2+} indicator fura red, it is possible to detect synchronous oscillations of cAMP and Ca^{2+} that are induced by GLP-1 and that stimulate pulsatile insulin secretion from β cells.[75] Furthermore, it is possible to demonstrate translocation of Cα-YFP from the PM to the nucleus in response to a sustained increase of cAMP, thereby demonstrating a likely role for Cα nuclear translocation in the PKA-dependent control of gene expression.[75] In additional studies using MIN6 cells and mouse β cells, Dyachok et al. also found that glucose, alone, has the capacity to induce oscillations of cAMP occurring immediately beneath the PM.[76] Because this action of glucose is blocked by the TMAC inhibitor 2′,5′-dideoxyadenosine (DDA, a P-site inhibitor), and is reduced by an L-type Ca^{2+} channel blocker (verapamil), it seems possible that β cell glucose metabolism initiates Ca^{2+} influx that activates AC-8 so that oscillations of cAMP are generated.[76]

3.5 BIOSENSORS BASED ON PKA-MEDIATED PHOSPHORYLATION

GPCR signaling pathways are linked to major alterations of PKA activity that potentiate GSIS from β cells.[27,31,37,85–95] Furthermore, PKA holoenzyme complexes containing RIIα regulatory subunits bound to A-kinase anchoring proteins mediate the action of GLP-1 to potentiate GSIS.[96,97] Recently, it was reported that PKA-mediated phosphorylation of the SNARE complex–associated protein Snapin is one means by which GLP-1 facilitates insulin exocytosis that is Ca^{2+}-dependent.[91] Collectively, these reports are consistent with the finding that GSIS is potentiated by 6-Bn-cAMP-AM, a cAMP analog that activates PKA holoenzyme complexes containing RII but not RI regulatory subunit isoforms (G.G. Holz, unpublished findings). To monitor PKA activity in islets, it is possible to perform immunochemical assays of cAMP-response element-binding protein phosphorylation, or biochemical assays of Kemptide phosphorylation.[25] However, a FRET-based assay using the biosensor AKAR3 allows real-time measurements of PKA activity in single β cells.[25] AKAR3 is a unimolecular A-kinase activity reporter designed by Allen and Zhang (Table 3.1).[72] It consists of an N-terminal CFP chromophore linked to a forkhead-associated phosphoamino acid-binding domain (FHA-D), a consensus PKA substrate motif (LRRATLVD), and a C-terminal YFP chromophore. Phosphorylation of AKAR3 by PKA leads to a conformational change caused by the binding of the FHA-D to the PKA substrate motif containing phosphothreonine. This molecular rearrangement allows increased FRET that is measurable as an increase of the 535/485 nm emission ratio when exciting at 440 nm. Importantly, the AKAR3 coding sequence can be incorporated into adenovirus for its expression in human β cells for live-cell imaging in real time.[25] As discussed in Section 3.13, AKAR3 can also be expressed in cell lines so that it can serve as a sensitive indicator of PKA activity when performing high-throughput assays using a plate-reading spectrofluorimeter equipped with dual emission monochromators.

3.6 BIOSENSORS BASED ON EPAC ACTIVATION

The discovery of the Epac class of cAMP-regulated guanine nucleotide exchange factors in 1998 provided a new explanation for the ability of cAMP to control cellular functions independently of PKA.[98,99] In terms of β cell function, these PKA-independent actions of cAMP were reported to be mediated by Epac2.[100–102] However, recent findings indicate that Epac1 also participates.[103] Epac-mediated actions of cAMP include the stimulation of insulin secretion,[24–28,33,34,90,104–109] the stimulation of Ca^{2+} signaling,[15,25,26,29,104,110–113] the inhibition of ATP-sensitive K^+ channel (K-ATP) function,[15,30,35,110] and the stimulation of glucokinase activity.[17,18] Much of what is known concerning Epac action in the β cells has been revealed through the use of Epac-selective cAMP analogs (ESCAs). These analogs activate Epac proteins but not PKA when they are used at appropriately low concentrations.[107,114–116] Recently, new small molecules that selectively inhibit Epac activation have become available,[117–119] and new strains of Epac KO mice have been reported.[120–122]

Because Epac1 and Epac2 are unimolecular proteins that undergo major conformational changes upon binding of cAMP, they are suitable for the construction of FRET-based biosensors that contain the isolated cyclic nucleotide-binding domain (CNBD) or the full-length (FL) protein (Table 3.1). Nikolaev et al. designed Epac1-camps, a biosensor composed of EYFP-$Epac1_{CNBD}$-ECFP,[74] whereas DiPilato et al. designed ICUE1 composed of ECFP-$Epac1_{FL}$-Citrine.[77] Using a similar strategy, Ponsioen et al. designed CFP-$Epac1_{FL}$-YFP.[78] Other variants exist, including those based on Epac2 such as EYFP-$Epac2_{CNBD}$-ECFP,[74] ECFP-$Epac2_{FL}$-EYFP,[82] or Cerulean-$Epac2_{FL}$-Venus.[81] Second generation cAMP biosensors with improved properties include those with reduced pH sensitivity,[83] reduced ion sensitivity,[79] and wider dynamic range.[80] Additional features of these Epac-based biosensors are the inclusion of a PM targeting motif for the detection of cAMP at the PM,[73] or fusion of the biosensor with phosphodiesterases (PDEs) to monitor PDE activity.[123]

3.7 MODELING OSCILLATIONS OF cAMP AND Ca^{2+}

Landa et al. were the first to demonstrate that a cAMP biosensor could detect the dynamic processes of cAMP production and degradation in β cells.[56] Furthermore, Landa et al. were the first to demonstrate the dynamic interplay of cAMP and Ca^{2+} in β cells, as studied using Epacl-camps in combination with fura-2.[56,61] These findings of Landa et al. are incorporated into a mathematical model of second messenger interplay developed by Fridlyand et al., in which Ca^{2+} stimulates AC-8 and cyclic nucleotide PDEs to control cAMP synthesis and degradation.[61] The key findings of Landa et al. were that sustained depolarization with KCl leads to sustained Ca^{2+} influx and a sustained increase of cAMP that is explained by Ca^{2+}-dependent activation of AC-8.[56] However, periodic depolarization generates oscillatory Ca^{2+} influx that not only activates AC-8 but also activates PDEs.[56] Thus, under such conditions of periodic depolarization, the levels of cAMP and Ca^{2+} oscillate in a synchronous but antiphasic manner because Ca^{2+}-stimulated PDE activity predominates over Ca^{2+}-stimulated AC-8 activity.[61]

One interesting feature of the model devised by Fridlyand et al. is its prediction that GLP-1 will determine the phase relationships of cAMP and Ca^{2+} oscillations under conditions in which β cells are exposed to glucose. Because high concentrations of glucose induce periodic depolarization of β cells, the model of Fridlyand et al. predicts that glucose, alone, will stimulate synchronous antiphasic oscillations of cAMP and Ca^{2+} due to the fact that PDE activity is elevated.[61] However, such antiphasic oscillations will quickly transition to synchronous in-phase oscillations under conditions in which β cells are simultaneously exposed to GLP-1 and glucose. This GLP-1-induced reversal of the oscillatory pattern is explained by synergistic actions of GLP-1 and glucose to activate AC-8 so that cAMP production will prevail over PDE-catalyzed cAMP degradation. More specifically, synergistic activation of AC-8 is explained by direct binding of G_s proteins and Ca^{2+}/calmodulin to AC-8.[61]

3.8 HIGH-THROUGHPUT DETECTION OF cAMP

Chepurny et al. were the first to report that a cAMP biosensor can be stably expressed in a mammalian cell line so that levels of cAMP can be monitored in assays of FRET using a sensitive plate-reading spectrofluorimeter (FlexStation 3, Molecular Devices, Sunnyvale, CA).[107] This assay was first developed using subclones of INS-1 cells stably expressing Epacl-camps. One subclone designated as C-10 was selected after screening multiple subclones for a ΔFRET in response to cAMP-elevating agents (forskolin, IBMX). It was then established that this approach was applicable to INS-1 cells expressing AKAR3, and also in human embryonic kidney cells (HEK293) expressing Epacl-camps, AKAR3, FL-Epacl, and FL-Epac2.[117,119] A significant advance is our optimization of this assay so that GLP-1R agonists can be screened in a high-throughput mode using HEK293 cells coexpressing the human GLP-1R and either Epacl-camps or AKAR3. It is expected that this FRET-based assay will be applicable to many cell lines that express GPCRs positively linked to cAMP production. Furthermore, this assay will be useful for screens of small molecules that act intracellularly as cAMP agonists or antagonists.

3.9 INS-1 CELL CULTURE AND TRANSFECTION

INS-1 cells are maintained in a humidified incubator (95% air, 5% CO_2) at 37°C in RPMI 1640 medium containing 10 mM Hepes, 11.1 mM glucose, 10% FBS, 100 U mL^{-1} penicillin G, 100 μg mL^{-1} streptomycin, 2.0 mM L-glutamine, 1.0 mM sodium pyruvate, and 50 μM 2-mercaptoethanol.[39] INS-1 cells are passaged by trypsinization and subcultured once a week. The INS-1 cell clones stably expressing Epacl-camps are generated by G418 selection after transfection of cells with the Epacl-camps coding sequence in pcDNA3.1.[74] A primary screen is performed to identify an INS-1 cell clone that is highly responsive to cAMP-elevating agents forskolin (2 μM) and IBMX (100 μM). Fluorescence microscopy is used to validate cytosolic expression of Epacl-camps in all cells of an individual clone.

3.10 SPECTROFLUORIMETRY

INS-1 cell clones expressing Epacl-camps are plated at 80% confluence on 96-well clear-bottomed assay plates (Costar 3904). The cells are grown in culture overnight prior to their use. Assays are performed using a FlexStation 3 microplate reader equipped with excitation and emission monochromators, and controlled using SoftMax Pro software (Molecular Devices). On the day of the experiment, the culture medium is replaced with 170 μL/well of a standard extracellular saline (SES) solution containing (in millimoles): 138 NaCl, 5.6 KCl, 2.6 $CaCl_2$, 1.2 $MgCl_2$, 11.1 glucose, 0.1% DMSO, and 10 Hepes (295 mOsm; pH 7.4). The excitation light is delivered at 435 nm (455 nm cutoff), and the emitted light is detected at 485 nm (CFP) or 535 nm (YFP). The excitation light source is a Xenon flash lamp, and the emission intensities are averages of 12 excitation flashes for each time point. Test solutions containing cAMP-elevating agents are first dissolved in SES solution containing 0.1% DMSO. The test solutions are then aliquoted into individual wells of V-bottomed 96-well plates (Greiner). An automated pipetting procedure is used to transfer 30 μL of each test solution to the assay plate containing the cells to be studied. The test solutions are injected into each well at a pipette height that corresponded to a fluid level of 150 μL, and the rate of injection was 31 μL/s. The CFP/YFP emission ratio is calculated for each well, and values for 8 to 12 wells are averaged for each time point. A running average algorithm embedded within SoftMax Pro is used to obtain time point-to-time point smoothing of the kinetic determinations of FRET over a time course of 400 s. Values of FRET ratios for each time point are quantified as the mean ± SEM. The time course of the change of FRET ratio is plotted after exporting these data to Origin v8.0 (OriginLab). Data are processed in a spreadsheet format for graphical display. Publication quality artwork is generated by exporting these data as Adobe Illustrator files for processing in a desktop publishing format.

3.11 LIVE-CELL IMAGING

Live-cell imaging is used to obtain independent confirmation of findings obtained in plate reader assays. Clonal C-10 INS-1 cells expressing Epacl-camps are plated onto glass coverslips and maintained in culture medium overnight. On the day of the experiment, the coverslips are mounted in an imaging chamber for perifusion with SES containing 0.1% DMSO and various test solutions. Imaging of cells is performed using a Nikon Ti inverted microscope equipped with a N.A. 1.45 TIRF objective (60×), a Photometrics Cascade 512b EMCCD camera (Roper Scientific), and a Chameleon-2 filter set (Chroma Technology Corp.) composed of a D440/20 excitation filter, a 455DCLP dichroic, and D485/40 (CFP) or D535/30 (YFP) emission filters.[25] The excitation light source is a DeltaRam X monochromator (Photon Technology Int.). Ratiometric analysis of the emitted light corresponding to fluorescence originating within a defined region of the cytoplasm was performed using Metafluor v.7.5 software (Molecular Devices).

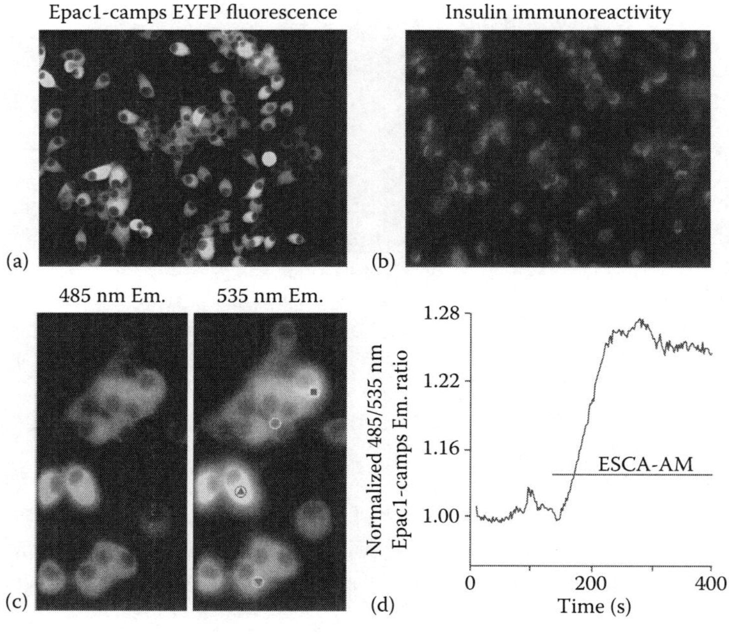

FIGURE 3.2 **(See color insert.)** INS-1 cell clone C-10 stably expresses Epacl-camps, as detected using an EYFP filter set (a). These cells also express insulin immunoreactivity that is detectable by fluorescence immunocytochemistry using an anti-insulin primary antiserum in combination with an Alexa Fluor 594 conjugated secondary antiserum (b). Live-cell imaging of C-10 cells using 440 nm excitation light allows detection of Epacl-camps in which the donor ECFP emission fluorescence is measured at 485 nm, whereas the acceptor EYFP emission fluorescence is measured at 535 nm (c). Activation of Epacl-camps by the Epac-selective cAMP analog 8-pCPT-2′-*O*-Me-cAMP-AM (ESCA-AM, bath application indicated by horizontal line) results in a decrease of FRET that is measured within defined regions of interest (circles over individual cells) as an increase of the 485/535 nm emission ratio (d).

3.12 Epac1-camps VALIDATION

Epac1-camps is stably expressed in monolayers of clone C-10 INS-1 cells that are insulin immunoreactive (Figure 3.2a and b). Activation of Epac1-camps is monitored by live-cell imaging of C-10 cells treated with 8-pCPT-2′-*O*-Me-cAMP-AM applied by bath perifusion (Figure 3.2c and d). This ESCA-AM ester is a prodrug that is highly membrane-permeable and that must undergo intracellular bioactivation so that it can activate Epac proteins.[107,116] Bioactivation of the ESCA-AM is catalyzed by cytosolic esterases that remove the AM moiety so that free 8-pCPT-2′-*O*-Me-cAMP is available.[115] When C-10 cell monolayers are tested in a high-throughput mode using a FlexStation 3 plate reader, Epac1-camps is also activated by 8-pCPT-2′-*O*-Me-cAMP-AM (Figure 3.3a–d). This action of the ESCA-AM is not mimicked by a vehicle solution composed of SES containing 0.1% DMSO (Figure 3.3a) or the negative control AM_3-ester (Figure 3.3d).

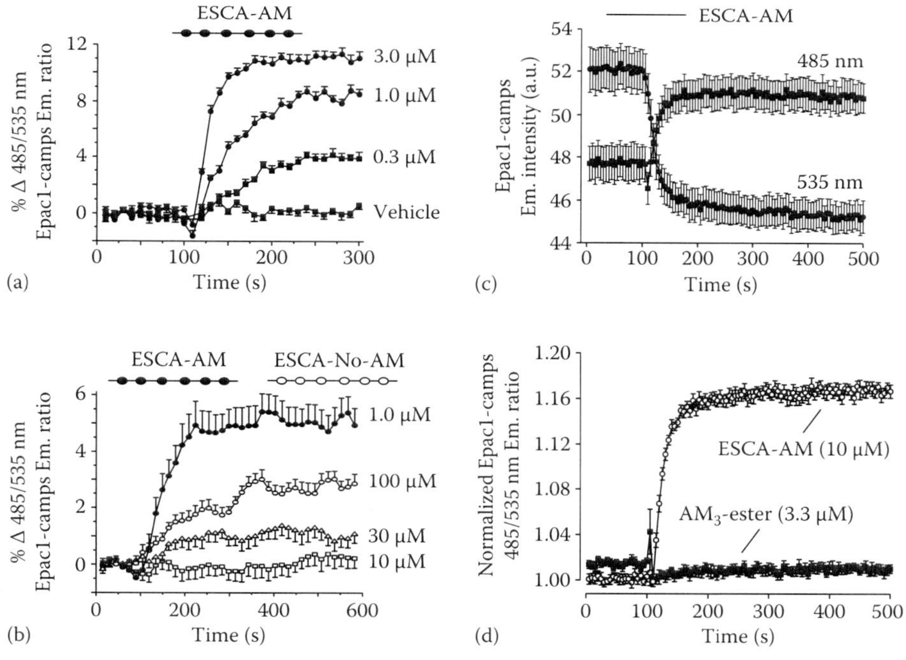

FIGURE 3.3 A FRET-based plate reader assay demonstrates that 8-pCPT-2′-*O*-Me-cAMP-AM (ESCA-AM) dose-dependently activates Epac1-camps in INS-1 cell clone C-10 (a). 8-pCPT-2′-*O*-Me-cAMP not conjugated to the AM-ester has less membrane permeability and a reduced capacity to activate Epac1-camps (b). Epac1-camps activation results in a ΔFRET that is measurable as a decrease of EYFP fluorescence (535 nm) and an increase of ECFP fluorescence (485 nm) (c). The ESCA-AM contains 1 mol of AM-ester per mole of prodrug, and it activates Epac1-camps when tested at a concentration of 10 μM. However, the negative control AM_3-ester contains 3 mol of AM-ester per mole of prodrug and it has no effect when tested at 3.3 μM (d). (Data shown in a and b were originally reported by Chepurny, O.G. et al., *Journal of Biological Chemistry* 284: 10728, 2009.)

8-pCPT-2′-*O*-Me-cAMP that is not conjugated to the AM-ester is at least 100-fold less potent as an activator of Epacl-camps in the FRET assay (Figure 3.3b). This finding is expected because 8-pCPT-2′-*O*-Me-cAMP has limited membrane permeability.[115] However, it can be argued that the action of 8-pCPT-2′-*O*-Me-cAMP-AM reported here is simply an artifact because bioactivation of the analog will generate acetic acid that lowers the cytosolic pH.[115] Because the acceptor chromophore in Epacl-camps is EYFP, it is possible that lowering the pH will quench EYFP fluorescence, thereby yielding an artifactual decrease of FRET unrelated to Epacl-camps activation.[124] This scenario is unlikely because extensive testing of C-10 cells demonstrates that unconjugated AM_3-ester has no effect at Epacl-camps (Figure 3.3d), despite the fact that it is a substrate for cytosolic esterases.[107,116]

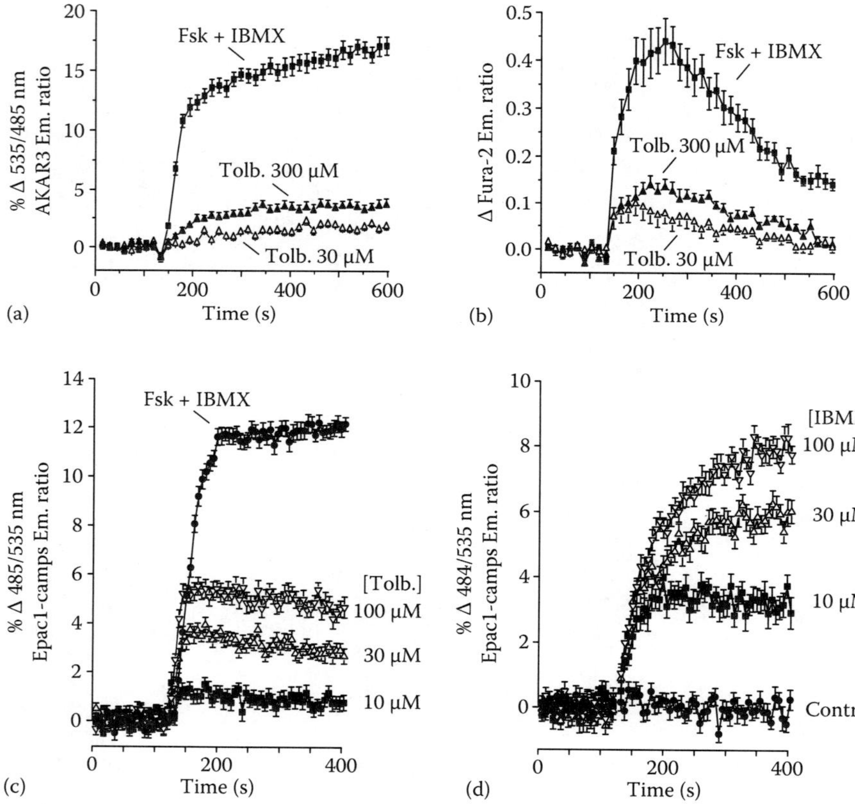

FIGURE 3.4 FRET-based plate reader assays demonstrate that AKAR3 is strongly activated by forskolin (2 μM) and IBMX (100 μM) in INS-1 cells virally transduced with this biosensor. However, tolbutamide produces a weaker effect (a). Fura-2-based plate reader assays performed according to Leech et al.[110] demonstrate that forskolin, IBMX, and tolbutamide also increase the $[Ca^{2+}]_i$ in INS-1 cells (b). When the FRET assay is performed using INS-1 cell clone C-10, Epacl-camps is also activated by forskolin, IBMX, and tolbutamide (c). Note that IBMX alone dose-dependently activates Eapcl-camps in C-10 cells (d).

3.13 AKAR3 VALIDATION

INS-1 cells virally transduced with AKAR3 using the methods of Chepurny et al.[25] provide an alternative means by which to monitor cAMP signaling because AKAR3 detects PKA-mediated phosphorylation. AKAR3 is activated by forskolin and IBMX (Figure 3.4a), and this activation is blocked by 10 μM of the PKA inhibitor H-89 (data not shown). Surprisingly, a high concentration (300 μM) of the sulfonylurea tolbutamide also activates AKAR3 (Figure 3.4a). This action of tolbutamide is accompanied by an increase of $[Ca^{2+}]_i$ in cells loaded with the Ca^{2+} indicator fura-2 (Figure 3.4b). Furthermore, we find that tolbutamide similarly activates Epac1-camps in the INS-1 cell clone C-10 (Figure 3.4c). Such findings seem to indicate that tolbutamide indirectly activates AKAR3 and Epac1-camps by raising the levels of cAMP. This concept is consistent with one prior report in which a high concentration of tolbutamide inhibits PDE activity in lysates derived from islets.[125] For our INS-1 cells, this might also be the case because Epac1-camps is activated by the PDE inhibitor IBMX (Figure 3.4d). Ongoing efforts seek to determine if sulfonylureas exert additional direct actions at cAMP biosensors,[81,82] even though this possibility is currently disputed.[126,127]

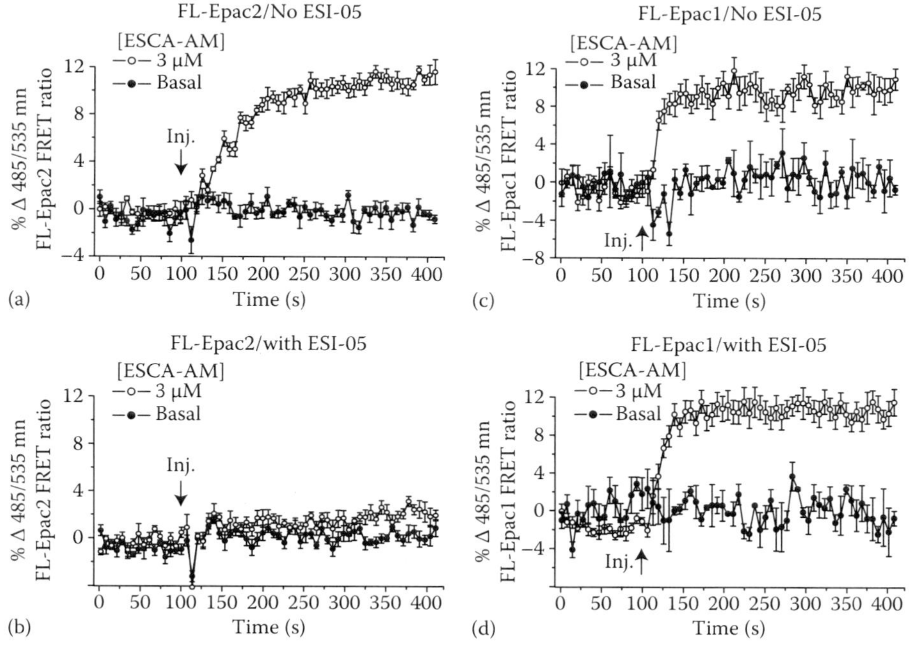

FIGURE 3.5 A full-length Epac2 biosensor (FL-Epac2) stably expressed in HEK293 cell clone C-1 is activated by 3 μM of the ESCA-AM (a), and this activation is blocked by pretreatment of cells with 10 μM of the Epac2-specific inhibitor ESI-05 (b). No such inhibitory action of ESI-05 is measured when a FL-Epac1 biosensor is activated by the ESCA-AM in HEK293 cell clone C-9 (c, d). (All data shown in a to d were originally reported by Tsalkova, T. et al., *Proceedings of the National Academy of Sciences of the United States of America* 109: 18613, 2012.)

3.14 STUDIES WITH HEK293 CELLS

HEK293 cell clones that stably express AKAR3, FL-Epacl, FL-Epac2, and Epacl-camps biosensors are new tools we created to enable high-throughput drug discovery efforts.[117] Evaluation of clones expressing FL-Epacl (C-9) or FL-Epac2 (C-1) biosensors reveals that these biosensors are activated by forskolin and IBMX, and also by the selective Epac activator 8-pCPT-2′-*O*-Me-cAMP-AM. Remarkably, we find that it is possible to inhibit Epac2 activation using ESI-05, a methylphenylsulfone first discovered by Cheng and coworkers.[117] For example, in the FRET assay, ESI-05 blocks FL-Epac2 but not FL-Epacl biosensor activation (Figure 3.5a–d). Independent confirmation of this finding is provided by live-cell imaging assays of HEK293 cells stably expressing the FL-Epac2 biosensor (Figure 3.6a and b). Recently, Rehmann[127] used an in vitro Rapl activation assay to independently confirm the selectivity with which ESI-05 blocks Epac2 activation.

Finally, we report that it is possible to monitor GPCR activation in the plate reader FRET assay using HEK293 cells virally transduced with cAMP biosensors. For example, we use Epacl-camps adenovirus to transduce HEK293 cells that stably

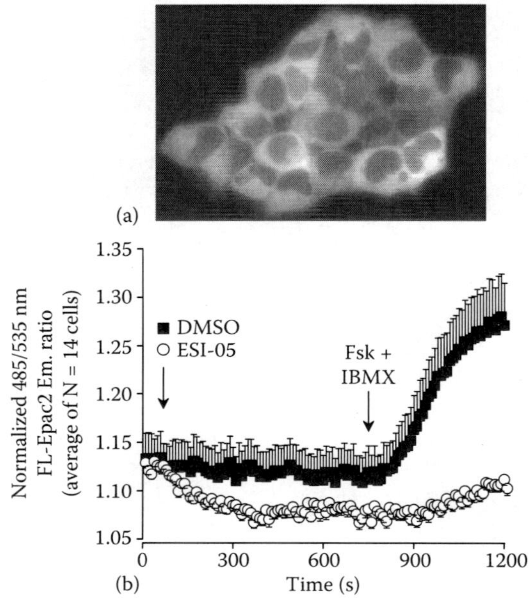

FIGURE 3.6 HEK293 cell clone C-1 expresses the FL-Epac2 biosensor, as imaged in a cluster of cells (a). Live-cell imaging demonstrates that 10 μM ESI-05 blocks activation of FL-Epac2 by 2 μM forskolin and 100 μM IBMX (b). Arrows indicate the times at which test substances were administered by bath perifusion. (All data shown in a and b were originally reported by Tsalkova, T. et al., *Proceedings of the National Academy of Sciences of the United States of America* 109: 18613, 2012.)

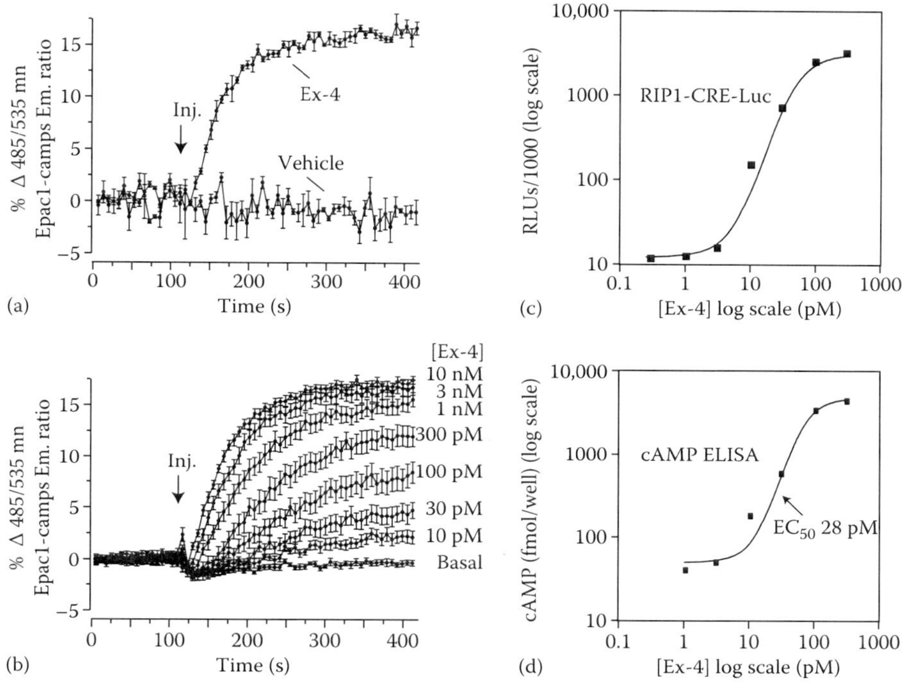

FIGURE 3.7 FRET-based plate reader assays of HEK293 cells stably expressing the GLP-1R and virally transduced with Epacl-camps demonstrate that the GLP-1R agonist exendin-4 (Ex-4, 10 nM) dose-dependently activates Epacl-camps (a, b). Luciferase assays demonstrate that Exendin-4 dose-dependently stimulates CRE-dependent gene expression, as measured in HEK293/GLP-1R cells transfected with RIP1-CRE-Luc (c). ELISA assays demonstrate the exendin-4 dose-dependently stimulates cAMP production in HEK293/GLP-1R cells (d).

express the human GLP-1R.[128] In our FRET assay, the GLP-1R agonist exendin-4 (Ex-4) dose-dependently activates Epacl-camps (Figure 3.7a and b). This action of exendin-4 is most likely cAMP-mediated because exendin-4 also stimulates cAMP-dependent gene transcription that is measurable in HEK293/GLP-1R cells transfected with RIP1-CRE-Luc,[128] a luciferase reporter that incorporates a cAMP response element (Figure 3.7c). Furthermore, an enzyme-linked immunosorbent assay (ELISA) reveals that exendin-4 increases the levels of cAMP in these cells (Figure 3.7d). A cAMP-dependent action of exendin-4 is also evident in HEK293/GLP-1R cells that are virally transduced with AKAR3. In these cells, AKAR3 activation occurs in response to forskolin and IBMX (Figure 3.8a) as well as in exendin-4 (Figure 3.8b). Such actions of forskolin, IBMX, and exendin-4 are nearly eliminated by the PKA inhibitor H-89 (Figure 3.8a, c, and d).

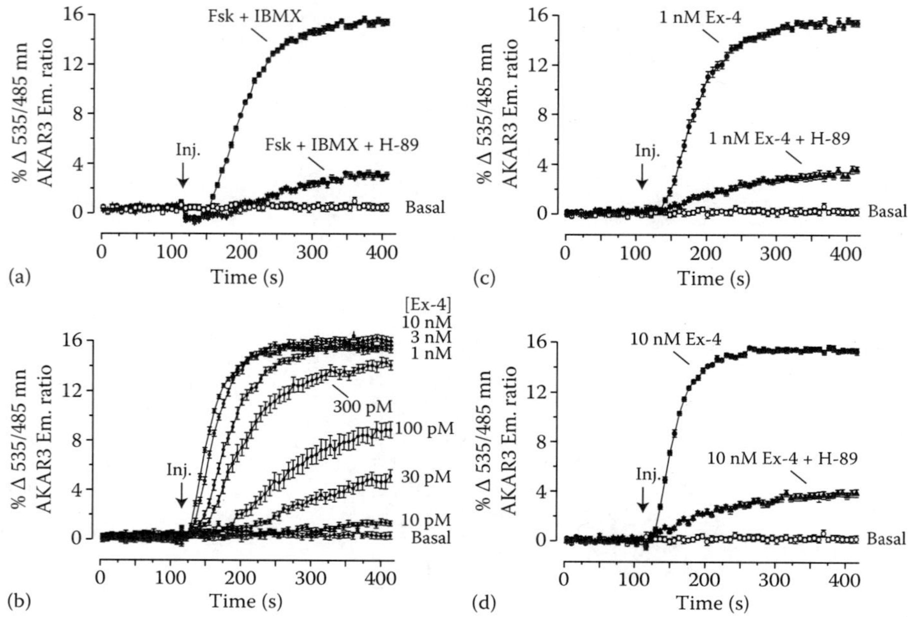

FIGURE 3.8 FRET-based plate reader assays of HEK293 cells stably expressing the GLP-1R and virally transduced with AKAR3 demonstrate that the PKA inhibitor H-89 (10 μM) blocks the actions of forskolin (Fsk, 2 μM) and IBMX (100 μM) to activate AKAR3 (a). The GLP-1R agonist exendin-4 (Ex-4) dose-dependently activates AKAR3 in HEK293/GLP-1R cells (b), and this action of Ex-4 is blocked by 10 μM of the PKA inhibitor H-89 (c, d).

3.15 CONCLUSION

Numerous cAMP biosensors are now available for the assessment of cAMP dynamics in living cells. To date, nearly all published studies using these biosensors have relied on live-cell imaging, as performed using EMCCD cameras interfaced with optics that allow wide-field fluorescence microscopy, confocal microscopy, and TIRF microscopy. Although such approaches allow cAMP to be monitored with fast temporal and high spatial resolution, the use of microscopy is not optimal for assays in which small molecule libraries are screened in a high-throughput mode. Ideally, high-throughput screening should instead be performed in a multiwell format using automated injection of test solutions. Unfortunately, transient transfection with cAMP biosensors does not afford sufficient efficiency to perform low-light level detection of FRET in monolayers of cells. To circumvent this limitation, we generated INS-1 and HEK293 cell clones that stably express AKAR3, Epac1-camps, FL-Epac1, and FL-Epac2 biosensors.[117] Using these clonal cell lines, we demonstrated that it is possible to use a plate-reading spectrofluorimeter in combination with automated injection protocols to perform accurate time-dependent measurements of FRET. Thus, all cAMP biosensors we tested are highly responsive to cAMP-elevating agents, and we demonstrated that the biosensors Epac1-camps and AKAR3 detect cAMP production in response to GLP-1R activation. Promising areas of future investigation include screens for GLP-1R agonists,[129–131] or screens for molecules that target PKA, Epac1, and Epac2[132–134] Potentially, small molecules might also be identified that target oscillatory circuits in which cAMP, PKA, and Ca^{2+} control insulin secretion.[135] A recent study indicates a prominent role for AC-5 in the control of human islet insulin secretion by glucose,[136] thus this AC isoform might also constitute a relevant target. We conclude that the future remains bright for fluorescence-based assays of cAMP signal transduction as it pertains to small molecule drug discovery.

AUTHOR CONTRIBUTIONS

G.G. Holz wrote the manuscript. C.A. Leech, O.G. Chepurny, and M.W. Roe edited the manuscript. G.G. Holz serves as guarantor of this work.

ACKNOWLEDGMENTS

This work was supported by American Diabetes Association Basic Science Awards to G.G. Holz (7-12-BS-077) and C.A. Leech (1-12-BS-109). M.W. Roe acknowledges the support of the National Institutes of Health (R01-DK092616). O.G. Chepurny acknowledges institutional support from SUNY Upstate Medical University.

REFERENCES

1. Seino, S., T. Shibasaki, and K. Minami. 2011. Dynamics of insulin secretion and the clinical implications for obesity and diabetes. *Journal of Clinical Investigation* 121: 2118.
2. Pessin, J.E., and A.R. Saltiel. 2000. Signaling pathways in insulin action: Molecular targets of insulin resistance. *Journal of Clinical Investigation* 106: 165.
3. Stumvoll, M., B.J. Goldstein, and T.W. van Haeften. 2005. Type 2 diabetes: Principles of pathogenesis and therapy. *Lancet* 365: 1333.
4. Ahrén, B. 2009. Islet G protein-coupled receptors as potential targets for treatment of type 2 diabetes. *Nature Reviews: Drug Discovery* 8: 369.
5. Furman, B., N. Pyne, P. Flatt, and F. O'Harte. 2004. Targeting beta-cell cyclic 3′5′ adenosine monophosphate for the development of novel drugs for treating type 2 diabetes mellitus. A review. *Journal of Pharmacy and Pharmacology* 56: 1477.
6. Henquin, J.C. 2000. Triggering and amplifying pathways of regulation of insulin secretion by glucose. *Diabetes* 49: 1751.
7. Winzell, M.S., and B. Ahrén. 2007. G-protein-coupled receptors and islet function—Implications for treatment of type 2 diabetes. *Pharmacology & Therapeutics* 116: 437.
8. Meier, J.J. 2012. GLP-1 receptor agonists for individualized treatment of type 2 diabetes mellitus. *Nature Reviews: Endocrinology* 8: 728.
9. Holz, G.G., and O.G. Chepurny. 2003. Glucagon-like peptide-1 synthetic analogs: New therapeutic agents for use in the treatment of diabetes mellitus. *Current Medicinal Chemistry* 10: 2471.
10. Nadkarni, P., O.G. Chepurny, and G.G. Holz. 2012. Regulation of glucose homeostasis by GLP-1. *Progress in Molecular Biology and Translational Science* 121: 23.

11. Thorens, B. 1992. Expression cloning of the pancreatic β cell receptor for the glucoincretin hormone glucagon-like peptide-1. *Proceedings of the National Academy of Sciences of the United States of America* 89: 8641.
12. Holz, G.G. 2004. Epac: A new cAMP-binding protein in support of glucagon-like peptide-1 receptor-mediated signal transduction in the pancreatic β-cell. *Diabetes* 53: 5.
13. Holz, G.G., G. Kang, M. Harbeck, M.W. Roe, and O.G. Chepurny. 2006. Cell physiology of cAMP sensor Epac. *Journal of Physiology* 577: 5.
14. Leech, C.A., O.G. Chepurny, and G.G. Holz. 2010. Epac2-dependent Rap1 activation and the control of islet insulin secretion by glucagon-like peptide-1. *Vitamins & Hormones* 84: 279.
15. Leech, C.A., I. Dzhura, O.G. Chepurny, G. Kang, F. Schwede, H.G. Genieser et al. 2011. Molecular physiology of glucagon-like peptide-1 insulin secretagogue action in pancreatic β cells. *Progress in Biophysics and Molecular Biology* 107: 236.
16. Holz, G.G., and J.F. Habener. 1992. Signal transduction crosstalk in the endocrine system: Pancreatic β-cells and the glucose competence concept. *Trends in Biochemical Sciences* 17: 388.
17. Ding, S.Y., A. Nkobena, C.A. Kraft, M.L. Markwardt, and M.A. Rizzo. 2011. Glucagon-like peptide 1 stimulates post-translational activation of glucokinase in pancreatic β cells. *Journal of Biological Chemistry* 286: 16768.
18. Park, J.H., S.J. Kim, S.H. Park, D.G. Son, J.H. Bae, H.K. Kim et al. 2012. Glucagon-like peptide-1 enhances glucokinase activity in pancreatic β-cells through the association of Epac2 with Rim2 and Rab3A. *Endocrinology* 153: 574.
19. Holz, G.G. 4th, W.M. Kühtreiber, and J.F. Habener. 1993. Pancreatic beta-cells are rendered glucose-competent by the insulinotropic hormone glucagon-like peptide-1(7–37). *Nature* 361: 362.
20. Dachicourt, N., P. Serradas, D. Bailbé, M. Kergoat, L. Doaré, and B. Portha. 1997. Glucagon-like peptide-1(7–36)-amide confers glucose sensitivity to previously glucose-incompetent β-cells in diabetic rats: In vivo and in vitro studies. *Journal of Endocrinology* 155: 369.
21. Cornu, M., H. Modi, D. Kawamori, R.N. Kulkarni, M. Joffraud, and B. Thorens. 2010. Glucagon-like peptide-1 increases β-cell glucose competence and proliferation by translational induction of insulin-like growth factor-1 receptor expression. *Journal of Biological Chemistry* 285: 10538.
22. Amisten, S., A. Salehi, P. Rorsman, P.M. Jones, and S.J. Persaud. 2013. An atlas and functional analysis of G-protein coupled receptors in human islets of Langerhans. *Pharmacology & Therapeutics* 139: 359.
23. Sharp, G.W. 1996. Mechanisms of inhibition of insulin release. *American Journal of Physiology—Cell Physiology* 271: C1781.
24. Kelley, G.G., O.G. Chepurny, F. Schwede, H.G. Genieser, C.A. Leech, M.W. Roe et al. 2009. Glucose-dependent potentiation of mouse islet insulin secretion by Epac activator 8-pCPT-2′-*O*-Me-cAMP-AM. *Islets* 1: 260.
25. Chepurny, O.G., G.G. Kelley, I. Dzhura, C.A. Leech, M.W. Roe, E. Dzhura et al. 2010. PKA-dependent potentiation of glucose-stimulated insulin secretion by Epac activator 8-pCPT-2′-*O*-Me-cAMP-AM in human islets of Langerhans. *American Journal of Physiology—Endocrinology and Metabolism* 298: E622.
26. Dzhura, I., O.G. Chepurny, C.A. Leech, M.W. Roe, E. Dzhura, X. Xu et al. 2011. Phospholipase Cε links Epac2 activation to the potentiation of glucose-stimulated insulin secretion from mouse islets of Langerhans. *Islets* 3: 121.
27. Renström, E., L. Eliasson, and P. Rorsman. 1997. Protein kinase A-dependent and -independent stimulation of exocytosis by cAMP in mouse pancreatic B-cells. *Journal of Physiology* 502: 105.

28. Eliasson, L., X. Ma, E. Renström, S. Barg, P.O. Berggren, J. Galvanovskis et al. 2003. SUR1 regulates PKA-independent cAMP-induced granule priming in mouse pancreatic B-cells. *Journal of General Physiology* 121: 181.
29. Kang, G., O.G. Chepurny, M.J. Rindler, L. Collis, Z. Chepurny, W.H. Li et al. 2005. A cAMP and Ca^{2+} coincidence detector in support of Ca^{2+}-induced Ca^{2+} release in mouse pancreatic β cells. *Journal of Physiology* 566: 173.
30. Kang, G., O.G. Chepurny, B. Malester, M.J. Rindler, H. Rehmann, J.L. Bos et al. 2006. cAMP sensor Epac as a determinant of ATP-sensitive potassium channel activity in human pancreatic β cells and rat INS-1 cells. *Journal of Physiology* 573: 595.
31. Hatakeyama, H., T. Kishimoto, T. Nemoto, H. Kasai, and N. Takahashi. 2006. Rapid glucose sensing by protein kinase A for insulin exocytosis in mouse pancreatic islets. *Journal of Physiology* 570: 271.
32. Hatakeyama, H., N. Takahashi, T. Kishimoto, T. Nemoto, and H. Kasai. 2007. Two cAMP-dependent pathways differentially regulate exocytosis of large dense-core and small vesicles in mouse β-cells. *Journal of Physiology* 582: 1087.
33. Shibasaki, T., H. Takahashi, T. Miki, Y. Sunaga, K. Matsumura, M. Yamanaka et al. 2007. Essential role of Epac2/Rap1 signaling in regulation of insulin granule dynamics by cAMP. *Proceedings of the National Academy of Sciences of the United States of America* 104: 19333.
34. Kwan, E.P., L. Xie, L. Sheu, T. Ohtsuka, and H.Y. Gaisano. 2007. Interaction between Munc13-1 and RIM is critical for glucagon-like peptide-1 mediated rescue of exocytotic defects in Munc13-1 deficient pancreatic β-cells. *Diabetes* 56: 2579.
35. Kang, G., C.A. Leech, O.G. Chepurny, W.A. Coetzee, and G.G. Holz. 2008. Role of the cAMP sensor Epac as a determinant of K_{ATP} channel ATP sensitivity in human pancreatic β-cells and rat INS-1 cells. *Journal of Physiology* 586: 1307.
36. Dzhura, I., O.G. Chepurny, G.G. Kelley, C.A. Leech, M.W. Roe, E. Dzhura et al. 2010. Epac2-dependent mobilization of intracellular Ca^{2+} by glucagon-like peptide-1 receptor agonist exendin-4 is disrupted in β-cells of phospholipase Cε knockout mice. *Journal of Physiology* 588: 4871.
37. Skelin, M., and M. Rupnik. 2011. cAMP increases the sensitivity of exocytosis to Ca^{2+} primarily through protein kinase A in mouse pancreatic beta cells. *Cell Calcium* 49: 89.
38. Skelin, M., M. Rupnik, and A. Cencic. 2010. Pancreatic beta cell lines and their applications in diabetes mellitus research. *Alternatives to Animal Experimentation* 27: 105.
39. Asfari, M., D. Janjic, P. Meda, G. Li, P.A. Halban, and C.B. Wollheim. 1992. Establishment of 2-mercaptoethanol-dependent differentiated insulin-secreting cell lines. *Endocrinology* 130: 167.
40. Ishihara, H., T. Asano, K. Tsukuda, H. Katagiri, K. Inukai, M. Anai et al. 1993. Pancreatic beta cell line MIN6 exhibits characteristics of glucose metabolism and glucose-stimulated insulin secretion similar to those of normal islets. *Diabetologia* 36: 1139.
41. Santerre, R.F., R.A. Cook, R.M. Crisel, J.D. Sharp, R.J. Schmidt, D.C. Williams et al. 1981. Insulin synthesis in a clonal cell line of simian virus 40-transformed hamster pancreatic beta cells. *Proceedings of the National Academy of Sciences of the United States of America* 78: 4339.
42. McClenaghan, N.H., and P.R. Flatt. 1999. Engineering cultured insulin-secreting pancreatic B-cell lines. *Journal of Molecular Medicine (Berlin, Germany)* 77: 235.
43. Ravassard, P., Y. Hazhouz, S. Pechberty, E. Bricout-Neveu, M. Armanet, P. Czernichow et al. 2011. A genetically engineered human pancreatic β cell line exhibiting glucose-inducible insulin secretion. *Journal of Clinical Investigation* 121: 3589.
44. Scharfmann, R., S. Pechberty, Y. Hazhouz, M. von Bülow, E. Bricout-Neveu, M. Grenier-Godard et al. 2014. Development of a conditionally immortalized human pancreatic β cell line. *Journal of Clinical Investigation* 124: 2087.

45. Li, H., A. Lam, A.M. Xu, K.S. Lam, and S.K. Chung. 2010. High dosage of exendin-4 increased early insulin secretion in differentiated beta cells from mouse embryonic stem cells. *Acta Pharmacologica Sinica* 31: 570.
46. Jiang, W., Y. Shi, D. Zhao, S. Chen, J. Yong, J. Zhang et al. 2007. In vitro derivation of functional insulin-producing cells from human embryonic stem cells. *Cell Research* 17: 333.
47. Bose, B., S.P. Shenoy, S. Konda, and P. Wangikar. 2012. Human embryonic stem cell differentiation into insulin secreting β-cells for diabetes. *Cell Biology International* 36: 1013.
48. Minami, K., H. Yano, T. Miki, K. Nagashima, C.Z. Wang, H. Tanaka et al. 2000. Insulin secretion and differential gene expression in glucose-responsive and -unresponsive MIN6 sublines. *American Journal of Physiology—Endocrinology and Metabolism* 279: E773.
49. Cheng, K., V. Delghingaro-Augusto, C.J. Nolan, N. Turner, N. Hallahan, S. Andrikopoulos et al. 2012. High passage MIN6 cells have impaired insulin secretion with impaired glucose and lipid oxidation. *PLoS One* 7: e40868. doi: 10.1371/journal.pone.0040868.
50. Chepurny, O.G., and G.G. Holz. 2002. Over-expression of the glucagon-like peptide-1 receptor on INS-1 cells confers autocrine stimulation of insulin gene promoter activity: A strategy for production of pancreatic β-cell lines for use in transplantation. *Cell and Tissue Research* 307: 191.
51. Nakashima, K., Y. Kanda, Y. Hirokawa, F. Kawasaki, M. Matsuki, and K. Kaku. 2009. MIN6 is not a pure beta cell line but a mixed cell line with other pancreatic endocrine hormones. *Endocrine Journal* 56: 45.
52. Janjic, D., P. Maechler, N. Sekine, C. Bartley, A.S. Annen, and C.B. Wolheim. 1999. Free radical modulation of insulin release in INS-1 cells exposed to alloxan. *Biochemical Pharmacology* 57: 639.
53. Hohmeier, H.E., H. Mulder, G. Chen, R. Henkel-Rieger, M. Prentki, and C.B. Newgard. 2000. Isolation of INS-1-derived cell lines with robust ATP-sensitive K^+ channel-dependent and -independent glucose-stimulated insulin secretion. *Diabetes* 49: 424.
54. Iwasaki, M., K. Minami, T. Shibasaki, T. Miki, J. Miyazaki, and S. Seino. 2010. Establishment of new clonal pancreatic β-cell lines (MIN6-K) useful for study of incretin/cyclic adenosine monophosphate signaling. *Journal of Diabetes Investigation* 1: 137.
55. Leech, C.A., M.A. Castonguay, and J.F. Habener. 1999. Expression of adenylyl cyclase subtypes in pancreatic β-cells. *Biochemical and Biophysical Research Communications* 254: 703.
56. Landa, L.R. Jr., M. Harbeck, K. Kaihara, O. Chepurny, K. Kitiphongspattana, O. Graf et al. 2005. Interplay of Ca^{2+} and cAMP signaling in the insulin-secreting MIN6 β-cell line. *Journal of Biological Chemistry* 280: 31294.
57. Guenifi, A., G.M. Portela-Gomes, L. Grimelius, S. Efendić, and S.M. Abdel-Halim. 2000. Adenylyl cyclase isoform expression in non-diabetic and diabetic Goto–Kakizaki (GK) rat pancreas. Evidence for distinct overexpression of type-8 adenylyl cyclase in diabetic GK rat islets. *Histochemistry and Cell Biology* 113: 81.
58. Delmeire, D., D. Flamez, S.A. Hinke, J.J. Cali, D. Pipeleers, and F. Schuit. 2003. Type VIII adenylyl cyclase in rat beta cells: Coincidence signal detector/generator for glucose and GLP-1. *Diabetologia* 46: 1383.
59. Roger, B., J. Papin, P. Vacher, M. Raoux, A. Mulot, M. Dubois et al. 2011. Adenylyl cyclase 8 is central to glucagon-like peptide 1 signalling and effects of chronically elevated glucose in rat and human pancreatic beta cells. *Diabetologia* 54: 390.
60. Everett, K.L., and D.M. Cooper. 2013. An improved targeted cAMP sensor to study the regulation of adenylyl cyclase 8 by Ca^{2+} entry through voltage-gated channels. *PLoS One* 8: e75942.

61. Fridlyand, L.E., M.C. Harbeck, M.W. Roe, and L.H. Philipson. 2007. Regulation of cAMP dynamics by Ca^{2+} and G protein-coupled receptors in the pancreatic β-cell: A computational approach. *American Journal of Physiology—Cell Physiology* 293: C1924.
62. Holz, G.G., E. Heart, and C.A. Leech. 2008. Synchronizing Ca^{2+} and cAMP oscillations in pancreatic β-cells: A role for glucose metabolism and GLP-1 receptors? Focus on "regulation of cAMP dynamics by Ca^{2+} and G protein-coupled receptors in the pancreatic β-cell: A computational approach." *American Journal of Physiology—Cell Physiology* 294: C4.
63. Takeda, Y., A. Amano, A. Noma, Y. Nakamura, S. Fujimoto, and N. Inagaki. 2011. Systems analysis of GLP-1 receptor signaling in pancreatic β-cells. *American Journal of Physiology—Cell Physiology* 301: C792.
64. Ramos, L.S., J.H. Zippin, M. Kamenetsky, J. Buck, and L.R. Levin. 2008. Glucose and GLP-1 stimulate cAMP production via distinct adenylyl cyclases in INS-1E insulinoma cells. *Journal of General Physiology* 132: 329.
65. Zippin, J.H., Y. Chen, S.G. Straub, K.C. Hess, A. Diaz, D. Lee et al. 2013. CO_2/HCO_3^- and calcium-regulated soluble adenylyl cyclase as a physiological ATP sensor. *Journal of Biological Chemistry* 288: 33283.
66. Rahman, N., J. Buck, and L.R. Levin. 2013. pH sensing via bicarbonate-regulated *soluble* adenylyl cyclase (sAC). *Frontiers in Physiology* 4: 343.
67. Chang, J.C., and R.P. Oude-Elferink. 2014. Role of the bicarbonate-responsive soluble adenylyl cyclase in pH sensing and metabolic regulation. *Frontiers in Physiology* 5: 42.
68. Parkkila, A.K., A.L. Scarim, S. Parkkila, A. Waheed, J.A. Corbett, and W.S. Sly. 1998. Expression of carbonic anhydrase V in pancreatic beta cells suggests role for mitochondrial carbonic anhydrase in insulin secretion. *Journal of Biological Chemistry* 273: 24620.
69. Sener, A., H. Jijakli, S. Zahedi Asl, P. Courtois, A.P. Yates, S. Meuris et al. 2007. Possible role of carbonic anhydrase in rat pancreatic islets: Enzymatic, secretory, metabolic, ionic, and electrical aspects. *American Journal of Physiology—Endocrinology and Metabolism* 292: E1624.
70. Adams, S.R., A.T. Harootunian, Y.J. Buechler, S.S. Taylor, and R.Y. Tsien. 1991. Fluorescence ratio imaging of cyclic AMP in single cells. *Nature* 349: 694.
71. Zaccolo, M., F. De Giorgi, C.Y. Cho, L. Feng, T. Knapp, P.A. Negulescu et al. 2000. A genetically encoded, fluorescent indicator for cyclic AMP in living cells. *Nature Cell Biology* 2: 25.
72. Allen, M.D., and J. Zhang. 1998. Subcellular dynamics of protein kinase A activity visualized by FRET-based reporters. *Biochemical and Biophysical Research Communications* 348: 716.
73. Depry, C., M.D. Allen, and J. Zhang. 2011. Visualization of PKA activity in plasma membrane microdomains. *Molecular BioSystems* 7: 52.
74. Nikolaev, V.O., M. Bünemann, L. Hein, A. Hannawacker, and M.J. Lohse. 2004. Novel single chain cAMP sensors for receptor-induced signal propagation. *Journal of Biological Chemistry* 279: 37215.
75. Dyachok, O., Y. Isakov, J. Sågetorp, and A. Tengholm. 2006. Oscillations of cyclic AMP in hormone-stimulated insulin-secreting β-cells. *Nature* 439: 349.
76. Dyachok, O., O. Idevall-Hagren, J. Sågetorp, G. Tian, A. Wuttke, C. Arrieumerlou et al. 2008. Glucose-induced cyclic AMP oscillations regulate pulsatile insulin secretion. *Cell Metabolism* 8: 26.
77. DiPilato, L.M., X. Cheng, and J. Zhang. 2004. Fluorescent indicators of cAMP and Epac activation reveal differential dynamics of cAMP signaling within discrete subcellular compartments. *Proceedings of the National Academy of Sciences of the United States of America* 101: 16513.

78. Ponsioen, B., J. Zhao, J. Riedl, F. Zwartkruis, G. van der Krogt, M. Zaccolo et al. 2004. Detecting cAMP-induced Epac activation by fluorescence resonance energy transfer: Epac as a novel cAMP indicator. *EMBO Reports* 5: 1176.
79. Salonikidis, P.S., M. Niebert, T. Ullrich, G. Bao, A. Zeug, and D.W. Richter. 2011. An ion-insensitive cAMP biosensor for long term quantitative ratiometric fluorescence resonance energy transfer (FRET) measurements under variable physiological conditions. *Journal of Biological Chemistry* 286: 23419.
80. Klarenbeek, J.B., J. Goedhart, M.A. Hink, T.W. Gadella, and K. Jalink. 2011. A mTurquoise-based cAMP sensor for both FLIM and ratiometric read-out has improved dynamic range. *PLoS One* 6: e19170.
81. Herbst, K.J., C. Coltharp, L.M. Amzel, and J. Zhang. 2011. Direct activation of Epac by sulfonylurea is isoform selective. *Chemistry and Biology* 18: 243.
82. Zhang, C.L., M. Katoh, T. Shibasaki, K. Minami, Y. Sunaga, H. Takahashi et al. 2009. The cAMP sensor Epac2 is a direct target of antidiabetic sulfonylurea drugs. *Science* 325: 607.
83. Everett, K.L., and D.M. Cooper. 2013. An improved targeted cAMP sensor to study the regulation of adenylyl cyclase 8 by Ca^{2+} entry through voltage-gated channels. *PLoS One* 8: e75942.
84. DeBernardi, M.A., and G. Brooker. 1996. Single cell Ca^{2+}/cAMP cross-talk monitored by simultaneous Ca^{2+}/cAMP fluorescence ratio imaging. *Proceedings of the National Academy of Sciences of the United States of America* 93: 4577.
85. Harris, T.E., S.J. Persaud, and P.M. Jones. 1997. Pseudosubstrate inhibition of cyclic AMP-dependent protein kinase in intact pancreatic islets: Effects on cyclic AMP-dependent and glucose-dependent insulin secretion. *Biochemical and Biophysical Research Communications* 232: 648.
86. Ding, W.G., and J. Gromada. 1997. Protein kinase A-dependent stimulation of exocytosis in mouse pancreatic β-cells by glucose-dependent insulinotropic polypeptide. *Diabetes* 46: 615.
87. Gromada, J., K. Bokvist, W.G. Ding, J.J. Holst, J.H. Nielsen, and P. Rorsman. 1998. Glucagon-like peptide 1 (7–36) amide stimulates exocytosis in human pancreatic β-cells by both proximal and distal regulatory steps in stimulus-secretion coupling. *Diabetes* 47: 57.
88. McQuaid, T.S., M.C. Saleh, J.W. Joseph, A. Gyulkhandanyan, J.E. Manning-Fox, J.D. MacLellan et al. 2006. cAMP-mediated signaling normalizes glucose-stimulated insulin secretion in uncoupling protein-2 overexpressing β-cells. *Journal of Endocrinology* 190: 669.
89. Goehring, A.S., B.S. Pedroja, S.A. Hinke, L.K. Langeberg, and J.D. Scott. 2007. MyRIP anchors protein kinase A to the exocyst complex. *Journal of Biological Chemistry* 282: 33155.
90. Idevall-Hagren, O., S. Barg, E. Gylfe, and A. Tengholm. 2010. cAMP mediators of pulsatile insulin secretion from glucose-stimulated single β-cells. *Journal of Biological Chemistry* 285: 23007.
91. Song, W.J., M. Seshadri, U. Ashraf, T. Mdluli, P. Mondal, M. Keil et al. 2011. Snapin mediates incretin action and augments glucose-dependent insulin secretion. *Cell Metabolism* 13: 308.
92. Li, N., B. Li, T. Brun, C. Deffert-Delbouille, Z. Mahiout, Y. Daali et al. 2012. NADPH oxidase NOX2 defines a new antagonistic role for reactive oxygen species and cAMP/PKA in the regulation of insulin secretion. *Diabetes* 61: 2842.
93. Nie, J., B.N. Lilley, Y.A. Pan, O. Faruque, X. Liu, W. Zhang et al. 2013. SAD-A potentiates glucose-stimulated insulin secretion as a mediator of glucagon-like peptide 1 response in pancreatic β cells. *Molecular and Cellular Biology* 33: 2527.
94. Kaihara, K.A., L.M. Dickson, D.A. Jacobson, N. Tamarina, M.W. Roe, L.H. Philipson et al. 2013. β-Cell-specific protein kinase A activation enhances the efficiency of glucose control by increasing acute-phase insulin secretion. *Diabetes* 62: 1527.

95. Cognard, E., C.G. Dargaville, D.L. Hay, and P.R. Shepherd. 2013. Identification of a pathway by which glucose regulates β-catenin signalling via the cAMP/protein kinase A pathway in β-cell models. *Biochemical Journal* 449: 803.
96. Lester, L.B., L.K. Langeberg, and J.D. Scott. 1997. Anchoring of protein kinase A facilitates hormone-mediated insulin secretion. *Proceedings of the National Academy of Sciences of the United States of America* 94: 14942.
97. Fraser, I.D., S.J. Tavalin, L.B. Lester, L.K. Langeberg, A.M. Westphal, R.A. Dean et al. 1998. A novel lipid-anchored A-kinase anchoring protein facilitates cAMP-responsive membrane events. *EMBO Journal* 17: 2261.
98. de Rooij, J., F.J. Zwartkruis, M.H. Verheijen, R.H. Cool, S.M. Nijman, A. Wittinghofer et al. 1998. Epac is a Rap1 guanine-nucleotide-exchange factor directly activated by cyclic AMP. *Nature* 396: 474.
99. Kawasaki, H., G.M. Springett, N. Mochizuki, S. Toki, M. Nakaya, M. Matsuda et al. 1998. A family of cAMP-binding proteins that directly activate Rap1. *Science* 282: 2275.
100. Ozaki, N., T. Shibasaki, Y. Kashima, T. Miki, K. Takahashi, H. Ueno et al. 2000. cAMP-GEFII is a direct target of cAMP in regulated exocytosis. *Nature Cell Biology* 2: 805.
101. Kashima, Y., T. Miki, T. Shibasaki, N. Ozaki, M. Miyazaki, H. Yano et al. 2001. Critical role of cAMP-GEFII—Rim2 complex in incretin-potentiated insulin secretion. *Journal of Biological Chemistry* 276: 46046.
102. Fujimoto, K., T. Shibasaki, N. Yokoi, Y. Kashima, M. Matsumoto, T. Sasaki et al. 2002. Piccolo, a Ca^{2+} sensor in pancreatic β-cells. Involvement of cAMP-GEFII.Rim2. Piccolo complex in cAMP-dependent exocytosis. *Journal of Biological Chemistry* 277: 50497.
103. Kai, A.K., A.K. Lam, Y. Chen, A.C. Tai, X. Zhang, A.K. Lai et al. 2013. Exchange protein activated by cAMP 1 (Epac1)-deficient mice develop β-cell dysfunction and metabolic syndrome. *FASEB Journal* 27: 4122.
104. Kang, G., J.W. Joseph, O.G. Chepurny, M. Monaco, M.B. Wheeler, J.L. Bos et al. 2003. Epac-selective cAMP analog 8-pCPT-2′-O-Me-cAMP as a stimulus for Ca^{2+}-induced Ca^{2+} release and exocytosis in pancreatic β-cells. *Journal of Biological Chemistry* 278: 8279.
105. Ying, Y., L. Li, W. Cao, D. Yan, Q. Zeng, X. Kong et al. 2012. The microtubule associated protein syntabulin is required for glucose-stimulated and cAMP-potentiated insulin secretion. *FEBS Letters* 586: 3674.
106. Idevall-Hagren, O., I. Jakobsson, Y. Xu, and A. Tengholm. 2013. Spatial control of Epac2 activity by cAMP and Ca^{2+}-mediated activation of Ras in pancreatic β cells. *Science Signaling* 6: ra29.1–11, S1–6.
107. Chepurny, O.G., C.A. Leech, G.G. Kelley, I. Dzhura, E. Dzhura, X. Li et al. 2009. Enhanced Rap1 activation and insulin secretagogue properties of an acetoxymethyl ester of an Epac-selective cyclic AMP analog in rat INS-1 cells: Studies with 8-pCPT-2′-O-Me-cAMP-AM. *Journal of Biological Chemistry* 284: 10728.
108. Song, W.J., P. Mondal, Y. Li, S.E. Lee, and M.A. Hussain. 2013. Pancreatic β-cell response to increased metabolic demand and to pharmacologic secretagogues requires EPAC2A. *Diabetes* 62: 2796.
109. Holz, G.G., O.G. Chepurny, and C.A. Leech. 2013. Epac2A makes a new impact in β-cell biology. *Diabetes* 62: 2665.
110. Leech, C.A., I. Dzhura, O.G. Chepurny, F. Schwede, H.G. Genieser, and G.G. Holz. 2010. Facilitation of β-cell K_{ATP} channel sulfonylurea sensitivity by a cAMP analog selective for the cAMP-regulated guanine nucleotide exchange factor Epac. *Islets* 2: 72.
111. Jarrard, R.E., Y. Wang, A.E. Salyer, E.P. Pratt, I.M. Soderling, M.L. Guerra et al. 2013. Potentiation of sulfonylurea action by an EPAC-selective cAMP analog in INS-1 cells: Comparison of tolbutamide and gliclazide and a potential role for EPAC activation of a 2-APB-sensitive Ca^{2+} influx. *Molecular Pharmacology* 83: 191.

112. Kang, G., O.G. Chepurny, and G.G. Holz. 2001. cAMP-regulated guanine nucleotide exchange factor II (Epac2) mediates Ca^{2+}-induced Ca^{2+} release in INS-1 pancreatic β-cells. *Journal of Physiology* 536: 375.
113. Liu, G., S.M. Jacobo, N. Hilliard, and G.H. Hockerman. 2006. Differential modulation of $Ca_v1.2$ and $Ca_v1.3$-mediated glucose-stimulated insulin secretion by cAMP in INS-1 cells: Distinct roles for exchange protein directly activated by cAMP 2 (Epac2) and protein kinase A. *Journal of Pharmacology and Experimental Therapeutics* 318: 152.
114. Enserink, J.M., A.E. Christensen, J. de Rooij, M. van Triest, F. Schwede, H.G. Genieser et al. 2002. A novel Epac-specific cAMP analogue demonstrates independent regulation of Rap1 and ERK. *Nature Cell Biology* 4: 901.
115. Holz, G.G., O.G. Chepurny, and F. Schwede. 2008. Epac-selective cAMP analogs: New tools with which to evaluate the signal transduction properties of cAMP-regulated guanine nucleotide exchange factors. *Cell Signaling* 20: 10.
116. Vliem, M.J., B. Ponsioen, F. Schwede, W.J. Pannekoek, J. Riedl, M.R. Kooistra et al. 2008. 8-pCPT-2′-O-Me-cAMP-AM: An improved Epac-selective cAMP analogue. *Chembiochem* 9: 2052.
117. Tsalkova, T., F.C. Mei, S. Li, O.G. Chepurny, C.A. Leech, T. Liu et al. 2012. Isoform-specific antagonists of exchange proteins directly activated by cAMP. *Proceedings of the National Academy of Sciences of the United States of America* 109: 18613.
118. Courilleau, D., M. Bisserier, J.C. Jullian, A. Lucas, P. Bouyssou, R. Fischmeister et al. 2012. Identification of a tetrahydroquinoline analog as a pharmacological inhibitor of the cAMP-binding protein Epac. *Journal of Biological Chemistry* 287: 44192.
119. Chen, H., T. Tsalkova, O.G. Chepurny, F.C. Mei, G.G. Holz, X. Cheng et al. 2013. Identification and characterization of small molecules as potent and specific EPAC2 antagonists. *Journal of Medicinal Chemistry* 56: 952.
120. Yang, Y., X. Shu, D. Liu, Y. Shang, Y. Wu, L. Pei et al. 2012. EPAC null mutation impairs learning and social interactions via aberrant regulation of miR-124 and Zif268 translation. *Neuron* 73: 774.
121. Zhao, K., R. Wen, X. Wang, L. Pei, Y. Yang, Y. Shang et al. 2013. EPAC inhibition of SUR1 receptor increases glutamate release and seizure vulnerability. *Journal of Neuroscience* 33: 8861.
122. Pereira, L., H. Cheng, D.H. Lao, L. Na, R.J. van Oort, J.H. Brown et al. 2013. Epac2 mediates cardiac β1-adrenergic-dependent sarcoplasmic reticulum Ca^{2+} leak and arrhythmia. *Circulation* 127: 913.
123. Herget, S., M.J. Lohse, and V.O. Nikolaev. 2008. Real-time monitoring of phosphodiesterase inhibition in intact cells. *Cell Signaling* 20: 1423.
124. Everett, K.L., and D.M. Cooper. 2012. cAMP measurements with FRET-based sensors in excitable cells. *Biochemical Society Transactions* 40: 179.
125. Goldfine, I.D., R. Perlman, and J. Roth. 1971. Inhibition of cyclic 3′,5′-AMP phosphodiesterase in islet cells and other tissues by tolbutamide. *Nature* 234: 295.
126. Tsalkova, T., A.V. Gribenko, and X. Cheng. 2011. Exchange protein directly activated by cyclic AMP isoform 2 is not a direct target of sulfonylurea drugs. *Assay Drug Development Technologies* 9: 88.
127. Rehmann, H. 2013. Epac-inhibitors: Facts and artefacts. *Scientific Reports* 3: 3032.
128. Chepurny, O.G., and G.G. Holz. 2007. A novel cyclic adenosine monophosphate responsive luciferase reporter incorporating a nonpalindromic cyclic adenosine monophosphate response element provides optimal performance for use in G protein coupled receptor drug discovery efforts. *Journal of Biomolecular Screening* 12: 740.
129. Holz, G.G., and J.F. Habener. 1998. Black widow spider α-latrotoxin: A presynaptic neurotoxin that shares structural homology with the glucagon-like peptide-1 family of insulin secretagogic hormones. *Comparative Biochemistry and Physiology: Biochemistry and Molecular Biology* 121: 177.

130. Holz, G.G. 2004. New insights concerning the glucose-dependent insulin secretagogue action of glucagon-like peptide-1 in pancreatic β-cells. *Hormone and Metabolic Research* 36: 787.
131. Clardy-James, S., O.G. Chepurny, C.A. Leech, G.G. Holz, and R.P. Doyle. 2013. Synthesis, characterization and pharmacodynamics of vitamin-B_{12}-conjugated glucagon-like peptide-1. *ChemMedChem* 8: 582.
132. Holz, G.G., C.A. Leech, and J.F. Habener. 2000. Insulinotropic toxins as molecular probes for analysis of glucagon-like peptide-1 receptor-mediated signal transduction in pancreatic β-cells. *Biochimie* 82: 915.
133. Chepurny, O.G., C.A. Leech, X. Cheng, and G.G. Holz. 2012. cAMP sensor Epac and gastrointestinal function. In: *Physiology of the Gastrointestinal Tract*, 5th Edition, edited by L. Johnson, F. Ghishan, J.M. Kaunitz, H. Said, and J. Wood, 1849–1861. Oxford: Elsevier.
134. Chen, H., C. Wild, X. Zhou, N. Ye, X. Cheng, and J. Zhou. 2013. Recent advances in the discovery of small molecules targeting exchange proteins directly activated by cAMP (EPAC). *Journal of Medicinal Chemistry* 57: 3651.
135. Ni, Q., A. Ganesan, N.N. Aye-Han, X. Gao, M.D. Allen, A. Levchenko et al. 2011. Signaling diversity of PKA achieved via a Ca^{2+}-cAMP-PKA oscillatory circuit. *Nature Chemical Biology* 7: 34.
136. Hodson, D.J., R.K. Mitchell, L. Marselli, T.J. Pullen, S.G. Brias, F. Semplici et al. 2014. ADCY5 couples glucose to insulin secretion in human islets. *Diabetes* 63: 3009.

4 Assessing Cyclic Nucleotide Recognition in Cells

Opportunities and Pitfalls for Selective Receptor Activation

Rune Kleppe, Lise Madsen, Lars Herfindal, Jan Haavik, Frode Selheim, and Stein Ove Døskeland

CONTENTS

4.1 BRIEF OVERVIEW OF CYCLIC NUCLEOTIDE RECEPTOR PROTEINS

The archetypical second messengers, cAMP and cGMP, act mainly via binding to the cyclic nucleotide (cN) binding domain (cNBD), first found in the catabolite activator protein of *Escherichia coli*. The cNs can also act via the cN binding GAF domain initially discovered in some cN phosphodiesterases (PDEs). An isolated case is provided by the *Dictyostelium* cAMP receptor of the seven transmembrane G protein–coupled receptor family, which responds to extracellular cAMP. In general, cN activation is driven by a higher cN affinity for the active rather than the inactive receptor conformation. The structural details of the molecular mechanisms involved in the activation process are emerging.[1,2]

Many cell types signal through both cAMP and cGMP, and additional binding site restraints seem to be required to ensure discrimination. The cAMP- and cGMP-dependent protein kinases (PKAs and PKGs) contain two tandem CNBDs. This arrangement limits cross-signaling because both cN sites must be occupied to ensure efficient activation.[3,4] Therefore, a 30-fold binding site preference for each site is theoretically sufficient to obtain a 900-fold preference for the two sites combined. Tandem CNBDs also permit intrachain allosteric interactions, as observed during the activation of PKA and PKG.[3,5]

Two genes, *PRKG1* and *PRKG2*, give rise to PKG type I (PKG-I) and type II (PKG-II). Their two CNBDs differ in cGMP affinity. A missense mutation (p.Arg177Gln) of the high-affinity site produces a constitutively active PKG-I and segregates with aortic disease in humans.[6] The PKG-Iα and PKG-Iβ isoforms are formed through alternative splicing and differ only in their far N-terminal region that contains the autoinhibitory motif. Despite having the same cGMP binding domains, PKG-Iα and PKG-Iβ have different cGMP activation constants (K_a = 0.1 and 0.9 μM, respectively) due to differential interaction between the CNBDs and the variable N-terminal.[7]

PKA is unique among protein kinases in that the autoinhibitory motif is located on a cAMP-binding regulatory subunit (PKA-R, where R = RIα, RIβ, RIIα, or RIIβ) that physically dissociates from the catalytic (PKA-C) subunit.[8–10] This arrangement produces a compact active kinase with enhanced ability to diffuse within the cell. The driving force for PKA dissociation is provided by the much tighter binding of cAMP to free PKA-R (K_d in the low nanomolar range) than PKA holoenzyme (K_d in the low micromolar range).[11] Structural studies reveal an extensive interaction surface between the R and C subunits, in particular for the CNBD-A domain.[12,13] This explains kinetic observations that cAMP binding to the B site (CNBD-B) precedes binding to the A site.[14,15]

Although PKA is present from yeast to man, the precursor of the mammalian cAMP-stimulated Rap exchange factors Epacl and Epac2 (RapGEF3 and 4)[16,17] first appeared in *Caenorhabditis elegans*, possibly as a result of fusion between a RapGEF and the two CNBDs of PKA-R. Apparently, two subsequent evolutionary steps have occurred. One is the degeneration (in Epac2) or loss (Epacl) of one of the tandem CNBDs. Another is the adjustment (lowering) of the cAMP affinity of CNBD from the low nanomolar range in PKA-R to the low micromolar range in Epac, which is closer to the cAMP level encountered in the intact cell.[11] This latter property of the

Epac CNBD has made it a useful component for engineering intracellular cAMP sensors.[18,19]

The CNBD-regulated ion channels, the cN-gated channels (CNGC) and the hyperpolarization-activated cN-gated channels (HCN1–4), are found in all phyla.[20,21] The HCNs are best described in the nervous system and in the heart, but they are expressed in other tissues as well. The cAMP binding site of the HNCs has been mapped and the structure elucidated.[22–24] Intriguingly, cyclic dinucleotides, previously held to signal mainly in prokaryotes, can bind to the peptide linking the cAMP site of HCN4 to the membrane pore and thereby counteract the effect of cAMP binding.[25] Genetic disruption of HCN1 has implicated its roles in spatial learning and memory in the entorhinal cortex[26] and hippocampus,[27] as well as motor learning in the cerebellum.[28] Disruption of the more ubiquitously expressed HCN2 can lead to arrhythmia and ataxia.[29,30] Several human mutations in HCN4 have been described and associated with bradycardia and arrhythmias.[31]

The GAF domains are bona fide cN receptors found in several of the mammalian PDE family members (PDE2, 5, 6, 10, and 11), often functioning as allosteric activating domains of the enzymes. They can bind cGMP and cAMP with different selectivity.[32] Binding of cGMP to the GAF domains mediates the allosteric stimulation of PDE2 and PDE5, leading to increased degradation of cGMP in PDE5, and of both cAMP and cGMP in PDE2.[33,34] The selectivity for cGMP and the conformational response to ligand binding has made these GAF domains effective components of FRET-based sensors for monitoring intracellular cGMP.[35]

The active sites of PDEs and of transport proteins, such as multidrug resistance–associated 4, whereas handling cNs, are not generally considered as cN receptors. The dual-specificity PDEs can, however, mediate cAMP/cGMP cross-talk through substrate competition, which is amplified by large differences in catalytic turnover rates between the cNs. Thus, cGMP-inhibited PDE3A degrade both cAMP and cGMP, but with 10-fold difference in the k_{cat} value.[36,37] The multitude of potential cN targets reflects the biological importance of the cNs, but is a challenge for the design of specific cN analogs and for understanding the complete cellular consequences of a cN signaling event.

4.2 OVERVIEW OF METHODS TO DETERMINE cN BINDING TO TARGET PROTEINS

4.2.1 cN Binding to Isolated Target Proteins

The use of labeled cN has advantages especially for time kinetic studies of receptor binding because the binding can be quenched by adding excess unlabeled cN. Furthermore, binding site interaction can be studied by monitoring the dissociation of bound labeled cN in the presence of various unlabeled cNs. The least intrusive labeling is by replacing the ^{1}H by the weak beta-emitter ^{3}H, as in [^{3}H]cAMP or [^{3}H]cGMP.

Bound labeled cN can be separated from free cN by membrane filtration. The precipitation of receptor-bound [^{3}H]cN in cold ammonium sulfate solution will often stabilize the bound cN, and has been used successfully to determine [^{3}H]cN bound to PKA-RI, PKA-RII,[38] PKG,[39] and Epac1.[40] Labeled cN can also be used to study binding interactions without separating bound and free ligand. In proximity scintillation counting assays, the solid phase scintillant distinguishes between bound and free ligand as it only responds to radiation from ^{3}H very close to the solid phase surface. The method has been used to determine the binding of [^{3}H]cAMP-complexed PKA-RI to solid phase anchored PKA-C in a solution containing excess [^{3}H]cAMP.[41] Another method to study equilibrium binding of small ligands, including labeled cNs, is through size exclusion gel filtration.[11]

Nuclear magnetic resonance (NMR) provides information about both cN binding stoichiometry and the conformational changes induced.[42–44] Present limitations are long incubation times, need for high protein concentrations, and weaker resolution for large proteins. These shortcomings can be overcome by the less precise method of mass spectrometry–based assessment of the rate of exchange of protein amide hydrogen with deuterium from heavy water. The method has revealed the differential exposure of residues of PKA-R upon binding of *R*p-cAMPS analogs and cAMP[45] and has shown that cAMP binding alters the exposure of the Epac2 C-terminal Rap interaction domain, the N-terminal anchoring domain, and the region between the two CNBDs.[46,47]

Native mass spectrometry is well suited to determine the stoichiometry of cN binding,[48,49] and is of particular value when several mass variants exist of either the cN or the receptor. Less direct methods rely on the altered apparent size of the binding protein upon cN binding, as deduced by, for example, alterations in fluorescence anisotropy and of scattering techniques.[50,51]

4.2.2 cN Binding to Target Proteins in Intact Cells

The use of light-emitting or light-quenching cN ligands or cN receptors has allowed elegant intact cell experiments yielding new information about cN signaling compartments. Fluorescently labeled cNs have revealed the dynamics of cN binding to ion channels,[52] and fluorescently tagged PKA-R and PKA-C subunits have been used to infer cAMP-induced PKA holoenzyme dissociation.[41,53] Such tools introduce extra molecules with the potential of perturbing the endogenous signaling, which should be taken into account when they are used instead of endogenous sensors of cN signaling.

The direct determination of the endogenous cAMP or cGMP bound to non-modified intracellular proteins in single live cells does not yet seem to be feasible. There are, however, methods to detect binding of endogenous cN to PKA-R in native cells or tissue snap-frozen in liquid nitrogen. The frozen sample is homogenized in a buffer (1 M ammonium sulfate in 15% glycerol at subzero temperatures) stabilizing the cAMP bound to PKA.[54,55] The amount of cAMP bound to the regulatory subunit of PKA type I (PKA-RI) and type II (PKA-RII) can be determined separately based on immune precipitation. In glucagon-stimulated hepatocytes, the total protein-associated cAMP was similar to the sum of cAMP bound to PKA-RI and PKA-RII,[54] indicating that the in vitro cAMP binding protein S-adensylhomocysteinase,[56,57] highly expressed in liver,[56] does not significantly bind cAMP in intact cells.

4.3 ESTIMATION AND MODELING OF FREE cN IN CELLS: THE EFFECT OF HIGH cN RECEPTOR DENSITY

4.3.1 Methods for Estimation of Free cN in Intact Cells

The level of $[cN]_{free}$ is more important than $[cN]_{total}$ (bound + free) to judge activation of cN effectors and cN turnover.[58,59] The $[cAMP]_{free}$ can be determined by FRET-coupled sensors based on Epac CNBD[18] or PKA subunits.[60] The Epac CNBD is preferred because PKA-based assays are prone to dilution of the FRET signal by endogenous PKA subunits and incomplete dissociation of PKA at saturating cAMP.[41] The $[cGMP]_{free}$ can be assessed by FRET sensors using the GAF of PDE or PKG CNBD.[61] Such sensors can be equipped with nuclear, mitochondrial, or surface membrane directing tags as well as tags directing them to submicroscopic compartments containing specific PKA-anchoring AKAP proteins.[62] The sensor itself should be present at sufficiently low abundance to not act as a cN sink. Also, proximity to endogenous fluorescence quenchers may affect the results. In cells where PKA is the major cAMP receptor, $[cAMP]_{free}$ can also be estimated by subtracting PKA-R bound cAMP from $[cAMP]_{total}$.[54]

4.3.2 The Effect of High Endogenous Platelet PKG Concentrations on $[cGMP]_{FREE}$: Implications for cGMP Metabolism and Target Activation

In cells such as human thrombocytes (blood platelets), the use of genetically engineered probes for sensing free cN is currently unfeasible. This is unfortunate because platelets have far higher concentrations of PKG (~7.3 μM PKG-Iβ monomers with 14.6 μM cGMP sites) and PKA subunits (~3.1 μM monomers, corresponding to 6.2 μM cAMP sites) compared with most, if not all, other cells,[63] and only a small proportion of total cN will be free. In such cells, the $[cGMP]_{free}$ can be estimated *in silico* based on the published binding constants for cGMP binding to PKG.

The high PKG, PKA, and PDE concentrations and the small size of the blood platelets resembles subcellular signaling compartments with coanchored PKA/PKG and PDEs. The platelet may therefore be an instructive model system for investigating the features of subcellular compartmentalization in cN biology. The modeling of $[cGMP]_{free}$ shown in Figure 4.1a is based on published values for the expression and cGMP affinity of the highly abundant platelet PKG-Iβ (see legend to Figure 4.1 for details).

Note that under basal conditions only a small fraction of total cGMP is free and available for PDE degradation due to massive cGMP sequestration by the *high affinity* site of PKG-Iβ (Figure 4.1a). The protection of bound cGMP from degradation explains why the thrombocyte $[cGMP]_{total}$ is far higher than predicted based on the assumed PDE activity and its basal soluble guanylyl cyclase activity.[64] The small increase of $[cGMP]_{free}$ predicted at moderate NO stimulation will be insufficient to activate the GAF domains of PDE2A and PDE5A, but may be sufficient to inhibit competitive cAMP degradation by PDE3 (Figure 4.1b). This agrees with the important role of PDE3A in regulating shape change and activation of NO-stimulated platelets.[36,65]

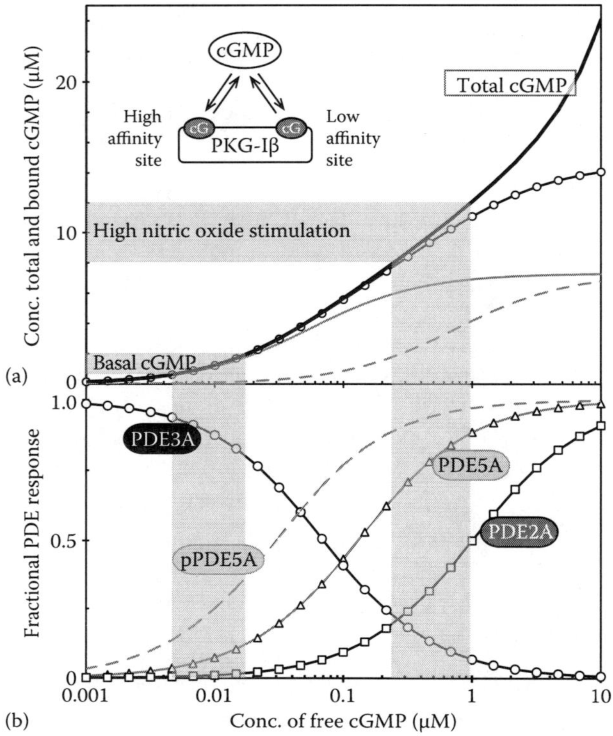

FIGURE 4.1 The effect of high cN receptor levels on signaling: the case of cGMP in blood platelets. (a) The estimated platelet total cGMP, which is typically measured experimentally, and the free cGMP levels in platelets, which is biologically more relevant is shown. The total amount of cGMP bound to PKG-Iβ (○), the amount cGMP bound to the high-affinity (gray solid line) and low-affinity (gray dashed line) sites of PKG as well as the total platelet cGMP level (free + bound; black solid line) are shown. Shaded areas illustrate reported levels of cGMP at basal and NO-stimulated conditions.[66] (b) The relative inhibition of PDE3A (○, at $[cAMP]_{free} = K_m$), the binding of cGMP to the GAF-domain of nonphosphorylated (△, K_d = 130 nM)[67] or PKG-phosphorylated PDE5A (pPDE5A, gray dashed line, K_d = 30 nM)[67] and the activation of PDE2A (□, $K_a \approx 1$ μM at 37°C)[68] are shown. Shaded areas illustrate $[cGMP]_{free}$ levels at basal and NO-stimulated conditions.

Negative feedback control of cGMP activation is reported in PKG-mediated phosphorylation of PDE5. This occurs only during cGMP-binding to the GAF-A domain of PDE5 and renders it more responsive to cGMP.[67,69] The model shows that PDE5, if phosphorylated by PKG during a preceding strong burst of cGMP, will be far more responsive to a small increase of $[cGMP]_{free}$. The physiological relevance of this feedback regulation is not evident unless cGMP sequestration is taken into account. This suggests that the autoinhibitory loop of cGMP activation of PDE5 only becomes activated upon very strong cGMP stimulation, as might occur if PDE5 is localized in a compartment close to a guanylate cyclase and PKG.[70]

4.4 cAMP/cGMP STRUCTURE ACTIVITY RELATIONSHIP FOR RECEPTOR BINDING

4.4.1 Role of the Cyclic Phosphate and Ribose Moiety

An intact cyclic phosphate is essential for the activation of PKA, PKG, and Epac. Only the cyclic ring oxygen S-substituted *R*p- and *S*p-diastereoisomers maintain reasonable affinity for PKA, PKG, and Epac, and to a lesser extent the HCN channels.[24] *S*p-cAMPS and *S*p-cGMPS are full PKA, Epac, and PKG activators whereas *R*p-cAMPS/*R*p-cGMPS are weak partial antagonists that compete with fully agonistic cN.[40,71,72]

Modification of the 2′-OH of the ribose severely compromises the cAMP binding affinity to PKA-R and HCN.[24] The first cN used to mimic cAMP, 2′-*O*-butyryl, *N*⁶-butyryl-cAMP (dibutyryl-cAMP), is inactive until converted to the PKA activator *N*⁶-(mono)-butyryl-cAMP by intracellular esterases. More recently, a number of 2′-modified cAMP analogs were compared for binding affinity for site B of PKA-RI and the homologous single cAMP site of Epac1.[11,40] Intact 2′-OH is required for cAMP to achieve full activation of PKA. In contrast, 2′-OH substituted analogs bind quite well to Epac, and 8-pCPT-2′-*O*-Me-cAMP is, if anything, a superactivator of Epac.[73] The 2′-*O*-methyl–modified analogs are therefore used extensively[40,74] for selective activation of Epac. The 2′-OH of cAMP contacts a conserved glutamate in the cN binding site of PKA that is replaced by glutamine in Epac1. Switching this Glu to Gln in PKA-RI lowers its cAMP affinity by several orders of magnitude so it becomes similar to that of Epac.[11] The Glu/Gln switch has been proposed to be responsible for the Epac preference of 2′-modified analogs,[73] but this is not supported by experimental studies of reciprocally switched PKA-RI and Epac1.[11] A Gln/Lys switch in Epac2 appears, however, to be responsible for a much lower potency of 8-CPT-2′-*O*-Me-cAMP as activator of Epac2 than Epac1.[75]

4.4.2 Role of Purine Ring Modifications

Purine ring–modified cN analogs are capable of fully activating either PKA-I,[76,77] PKA-II,[4] or PKG at saturating concentrations.[78] Therefore, cross-activation of PKG by cAMP may occur if the cAMP level is very high, which can occur in smooth muscles.[79] Cyclic AMP and cGMP may, however, induce subtly different PKG conformations because PKG autophosphorylation is favored more by bound cAMP than cGMP.[3,80]

Both cN analog mapping and the x-ray structure of cAMP bound to PKA-R indicate that the purine moiety of cAMP binds in a nonpolar pocket without forming hydrogen bonds.[77,81,82] The pocket is malleable because cN analogs with bulky adenine *N*⁶-substituents bind well to site AI of PKA-RI even if a steric clash is predicted by the crystal structure of the complex of cAMP and PKA-RI.[13,82,83] Functional and structural data therefore indicate that the purine moiety of cNs with intact cyclic phosphate serves to provide enhanced binding affinity and binding site discrimination within the PKA/PKG family. Puzzlingly, the (partial) activation of PKA becomes purine ring–dependent for *R*p-stereoisomers of cAMP. Thus, although *R*p-8-Br-cAMPS has no partial agonism and is a strong competitive inhibitor of cAMP-activated PKA-I in both in vitro and intact cells, the compound *R*p-*N*⁶-phenyl-cAMPS is a strong partial agonist that is unable to counteract PKA activation in intact cells.[71] This indicates that *R*p-modified cyclic phosphate ring affects the purine binding pocket differently in free PKA-R and the PKA-C complexed PKA holoenzyme.

An unmodified 6-position is also required for Epac activation,[40] possibly an adaption to prevent cross-activation by cGMP in compartments where Epac is anchored close to a guanylyl cyclase. PKA activating 6-modified cAMP analogs like N^6-benzoyl-cAMP and N^6-butyryl-cAMP do not activate Epac.[40] The recently developed non–cN-based Epac inhibitors that target the cN site and Epac2-selective inhibitors that bind an allosteric site between the cAMP binding domains of Epac2, suggest that cN receptor targeting extends beyond the cN-modified compounds (see also Chapter 1).[47,84]

4.5 PROBING OF cN FUNCTIONS IN INTACT CELLS WITH cN ANALOGS

The use of cN analogs has several advantages compared with the genetic modification of signaling pathways for cell studies: (1) allows rapid onset and cessation of signal perturbation, which avoids any compensatory regulation as a consequence of altered gene expression; (2) it allows graded stimulation through concentration variation; and (3) it is less costly and time-consuming in terms of labor.

4.5.1 Preferential Activation of PKA-I and PKA-II by cAMP Analog with Complementary Site A/B Preference

It was realized early on that modification of the purine ring of cAMP or cGMP could yield cN analogs that would be able to discriminate between the A and B sites of PKA-RI,[77] PKA-RII,[4] and PKG.[78] Recently, more discriminative analogs have been developed, such as N^6-benzoyl-8-piperidino-cAMP, which not only has improved selectivity for site AI of PKA-I (Table 4.1) but is also unable to activate Epac.[85–87] Because both site A and B must be occupied by cN for significant cellular activation, a pair of analogs with complementary site A/B selectivity for either PKA-I or PKA-II (Tables 4.1 and 4.2) will produce PKA-I and PKA-II synergism (Table 4.3). The use of such PKA-I or PKA-II directed analog combinations is illustrated in Figure 4.2. It shows PKA-I directed cN pair synergy for 3T3 cell preadipocyte differentiation, suggesting that PKA-I is the key mediator of the PKA-induced inhibition of Rho-kinase signaling during adipocyte differentiation.[88]

In another example (Figure 4.3), PKA-I directed cAMP analog pairs synergized to induce apoptosis in IPC-81 leukemia cells.[87] The degree of analog pair synergy was in accord with predictions based on the cN analog binding site preference (Tables 4.1 through 4.3). Note that the Epac activator 8-CPT-2′-*O*-Me-cAMPS was unable to induce apoptosis (Figure 4.3).

4.5.2 Precautions for the Use of cN Analog Probes in Intact Cells

The major disadvantages of cN analogs are off-target effects, either from the original cN analog or from its degradation products. Examples of the latter are the highly toxic

TABLE 4.1
N^6-Bz-8-Pip-cAMP Has Superior Selectivity for cAMP Binding Site AI of PKA-I

Compound	Site AI	Site BI	Site AII	Site BII	AI/BI	AI/AII	AI/BII
cAMP	1.0	1.0	1.0	1.0	1.0	1.0	1.0
N^6-Bz-8-Pip-cAMP	**1.7**	**0.018**	**0.0011**	**0.065**	**94**	**1550**	**26**
N^6-Bz-cAMP	0.50	0.26	3.8	0.0037	2	1/7.6	135
8-Pip-cAMP	1.2	0.022	0.037	2.4	55	32	1/2
8-CPT-cAMP	3.9	1.7	0.054	19	2.3	72	1/4.9
N^6-MB-cAMP	3.9	0.78	0.74	0.071	5	5.3	55
N^6-MBC-cAMP	0.50	0.086	13	0.066	5.8	1/26	7.6

Source: Huseby, S., Gausdal, G., Keen, T.J. et al. *Cell Death Dis* 2 (2011): e237.

Note: The dissociation constants of the analogs for binding to site A and B of RI and RII, relative to cAMP (K_d(cAMP)/K_d(analog)). The selectivity for site A of RI relative to the other sites is also shown (three right columns). The highly site AI-selective analog N^6-Bz-8-Pip-cAMP is highlighted (boldfaced). It also shows data for other analogs previously used to preferentially occupy site AI, as well as for 6-MBC-cAMP, used to selectively occupy site AII of PKA-II. N^6-Bz-8-Pip-cAMP is the most site AI/AII-selective (1550×) and AI/BI-selective (94×) compound on record.

TABLE 4.2
The 2-Cl-Substituted 8-AHA-cAMP Has Superior Site BI Specificity

Compound	Site AI	Site BI	Site AII	Site BII	BI/AI	BI/AII	BI/BII
cAMP	1.0	1.0	1.0	1.0	1.0	1.0	1.0
2-Cl-8-AHA-cAMP	**0.0052**	**3.9**	**0.0012**	**0.50**	**750**	**3250**	**7.8**
8-AHA-cAMP	0.73	3.6	0.028	0.31	4.9	128	12
8-NH-CH_3-cAMP	0.09	1.4	0.026	1.6	1.6	54	0.88
8-CPT-cAMP	3.9	1.7	0.054	19	0.44	31	1/11
Sp-5,6-DCl-cBIMPS	0.22	0.13	0.034	14	0.60	3.8	1/11

Source: Huseby, S., Gausdal, G., Keen, T.J. et al. *Cell Death Dis* 2 (2011): e237.

Note: The dissociation constants of the analogs for binding to site A and B of RI and RII, relative to cAMP (K_d(analog)/K_d(cAMP)). The selectivity for site A of RI relative to the other sites is also shown (three right columns). The new analog 2-Cl-8-AHA-cAMP, which is highly selective for site BI of PKA-I, is highlighted (boldfaced). The other compounds are listed because they have been used previously for selective PKA isozyme activation, combined with site AI- or site AII-preferring analogs. For further details, see the legend to Table 4.1.

compounds 8-Cl-adenosine and 7-deaza-adenosine (tubercidine), which are metabolites of 8-Cl-cAMP and 7-deaza-cAMP (7-CH-cAMP). These cNs should therefore be considered protoxic substances, which is unfortunate because 7-CH-cAMP is a highly potent activator of HCN channels.[24] The use of 8-Cl-cAMP is unnecessary because 8-Br-cAMP shares the cN binding properties of 8-Cl-cAMP without being toxic.[89] A more benign, but cumbersome, side effect is the ability of cN analogs to inhibit the PDE catalyzed degradation of endogenous cAMP or cGMP[90] (Chapter 2).

TABLE 4.3
The Expected Synergy for PKA-I and PKA-II Activation by Selected cAMP Analog Pairs

cAMP Analog Pair ($x + y$)	Predicted Synergy PKA-I $\frac{\sqrt{(AI^x + AI^y)(BI^x + BI^y)}}{\sqrt{(AI^x)(BI^x)} + \sqrt{(AI^y)(BI^y)}} - 1$	Predicted Synergy PKA-II $\frac{\sqrt{(AII^x + AII^y)(BII^x + BII^y)}}{\sqrt{(AII^x)(BII^x)} + \sqrt{(AII^y)(BII^y)}} - 1$
N^6-Bz-8-Pip-cAMP + 2-Cl-8-AHA-cAMP	7.2	0.09
N^6-Bz-8-Pip-cAMP + *Sp*-5,6-DCl-cBIMPS	2.6	0.14
N^6-Bz-8-Pip-cAMP + *N*^{6}-MB-cAMP	0.06	0.34
6-MBA-cAMP + *Sp*-5,6-DCl-cBIMPS	0.29	7.3

Source: Huseby, S., Gausdal, G., Keen, T.J. et al. *Cell Death Dis* 2 (2011): e237.

Note: The table shows how much more efficiently PKA-I or PKA-II can be activated by a mixture of two cAMP analogs (x,y) compared with that expected from simple additivity. The formulas are based on the analog affinity relative to cAMP, as given in Tables 4.1 and 4.2 for the cAMP binding sites of PKA-I (abbreviated AI^x, AI^y, and BI^x, BI^y) and PKA-II (AII^x, AII^y and BII^x, BII^y). The upper term of the formulas, $[(A^x + A^y) \times (B^x + B^y)]^{0.5}$, represent the expected potency of the $x + y$ mixture, whereas the lower term, $(A^x \times B^x)^{0.5} + (A^y \times B^y)^{0.5}$, shows the potency expected from simple additivity. If there is no synergy the ratio between these terms is 1 (see Christensen et al.[85] for further details), which becomes 0 after subtraction of 1, as done here. The combination of *N*^{6}-Bz-8-Pip-cAMP and 2-Cl-8-AHA-cAMP produces stronger PKA-I synergism than previously available analog combinations,[85] and is devoid of PKA-II synergy. The combination 6-MBA-cAMP + *Sp*-5,6-DCl-cBIMPS (row 4) produces strong PKA-II synergy.

A generally neglected possibility is that cN analogs can induce PKA-dependent phosphorylation of PDE4, whose subsequent activation decreases cellular cAMP.[91]

The 8-substitution of the Epac-selective compound 2′-*O*-Me-cAMP with a sulfur joined to an aromatic ring causes a large increase of Epac binding activity[40] and has made 8-pCPT-2′-*O*-Me-cAMP an extremely popular probe for Epac activation effects. Unfortunately, this compound has more off-target effects than the less potent Epac activator 8-Br-2′-*O*-Me-cAMP. They range from interference with receptor signaling[92,93] to disturbance of membrane transporters[94] and inhibition of PDEs.[90]

The essential nature of the cyclic phosphate and ribose moiety for cN receptor activation has made them targets for the production of inactive compounds with an enhanced ability to penetrate into cells, where they become activated upon contact with intracellular enzymes. The first such prodrug-like compound, db-cAMP, is converted intracellularly to the PKA activator *N*^{6}-(mono)butyryl-cAMP + butyrate. Because butyrate is bioactive, db-cAMP has been replaced largely by cN cyclic phosphate esters (see Chapter 2).

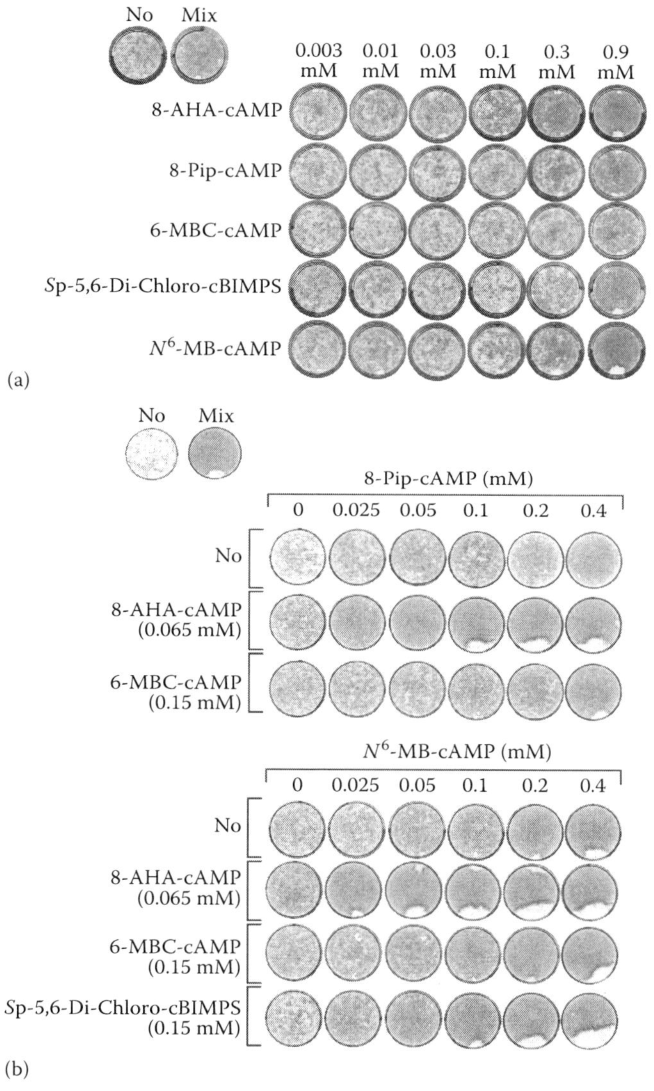

FIGURE 4.2 Using cAMP analogs to elucidate PKA isozyme contribution in adipocyte differentiation. (a) The amount of fat accumulation (assessed by staining with Congo red) in cells incubated with vehicle, the PDE inhibitor isobutylmethylxanthine (MIX), or various concentrations of 8-AHA-cAMP (site preference BI > BII >> AI,AII), *S*p-5,6-Di-Chloro-cBIMPS (BII > BI >> AI,AII), N^6-MB-cAMP (AI > AII >> BI,BII), 6-MBC-cAMP (AII > AI >> BI,BII), or 8-piperidino-cAMP (AI = BII > BI > AII; selects BOTH site AI and BII) are shown. (b, upper panels) 8-piperidino-cAMP (AI, BII) synergizes with the site BI-preferring analog 8-AHA-cAMP rather than with the BII-preferring analog 6-MBC-cAMP. This indicates that PKA-I can mediate adipogenesis. (b, lower panels) N^6-MB-cAMP synergizes more strongly with 8-AHA-cAMP than with either *S*p-5,6-Di-Chloro-cBIMPS or 6-MBC-cAMP, again confirming the PKA-I mediation of adipocyte differentiation.

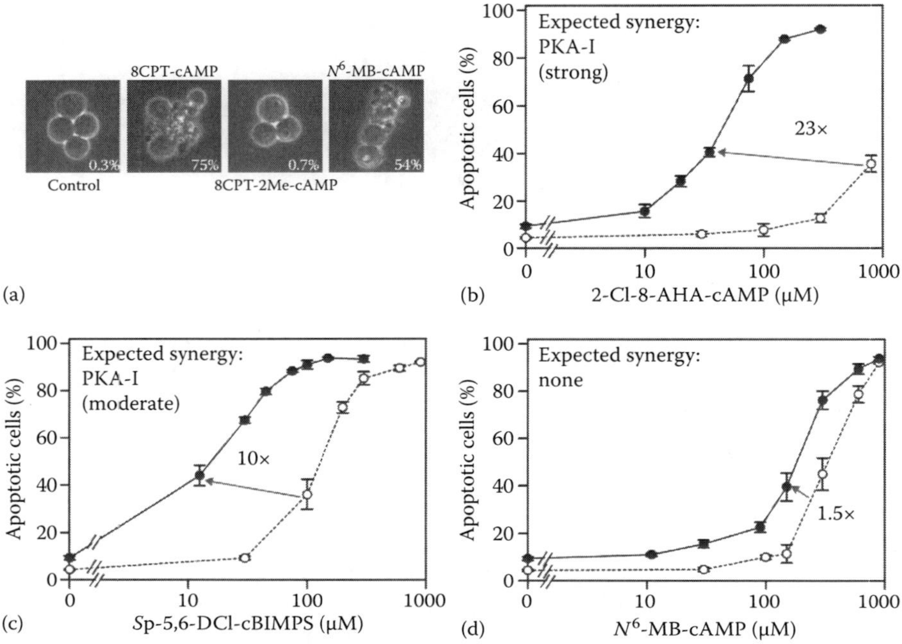

FIGURE 4.3 Synergistic activation of PKA-I by site A and site B analog pairs to facilitate leukemia cell death. (a) Apoptosis of IPC leukemia cells after 5 h of incubation with nonselective (8CPT-cAMP, 200 μM) and PKA-selective (*N*⁶-MB-cAMP, 700 μM) cAMP analogs, whereas cells incubated with Epac-selective activator (8CPT-2Me-cAMP, 700 μM) or vehicle were unaffected. (b) Induction of apoptosis using the 2-Cl-8-AHA-cAMP analog (specific for PKA-I site B) alone (○) or in combination with *N*⁶-Bnz-8-Pip-cAMP (specific for PKA-I site A, 300 μM; ●). This analog pair is synergistic for PKA-I, but not PKA-II (Tables 4.1 through 4.3). The analogs *S*p-5,6-DCl-cBIMPS (c) and *N*⁶-MB-cAMP (d) were moderately or marginally more efficient when combined with *N*⁶-Bnz-8-Pip-cAMP (300 μM; ●), than when used alone (○).

It is strongly recommended that several chemically distinct cN analogs with similar target effects be used before concluding that a target is regulated by cN in intact cells. The demonstration of PKA-I or PKA-II directed synergy (Figures 4.2 and 4.3) is strong evidence of PKA involvement. In addition, known or expected cN analog metabolites should be tested (see the Biolog web site; http://www.biolog.de/). The use of *prodrug/caged* cN analogs is advised for cells with poor cN permeability or if extracellular PDE degradation or effects on external receptors or channel proteins is an issue.

4.6 POSSIBLE THERAPEUTIC IMPLICATIONS OF cN RECEPTOR TARGETING

Pharmacologically directed targeting of specific cN binding sites of human cN receptors is attractive, as it avoids the broad actions expected from the manipulation of cellular cAMP and cGMP levels. On the other hand, the use of cAMP analogs in intact mammals has its own challenges. The cN analogs are exposed to PDE in blood, are readily taken up and presumably metabolized in parenchymal hepatocytes, and cleared rapidly from the circulation through renal secretion.[95] Therefore, it seems that the in vivo use of cN analogs is currently limited to short-term modulation of cN-dependent processes, for which a rapid clearing of active substance(s) may be desirable. To resolve their poor pharmacokinetics, we envision two possible solutions: (1) a caging strategy, possibly using a nanocarrier formulation, or (2) novel chemistries that can replace the strong selectivity for the ribose 3,5-cyclic-monophosphate moiety, which is important for cN receptor activation.

Hence, cAMP analogs currently seem to be most useful in acute respiratory or cardiovascular disease conditions or for the manipulation of circulating white blood cells or platelets. In addition to intravenous delivery, an aerosol-based vehicle could facilitate uptake and improve the pharmacokinetics in pulmonary tissue, as demonstrated for PDE4 inhibitors.[96]

The combination of cN analogs with agents targeting G protein–coupled receptors, PDEs, or novel non–cN site-directed compounds[47] (see also Chapter 1), is expected to enhance the biological potency and target specificity against cN receptors. The recent progress in structural elucidation provides a good starting point for the discovery of cN receptor targeting by compounds acting outside of the cN binding sites.

4.7 FUTURE PERSPECTIVES

The use of sophisticated intracellular probes and advances in microscopy have provided insight into intricate signaling compartments, some so small that they cannot be directly visualized using current microscopy techniques, although they are detectable using AKAP-anchored cAMP sensors.[62,97] The quantitative understanding of the intricate temporal and spatial cross-signaling in such compartments is still in its infancy, and too little is known about the effect of proximity on signaling, both conceptually and experimentally. Furthermore, quantitative systems biology of cN signaling is hampered by the lack of conventional biochemical kinetic data obtained under physiologically relevant assay conditions. Hopefully, we will soon see rapid progresses in these areas to reveal even more of the complexity and marvels of cN signaling.

REFERENCES

1. Bruystens, J.G., Wu, J., Fortezzo, A., Kornev, A.P., Blumenthal, D.K., and Taylor, S.S. PKA RIalpha homodimer structure reveals an intermolecular interface with implications for cooperative cAMP binding and Carney complex disease. *Structure* 22 (2014): 59–69.
2. Huang, G.Y., Kim, J.J., Reger, A.S. et al. Structural basis for cyclic-nucleotide selectivity and cGMP-selective activation of PKG I. *Structure* 22 (2014): 116–124.
3. Corbin, J.D., and Doskeland, S.O. Studies of two different intrachain cGMP-binding sites of cGMP-dependent protein kinase. *J Biol Chem* 258 (1983): 11391–11397.
4. Ogreid, D., Ekanger, R., Suva, R.H. et al. Activation of protein kinase isozymes by cyclic nucleotide analogs used singly or in combination. Principles for optimizing the isozyme specificity of analog combinations. *Eur J Biochem* 150 (1985): 219–227.

5. Doskeland, S.O. Evidence that rabbit muscle protein kinase has two kinetically distinct binding sites for adenosine 3′;5′-cyclic monophosphate. *Biochem Biophys Res Commun* 83 (1978): 542–549.
6. Guo, D.C., Regalado, E., Casteel, D.E. et al. Recurrent gain-of-function mutation in PRKG1 causes thoracic aortic aneurysms and acute aortic dissections. *Am J Hum Genet* 93 (2013): 398–404.
7. Ruth, P., Landgraf, W., Keilbach, A., May, B., Egleme, C., and Hofmann, F. The activation of expressed cGMP-dependent protein kinase isozymes I alpha and I beta is determined by the different amino-termini. *Eur J Biochem* 202 (1991): 1339–1344.
8. Moon, T.M., Osborne, B.W., and Dostmann, W.R. The switch helix: A putative combinatorial relay for interprotomer communication in cGMP-dependent protein kinase. *Biochim Biophys Acta* 1834 (2013): 1346–1351.
9. Beavo, J.A., Bechtel, P.J., and Krebs, E.G. Activation of protein kinase by physiological concentrations of cyclic AMP. *Proc Natl Acad Sci U S A* 71 (1974): 3580–3583.
10. Taylor, S.S., Ilouz, R., Zhang, P., and Kornev, A.P. Assembly of allosteric macromolecular switches: Lessons from PKA. *Nat Rev Mol Cell Biol* 13 (2012): 646–658.
11. Dao, K.K., Teigen, K., Kopperud, R. et al. Epac1 and cAMP-dependent protein kinase holoenzyme have similar cAMP affinity, but their cAMP domains have distinct structural features and cyclic nucleotide recognition. *J Biol Chem* 281 (2006): 21500–21511.
12. Kim, C., Cheng, C.Y., Saldanha, S.A., and Taylor, S.S. PKA-I holoenzyme structure reveals a mechanism for cAMP-dependent activation. *Cell* 130 (2007): 1032–1043.
13. Wu, J., Brown, S.H., von Daake, S., and Taylor, S.S. PKA type II alpha holoenzyme reveals a combinatorial strategy for isoform diversity. *Science* 318 (2007): 274–279.
14. Ogreid, D., and Doskeland, S.O. The kinetics of association of cyclic AMP to the two types of binding sites associated with protein kinase II from bovine myocardium. *FEBS Lett* 129 (1981): 287–292.
15. Ogreid, D., and Doskeland, S.O. The kinetics of the interaction between cyclic AMP and the regulatory moiety of protein kinase II. Evidence for interaction between the binding sites for cyclic AMP. *FEBS Lett* 129 (1981): 282–286.
16. de Rooij, J., Rehmann, H., van Triest, M., Cool, R.H., Wittinghofer, A., and Bos, J.L. Mechanism of regulation of the Epac family of cAMP-dependent RapGEFs. *J Biol Chem* 275 (2000): 20829–20836.
17. Rehmann, H., Prakash, B., Wolf, E. et al. Structure and regulation of the cAMP-binding domains of Epac2. *Nat Struct Biol* 10 (2003): 26–32.
18. Ponsioen, B., Zhao, J., Riedl, J. et al. Detecting cAMP-induced Epac activation by fluorescence resonance energy transfer: Epac as a novel cAMP indicator. *EMBO Rep* 5 (2004): 1176–1180.
19. Agarwal, S.R., Yang, P.C., Rice, M. et al. Role of membrane microdomains in compartmentation of cAMP signaling. *PLoS One* 9 (2014): e95835.
20. Kaupp, U.B., and Seifert, R. Cyclic nucleotide-gated ion channels. *Physiol Rev* 82 (2002): 769–824.
21. Peuker, S., Cukkemane, A., Held, M., Noe, F., Kaupp, U.B., and Seifert, R. Kinetics of ligand–receptor interaction reveals an induced-fit mode of binding in a cyclic nucleotide-activated protein. *Biophys J* 104 (2013): 63–74.
22. Wicks, N.L., Wong, T., Sun, J., Madden, Z., and Young, E.C. Cytoplasmic cAMP-sensing domain of hyperpolarization-activated cation (HCN) channels uses two structurally distinct mechanisms to regulate voltage gating. *Proc Natl Acad Sci U S A* 108 (2011): 609–614.
23. Xu, X., Marni, F., Wu, S. et al. Local and global interpretations of a disease-causing mutation near the ligand entry path in hyperpolarization-activated cAMP-gated channel. *Structure* 20 (2012): 2116–2123.

24. Moller, S., Alfieri, A., Bertinetti, D. et al. Cyclic nucleotide mapping of hyperpolarization-activated cyclic nucleotide-gated (HCN) channels. *ACS Chem Biol* 9 (2014): 1128–1137.
25. Lolicato, M., Bucchi, A., Arrigoni, C. et al. Cyclic dinucleotides bind the C-linker of HCN4 to control channel cAMP responsiveness. *Nat Chem Biol* 10 (2014): 457–462.
26. Giocomo, L.M., Hussaini, S.A., Zheng, F., Kandel, E.R., Moser, M.B., and Moser, E.I. Grid cells use HCN1 channels for spatial scaling. *Cell* 147 (2011): 1159–1170.
27. Hussaini, S.A., Kempadoo, K.A., Thuault, S.J., Siegelbaum, S.A., and Kandel, E.R. Increased size and stability of CA1 and CA3 place fields in HCN1 knockout mice. *Neuron* 72 (2011): 643–653.
28. Nolan, M.F., Malleret, G., Lee, K.H. et al. The hyperpolarization-activated HCN1 channel is important for motor learning and neuronal integration by cerebellar Purkinje cells. *Cell* 115 (2003): 551–564.
29. Chung, W.K., Shin, M., Jaramillo, T.C. et al. Absence epilepsy in apathetic, a spontaneous mutant mouse lacking the h channel subunit, HCN2. *Neurobiol Dis* 33 (2009): 499–508.
30. Ludwig, A., Budde, T., Stieber, J. et al. Absence epilepsy and sinus dysrhythmia in mice lacking the pacemaker channel HCN2. *EMBO J* 22 (2003): 216–224.
31. DiFrancesco, D. Funny channel gene mutations associated with arrhythmias. *J Physiol* 591 (2013): 4117–4124.
32. Heikaus, C.C., Pandit, J., and Klevit, R.E. Cyclic nucleotide binding GAF domains from phosphodiesterases: structural and mechanistic insights. *Structure* 17 (2009): 1551–1557.
33. Martinez, S.E., Wu, A.Y., Glavas, N.A. et al. The two GAF domains in phosphodiesterase 2A have distinct roles in dimerization and in cGMP binding. *Proc Natl Acad Sci U S A* 99 (2002): 13260–13265.
34. Rybalkin, S.D., Rybalkina, I.G., Shimizu-Albergine, M., Tang, X.B., and Beavo, J.A. PDE5 is converted to an activated state upon cGMP binding to the GAF A domain. *EMBO J* 22 (2003): 469–478.
35. Nikolaev, V.O., Gambaryan, S., and Lohse, M.J. Fluorescent sensors for rapid monitoring of intracellular cGMP. *Nat Methods* 3 (2006): 23–25.
36. Jensen, B.O., Selheim, F., Doskeland, S.O., Gear, A.R., and Holmsen, H. Protein kinase A mediates inhibition of the thrombin-induced platelet shape change by nitricoxide. *Blood* 104 (2004): 2775–2782.
37. Harrison, S.A., Reifsnyder, D.H., Gallis, B., Cadd, G.G., and Beavo, J.A. Isolation and characterization of bovine cardiac muscle cGMP-inhibited phosphodiesterase: A receptor for new cardiotonic drugs. *Mol Pharmacol* 29 (1986): 506–514.
38. Doskeland, S.O., and Ogreid, D. Ammonium sulfate precipitation assay for the study of cyclic nucleotide binding to proteins. *Methods Enzymol* 159 (1988): 147–150.
39. Doskeland, S.O., Vintermyr, O.K., Corbin, J.D., and Ogreid, D. Studies on the interactions between the cyclic nucleotide-binding sites of cGMP-dependent protein kinase. *J Biol Chem* 262 (1987): 3534–3540.
40. Christensen, A.E., Selheim, F., de Rooij, J. et al. cAMP analog mapping of Epac1 and cAMP kinase. Discriminating analogs demonstrate that Epac and cAMP kinase act synergistically to promote PC-12 cell neurite extension. *J Biol Chem* 278 (2003): 35394–35402.
41. Kopperud, R., Christensen, A.E., Kjarland, E., Viste, K., Kleivdal, H., and Doskeland, S.O. Formation of inactive cAMP-saturated holoenzyme of cAMP-dependent protein kinase under physiological conditions. *J Biol Chem* 277 (2002): 13443–13448.
42. Byeon, I.J., Dao, K.K., Jung, J. et al. Allosteric communication between cAMP binding sites in the RI subunit of protein kinase A revealed by NMR. *J Biol Chem* 285 (2010): 14062–14070.

43. Mazhab-Jafari, M.T., Das, R., Fotheringham, S.A., SilDas, S., Chowdhury, S., and Melacini, G. Understanding cAMP-dependent allostery by NMR spectroscopy: Comparative analysis of the EPAC1 cAMP-binding domain in its apo and cAMP-bound states. *J Am Chem Soc* 129 (2007): 14482–14492.
44. McNicholl, E.T., Das, R., SilDas, S., Taylor, S.S., and Melacini, G. Communication between tandem cAMP binding domains in the regulatory subunit of protein kinase A-Ialpha as revealed by domain-silencing mutations. *J Biol Chem* 285 (2010): 15523–15537.
45. Badireddy, S., Yunfeng, G., Ritchie, M. et al. Cyclic AMP analog blocks kinase activation by stabilizing inactive conformation: Conformational selection highlights a new concept in allosteric inhibitor design. *Mol Cell Proteomics* 10 (2011): M110.004390.
46. Li, S., Tsalkova, T., White, M.A. et al. Mechanism of intracellular cAMP sensor Epac2 activation: cAMP-induced conformational changes identified by amide hydrogen/deuterium exchange mass spectrometry (DXMS). *J Biol Chem* 286 (2011): 17889–17897.
47. Tsalkova, T., Mei, F.C., Li, S. et al. Isoform-specific antagonists of exchange proteins directly activated by cAMP. *Proc Natl Acad Sci U S A* 109 (2012): 18613–18618.
48. Scholten, A., Poh, M.K., van Veen, T.A., van Breukelen, B., Vos, M.A., and Heck, A.J. Analysis of the cGMP/cAMP interactome using a chemical proteomics approach in mammalian heart tissue validates sphingosine kinase type 1-interacting protein as a genuine and highly abundant AKAP. *J Proteome Res* 5 (2006): 1435–1447.
49. Aye, T.T., Mohammed, S., van den Toorn, H.W. et al. Selectivity in enrichment of cAMP-dependent protein kinase regulatory subunits type I and type II and their interactors using modified cAMP affinity resins. *Mol Cell Proteomics* 8 (2009): 1016–1028.
50. Vigil, D., Blumenthal, D.K., Heller, W.T. et al. Conformational differences among solution structures of the type Ialpha, IIalpha and IIbeta protein kinase A regulatory subunit homodimers: Role of the linker regions. *J Mol Biol* 337 (2004): 1183–1194.
51. Li, F., Gangal, M., Jones, J.M. et al. Consequences of cAMP and catalytic-subunit binding on the flexibility of the A-kinase regulatory subunit. *Biochemistry* 39 (2000): 15626–15632.
52. Biskup, C., Kusch, J., Schulz, E. et al. Relating ligand binding to activation gating in CNGA2 channels. *Nature* 446 (2007): 440–443.
53. Harootunian, A.T., Adams, S.R., Wen, W., Meinkoth, J.L., Taylor, S.S., and Tsien, R.Y. Movement of the free catalytic subunit of cAMP-dependent protein kinase into and out of the nucleus can be explained by diffusion. *Mol Biol Cell* 4 (1993): 993–1002.
54. Ekanger, R., Sand, T.E., Ogreid, D., Christoffersen, T., and Doskeland, S.O. The separate estimation of cAMP intracellularly bound to the regulatory subunits of protein kinase I and II in glucagon-stimulated rat hepatocytes. *J Biol Chem* 260 (1985): 3393–3401.
55. Ekanger, R., Vintermyr, O.K., Houge, G. et al. The expression of cAMP-dependent protein kinase subunits is differentially regulated during liver regeneration. *J Biol Chem* 264 (1989): 4374–4382.
56. Ueland, P.M., and Doskeland, S.O. An adenosine 3′:5′-monophosphate-adenosine binding protein from mouse liver. A study on its interaction with adenosine 3′:5′-monophosphate and adenosine. *J Biol Chem* 253 (1978): 1667–1676.
57. de Gunzburg, J., Hohman, R., Part, D., and Veron, M. Evidence that a cAMP binding protein from *Dictyostelium discoideum* carries S-adenosyl-L-homocysteine hydrolase activity. *Biochimie* 65 (1983): 33–41.
58. Kotera, J., Grimes, K.A., Corbin, J.D., and Francis, S.H. cGMP-dependent protein kinase protects cGMP from hydrolysis by phosphodiesterase-5. *Biochem J* 372 (2003): 419–426.
59. Gopal, V.K., Francis, S.H., and Corbin, J.D. Allosteric sites of phosphodiesterase-5 (PDE5). A potential role in negative feedback regulation of cGMP signaling in corpus cavernosum. *Eur J Biochem* 268 (2001): 3304–3312.

60. Adams, S.R., Harootunian, A.T., Buechler, Y.J., Taylor, S.S., and Tsien, R.Y. Fluorescence ratio imaging of cyclic AMP in single cells. *Nature* 349 (1991): 694–697.
61. Nikolaev, V.O., and Lohse, M.J. Novel techniques for real-time monitoring of cGMP in living cells. *Handb Exp Pharmacol* (2009): 229–243.
62. Di Benedetto, G., Zoccarato, A., Lissandron, V. et al. Protein kinase A type I and type II define distinct intracellular signaling compartments. *Circ Res* 103 (2008): 836–844.
63. Eigenthaler, M., Nolte, C., Halbrugge, M., and Walter, U. Concentration and regulation of cyclic nucleotides, cyclic-nucleotide-dependent protein kinases and one of their major substrates in human platelets. Estimating the rate of cAMP-regulated and cGMP-regulated protein phosphorylation in intact cells. *Eur J Biochem* 205 (1992): 471–481.
64. Mo, E., Amin, H., Bianco, I.H., and Garthwaite, J. Kinetics of a cellular nitric oxide/cGMP/phosphodiesterase-5 pathway. *J Biol Chem* 279 (2004): 26149–26158.
65. Manns, J.M., Brenna, K.J., Colman, R.W., and Sheth, S.B. Differential regulation of human platelet responses by cGMP inhibited and stimulated cAMP phosphodiesterases. *Thromb Haemost* 87 (2002): 873–879.
66. Jensen, B.O., Kleppe, R., Kopperud, R. et al. Dipyridamole synergizes with nitric oxide to prolong inhibition of thrombin-induced platelet shape change. *Platelets* 22 (2011): 8–19.
67. Corbin, J.D., Turko, I.V., Beasley, A., and Francis, S.H. Phosphorylation of phosphodiesterase-5 by cyclic nucleotide-dependent protein kinase alters its catalytic and allosteric cGMP-binding activities. *Eur J Biochem* 267 (2000): 2760–2767.
68. Wada, H., Osborne, J.C., Jr., and Manganiello, V.C. Effects of temperature on allosteric and catalytic properties of the cGMP-stimulated cyclic nucleotide phosphodiesterase from calf liver. *J Biol Chem* 262 (1987): 5139–5144.
69. Turko, I.V., Francis, S.H., and Corbin, J.D. Binding of cGMP to both allosteric sites of cGMP-binding cGMP-specific phosphodiesterase (PDE5) is required for its phosphorylation. *Biochem J* 329 (Pt 3) (1998): 505–510.
70. Wilson, L.S., Elbatarny, H.S., Crawley, S.W., Bennett, B.M., and Maurice, D.H. Compartmentation and compartment-specific regulation of PDE5 by protein kinase G allows selective cGMP-mediated regulation of platelet functions. *Proc Natl Acad Sci U S A* 105 (2008): 13650–13655.
71. Gjertsen, B.T., Mellgren, G., Otten, A. et al. Novel (Rp)-cAMPS analogs as tools for inhibition of cAMP-kinase in cell culture. Basal cAMP-kinase activity modulates interleukin-1 beta action. *J Biol Chem* 270 (1995): 20599–20607.
72. Van Haastert, P.J., Van Driel, R., Jastorff, B., Baraniak, J., Stec, W.J., and De Wit, R.J. Competitive cAMP antagonists for cAMP-receptor proteins. *J Biol Chem* 259 (1984): 10020–10024.
73. Bos, J.L. Epac proteins: Multi-purpose cAMP targets. *Trends Biochem Sci* 31 (2006): 680–686.
74. Enserink, J.M., Christensen, A.E., de Rooij, J. et al. A novel Epac-specific cAMP analogue demonstrates independent regulation of Rap1 and ERK. *Nat Cell Biol* 4 (2002): 901–906.
75. Schwede F., Bertyinetti, D., Lanerijs, C.N. et al. Structure-guided design of selective Epac1 and Epac2 agonists. *PLoS Biol* 13 (2015): e1002038.
76. de Wit, R.J., Hoppe, J., Stec, W.J., Baraniak, J., and Jastorff, B. Interaction of cAMP derivatives with the 'stable' cAMP-binding site in the cAMP-dependent protein kinase type I. *Eur J Biochem* 122 (1982): 95–99.
77. Doskeland, S.O., Ogreid, D., Ekanger, R., Sturm, P.A., Miller, J.P., and Suva, R.H. Mapping of the two intrachain cyclic nucleotide binding sites of adenosine cyclic 3′,5′-phosphate dependent protein kinase I. *Biochemistry* 22 (1983): 1094–1101.

78. Corbin, J.D., Ogreid, D., Miller, J.P., Suva, R.H., Jastorff, B., and Doskeland, S.O. Studies of cGMP analog specificity and function of the two intrasubunit binding sites of cGMP-dependent protein kinase. *J Biol Chem* 261 (1986): 1208–1214.
79. Jiang, H., Colbran, J.L., Francis, S.H., and Corbin, J.D. Direct evidence for cross-activation of cGMP-dependent protein kinase by cAMP in pig coronary arteries. *J Biol Chem* 267 (1992): 1015–1019.
80. Smith, J.A., Francis, S.H., Walsh, K.A., Kumar, S., and Corbin, J.D. Autophosphorylation of type Ibeta cGMP-dependent protein kinase increases basal catalytic activity and enhances allosteric activation by cGMP or cAMP. *J Biol Chem* 271 (1996): 20756–20762.
81. Wu, J., Brown, S., Xuong, N.H., and Taylor, S.S. RIalpha subunit of PKA: A cAMP-free structure reveals a hydrophobic capping mechanism for docking cAMP into site B. *Structure* 12 (2004): 1057–1065.
82. Wu, J., Jones, J.M., Nguyen-Huu, X., Ten Eyck, L.F., and Taylor, S.S. Crystal structures of RIalpha subunit of cyclic adenosine 5′-monophosphate (cAMP)-dependent protein kinase complexed with (Rp)-adenosine 3′,5′-cyclic monophosphothioate and (Sp)-adenosine 3′,5′-cyclic monophosphothioate, the phosphothioate analogues of cAMP. *Biochemistry* 43 (2004): 6620–6629.
83. Schwede, F., Christensen, A., Liauw, S. et al. 8-Substituted cAMP analogues reveal marked differences in adaptability, hydrogen bonding, and charge accommodation between homologous binding sites (AI/AII and BI/BII) in cAMP kinase I and II. *Biochemistry* 39 (2000): 8803–8812.
84. Chen, H., Tsalkova, T., Mei, F.C., Hu, Y., Cheng, X., and Zhou, J. 5-Cyano-6-oxo-1,6-dihydro-pyrimidines as potent antagonists targeting exchange proteins directly activated by cAMP. *Bioorg Med Chem Lett* 22 (2012): 4038–4043.
85. Christensen, A.E., Viste, K., and Doskeland, S.O. Cyclic nucleotide analogs as tools to investigate cyclic nucleotide signaling. In: *Handbook of Cell Signaling*, edited by E.A. Dennis and R.A. Bradshaw (Elsevier, Oxford, 2009), 1556–1562.
86. Gausdal, G., Wergeland, A., Skavland, J. et al. Cyclic AMP can promote APL progression and protect myeloid leukemia cells against anthracycline-induced apoptosis. *Cell Death Dis* 4 (2013): e516.
87. Huseby, S., Gausdal, G., Keen, T.J. et al. Cyclic AMP induces IPC leukemia cell apoptosis via CRE-and CDK-dependent Bim transcription. *Cell Death Dis* 2 (2011): e237.
88. Petersen, R.K., Madsen, L., Pedersen, L.M. et al. Cyclic AMP (cAMP)-mediated stimulation of adipocyte differentiation requires the synergistic action of Epac- and cAMP-dependent protein kinase-dependent processes. *Mol Cell Biol* 28 (2008): 3804–3816.
89. Boe, R., Gjertsen, B.T., Doskeland, S.O., and Vintermyr, O.K. 8-Chloro-cAMP induces apoptotic cell death in a human mammary carcinoma cell (MCF-7) line. *Br J Cancer* 72 (1995): 1151–1159.
90. Poppe, H., Rybalkin, S.D., Rehmann, H. et al. Cyclic nucleotide analogs as probes of signaling pathways. *Nat Methods* 5 (2008): 277–278.
91. Gettys, T.W., Blackmore, P.F., Redmon, J.B., Beebe, S.J., and Corbin, J.D. Short-term feedback regulation of cAMP by accelerated degradation in rat tissues. *J Biol Chem* 262 (1987): 333–339.
92. Herfindal, L., Nygaard, G., Kopperud, R., Krakstad, C., Doskeland, S.O., and Selheim, F. Off-target effect of the Epac agonist 8-pCPT-2′-*O*-Me-cAMP on P2Y12 receptors in blood platelets. *Biochem Biophys Res Commun* 437 (2013): 603–608.
93. Sand, C., Grandoch, M., Borgermann, C. et al. 8-pCPT-conjugated cyclic AMP analogs exert thromboxane receptor antagonistic properties. *Thromb Haemost* 103 (2010): 662–678.

94. Herfindal, L., Krakstad, C., Myhren, L. et al. Introduction of aromatic ring–containing substituents in cyclic nucleotides is associated with inhibition of toxin uptake by the hepatocyte transporters OATP 1B1 and 1B3. *PLoS One* 9 (2014): e94926.
95. Coulson, R., Baraniak, J., Stec, W.J., and Jastorff, B. Transport and metabolism of *N*^{6}- and C8-substituted analogs of adenosine 3′,5′-cyclic monophosphate and adenosine 3′5′-cyclic phosphorothioate by the isolated perfused rat kidney. *Life Sci* 32 (1983): 1489–1498.
96. Chapman, R.W., House, A., Richard, J. et al. Pharmacology of a potent and selective inhibitor of PDE4 for inhaled administration. *Eur J Pharmacol* 643 (2010): 274–281.
97. Lefkimmiatis, K., and Zaccolo, M. cAMP signaling in subcellular compartments. *Pharmacol Ther* 143 (2014): 295–304.

5 Monitoring Cyclic Nucleotides Using Genetically Encoded Fluorescent Reporters

Kirill Gorshkov and Jin Zhang

CONTENTS

5.1 INTRODUCTION

5.1.1 Spatiotemporal Regulation of Cyclic Nucleotide Signaling

3′,5′-cyclic adenosine monophosphate (cAMP) and 3′,5′-cyclic guanosine monophosphate (cGMP) are ubiquitous second messengers that transduce a wide variety of stimuli into intracellular signals. cAMP and cGMP are involved in many physiological functions including memory formation and vasodilation.[1–4] To achieve these diverse biological functions, the production and degradation of cyclic nucleotides are spatiotemporally regulated inside cells. cAMP and cGMP are generated by enzymes called adenylyl cyclases (AC) and guanylyl cyclases (GC) and degraded by phosphodiesterases (PDE). Restricting the action of cyclic nucleotides to small domains results in localized stimulation of downstream effector proteins such as protein kinase A (PKA) and protein kinase G (PKG). The tight control of cAMP signaling is mediated in part by the A-kinase anchoring protein (AKAP), a class of scaffold proteins involved in bringing together ACs, cAMP-specific PDEs, PKA, and its substrate.[5] cGMP is also highly regulated in space and time by GC, PKG, and cGMP-specific PDE interactions.[6] To investigate the spatiotemporal regulation of cyclic nucleotide signaling, researchers have developed many genetically encoded fluorescent reporters in an effort to monitor real-time changes in cyclic nucleotide concentration at the subcellular level. These biosensors have provided new insights into cyclic nucleotide signaling networks using real-time imaging in living cells.[7]

5.1.2 Fluorescence and Fluorescent Biosensor Basics

5.1.2.1 Fluorescence Concepts and Considerations

Fluorescence microscopy is a cornerstone of modern biological investigation from the nano- to the whole organism level. Fluorescent protein (FP) technology revolutionized the field of cell and developmental biology by providing a genetically encodable fluorescent tag. Before the cloning of *Aqueorea victoria* GFP, cellular imaging required the preparation and microinjection of macromolecules conjugated with fluorescent dyes. Genetic encodability relieved researchers from the cumbersome protocols of labeling macromolecules by letting the cells do the work. Given the fact that the FP primary amino acid sequence contains all the information needed for fluorescence, there is tremendous versatility in the use of FPs to precisely monitor the signaling dynamics of living cells. In addition, the ability to add a specific subcellular localization sequence to a fusion construct increases the spatial resolution to *submicroscopic* levels.[8] Extensive studies involving fluorescent fusion proteins have proven that FPs do not, in most cases, affect a protein's function or movement.

FPs, including green fluorescent protein (GFP) from *A. victoria* (GFP) and dsRed, have chromophores buried within their tertiary structures that can absorb and emit light.[9] The chromophore is produced autocatalytically in an aerobic environment through a process known as maturation. Generally, an 11-stranded β-barrel makes up the tertiary structure, which protects the chromophore from the solvent. The specific microenvironment of each chromophore is shaped by the surrounding amino acids and contributes to the different spectral properties exhibited by different fluorophores.

There are dozens of enhanced FPs to choose from with different properties including the excitation and emission spectra, fluorescence lifetime, extinction coefficient, quantum yield, photostability, and maturation speed.[10] The seven major variants of FPs are divided by their emission maxima as follows: blue (BFP 440–470 nm), cyan (CFP; 471–500 nm), green (GFP; 501–520 nm), yellow (YFP; 521–550 nm), orange (OFP; 551–575 nm), red (RFP; 576–610 nm), and far-red RFP (FRFP; 611–660 nm). It is important to choose the appropriate FP for the desired application. For a more comprehensive overview of FPs, please refer to the 2011 review by Sample et al.[10] in *Chemical Society Reviews*.

FP technology has long been used to passively report on cellular status. FP-based biosensors with the ability to track the dynamics of cellular activity have now exponentially enhanced our views of cellular and subcellular organization of macromolecules and their activities.[11] These kinds of indicators have been designed to detect a variety of intracellular signaling events including enzymatic action and protein–protein interactions as well as changes in ion or molecule concentration. Many of these biosensors utilize a phenomenon known as Förster resonance energy transfer (FRET). FRET is a nonradiative transfer of energy from an excited donor molecule to an acceptor molecule. For the purposes of this chapter, FRET refers to energy transfer between two FPs unless otherwise specified. This phenomenon depends on the spectral overlap of donor emission and acceptor excitation as well as their relative distance and orientation. FRET efficiency (E) is inversely proportional to the sixth power of the distance between donor and acceptor. Therefore, FRET decreases steeply with increased distance and typically occurs when a donor and acceptor pair are separated by approximately 2 to 10 nm.[12] In this way, FRET can be used as a molecular ruler for measuring macromolecules and their interactions.

Several processes happen during FRET: (1) the donor fluorescence is quenched, (2) the lifetime fluorescence of the donor excited state decreases, and (3) the emission of the acceptor fluorescence increases. Acceptor emission upon donor excitation is called sensitized FRET, and will be used in the protocol following this introduction. The FP toolbox for designing FRET biosensors is diverse and expanding. To date, CFP and YFP are the most widely used fluorophores for FRET because the CFP emission spectrum overlaps with the YFP absorbance spectrum and their excitation spectra are well separated.

In addition to energy transfer between two FPs, bioluminescence resonance energy transfer (BRET) involves resonance energy transfer between a bioluminescent donor such as luciferase and an acceptor.[13] Luciferases catalyze enzymatic reactions on exogenously added substrates producing light. When the donor luciferase and acceptor FP are close enough and oriented properly, nonradiative energy transfer can occur and the FP emits a photon at its own emission wavelength.[14]

5.1.2.2 Fluorescent Biosensor Design

Genetically encoded biosensors have been developed to report on the enzymatic activities, levels of second messengers, protein–protein interactions, and other cellular processes. To accomplish this, the biosensor contains a sensing and reporting element that work in concert. The sensing element is made up of one or more endogenous protein domains or engineered sequences to detect the biochemical activity or molecules. For example, the A-kinase activity reporter, or AKAR, sensing element contains a PKA substrate sequence and the phospho-amino acid binding domain

FHA1.[8] FHA1 binds to the substrate when it is phosphorylated by PKA. Together, they act as a molecular switch to detect PKA activity. Therefore, PKA phosphorylation of the substrate induces a conformational change in the biosensor. The reporting element for AKAR consists of two FPs that flank the sensing element and increase FRET when the molecular switch is activated. Another example is the calcium sensor GCaMP, whose sensing unit consists of the calcium-binding protein calmodulin (CaM) and a peptide called M13, which binds to calcium-bound calmodulin. The reporting unit consists of circularly permuted GFP (cpGFP), whose β-barrel has an opening that allows the solvent to quench the chromophore.[15] In the absence of calcium, the conformation of GCaMP is such that the barrel of cpGFP is accessible to solvent molecules resulting in low fluorescence. In the presence of calcium, CaM wraps around M13, occludes the barrel opening, and increases fluorescence because the chromophore is protected from solvent. These two examples highlight the modular design of FP-based biosensors using molecular switches. The 2011 review by Newman et al. contains a very thorough discussion of FPs and biosensors for many different biochemical activities and should be referred to for further information.[13]

5.1.2.3 Fluorescent Biosensors for Cyclic Nucleotides

Genetically encoded biosensors for cyclic nucleotides vary in their FPs and cyclic nucleotide–binding domains (CNBD). These variations result in differences in dynamic range, sensitivity, and specificity for sensing cAMP and cGMP. Generally, the CNBDs make up a sensing unit. The reporting unit can be composed of a single FP such as the cGMP sensor FlincG, or a FRET pair that flanks the CNBD (see Table 5.1; Figure 5.1a). Upon binding of a cyclic nucleotide, the CNBD undergoes a conformational change conducive to changes in fluorescent readouts. In FRET-based biosensors such as ICUE3 and cGES-DE5, the conformational changes induced by binding of the cyclic nucleotides are translated into changes in FRET by affecting the relative distance and orientation between the donor and acceptor FPs. In a single FP-based biosensor such as FlincG, binding of cGMP to the sensing element consisting of two regulatory fragments from PKG that are arranged in tandem with each other[16] induces a conformational change in the biosensor, resulting in an increase in fluorescence (Figure 5.1a,d). The specifics of ICUE3 and cGES-DE5 will be discussed in Section 5.3.1. Several BRET-based sensors have been developed for cyclic nucleotides including a PKA-based sensor,[17] an Epac1-based sensor named CAMYEL,[18] and a cGMP sensor GAF-BRET.[19]

These biosensors allow the monitoring of cyclic nucleotide dynamics in living cells with high spatiotemporal resolution. They can be targeted to many locations including the nucleus, mitochondria, ER, lysosome, plasma membrane, and cytosol using specific localization sequences placed at the N-terminus or C-terminus.[20] Indeed, they can even be targeted to specific proteins and structures such as ion channels, scaffolding proteins, and membrane rafts. The cAMP and cGMP biosensors described in Table 5.1 have been used in studies ranging from in vitro characterization to in vivo analysis in transgenic animals.[21] The next section will detail a protocol for imaging cAMP and cGMP in live cells using an epifluorescence microscope. It also includes a short protocol for calibrating the FRET ratio to cyclic nucleotide concentration.

TABLE 5.1
Examples of Biosensors Used to Monitor cAMP and cGMP

	Reporting Unit	Sensing Unit	EC_{50}	Response	Advantages/Disadvantages	References
FRET-Based cAMP Sensors						
RII-CFP and C-YFP (PKA based)	CFP/YFP	PKA-RII and CAT	~0.3 µM	FRET ↓	Bimolecular, high affinity	Zaccolo and Pozzan[22]
ICUE3 (Epac based)	ECFP/cpVen-L194	Epac1$^{149-881}$ Epac1$^{157-316}$	~12.5 µM	FRET ↓	Unimolecular, large dynamic range	DiPilato and Zhang[23]
Epac1/2-cAMPS (Epac based)	CFP/YFP	Epac2$^{285-443}$	2.4/0.9 µM	FRET ↓	Unimolecular, high affinity	Nikolaev et al.[24]
HCN2-cAMPS (CNGC based)	CFP/YFP	HCN2$^{467-638}$	6.0 µM	FRET ↓	Used for cells with high basal (cAMP)	Nikolaev et al.[25]
FRET-Based cGMP Sensors						
Cygnet1/2 (PKG based)	CFP/YFP	PKGIα Δ1–77/ Δ1–77, T516A	1.5/1.9 µM	FRET ↓	Unimolecular, relatively low sensitivity, and temporal resolution	Honda et al.[26]
cGES-DE2/5 (PDE based)	EYFP/ECFP	PDE5A GAF-A domain	0.9/1.5 µM	FRET ↑	Unimolecular, high affinity, small size, small dynamic range	Nikolaev et al.[27]
Single Fluorescent Protein–Based Sensor						
δ-FlincG (PKG based)	cpGFP	PKGIα$^{77-356}$	0.15 µM	Intensity ↑	Good dynamic range, rapid kinetics	Nausch et al.[16]

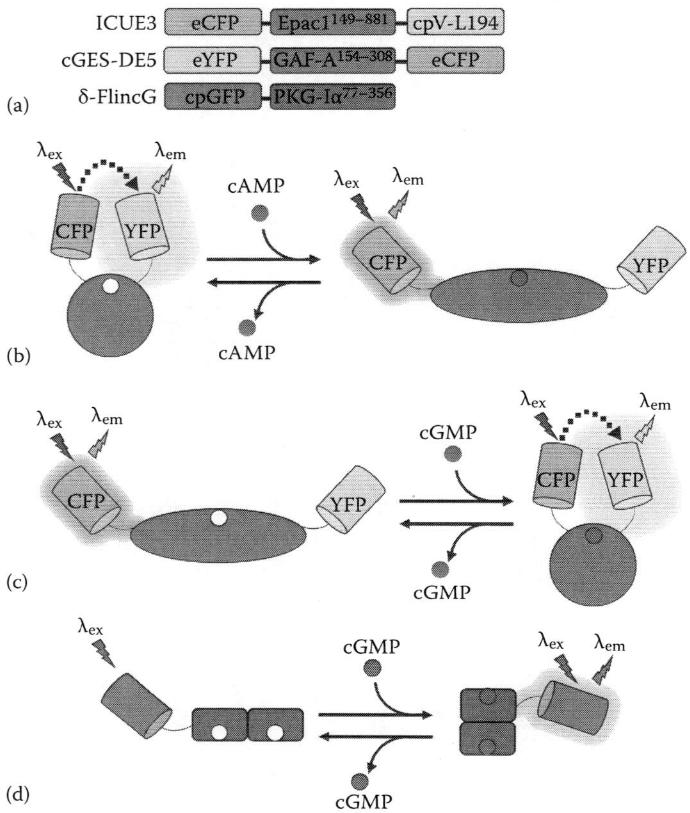

FIGURE 5.1 **(See color insert.)** The design of ICUE3, cGES-DE5, and δ-FlincG. (a) ICUE3 incorporates truncated Epac1$^{149-881}$ positioned between eCFP (FRET donor) and cpVenus L194 (FRET acceptor). cGES-DE5 incorporates truncated GAF-A$^{154-308}$ domain of PDE5A. δ-FlincG uses cpGFP as the FP and PKGIα$^{77-356}$. (b) General schematic of ICUE3. A cAMP-induced conformational change decreases the proximity of the donor and acceptor FPs. Donor excitation causes nonradiative energy transfer to the acceptor FP and emits light. (c) General schematic of cGES-DE5. A cGMP-induced conformational change increases the proximity of the donor and acceptor FPs and increases FRET. (d) General schematic for δ-FlincG. δ-FlincG is a single FP sensor that uses tandem cGMP binding domains from PKG-I$^{77-356}$. When cGMP binds, the domains undergo a conformational change resulting in fluorescence from cpGFP.

5.2 MATERIALS AND METHODS FOR IMAGING cAMP AND USING FRET BIOSENSORS

5.2.1 Cell Culture and Transfection Materials*

1. African green monkey kidney fibroblasts (Cos-7; see Note 1)
2. Dulbecco's phosphate-buffered saline (DPBS)—without Mg^{2+} and Ca^{2+} (Gibco)

* Section 5.2.5 for all notes.

3. Growth medium—Dulbecco's modified Eagle's medium (DMEM; Gibco/BRI, Bethesda, MD) supplemented with 10% fetal bovine serum (FBS, Sigma) and 1% penicillin-streptomycin (Sigma-Aldrich; DMEM-Cos-7; see Note 2)
4. T-25 cm^2 flask for propagation and 35 mm glass-bottomed imaging dishes for experiments (MatTEK, Ashland, MA)
5. Humidified incubator at 5% atmospheric CO_2 and 37°C
6. Solution of trypsin 0.25% for COS-7 and ethylenediamine tetraacetic acid (EDTA, 0.53 mM; Invitrogen, Carlsbad, CA)
7. Lipofectamine 2000 (Invitrogen)
8. OPTI-MEM I Reduced Serum Medium (Opti-MEM, Gibco)
9. ICUE3 or cGES-DE5 plasmid DNA (see Note 3)

5.2.2 Imaging Equipment and Materials

1. Microscope: Axiovert 200 M inverted microscope using a 40×/1.3 NA oil immersion objective lens equipped with an Aqua Stop to prevent liquid from running down the objective (Zeiss, Thornwood, NY). Images are captured using a MicroMAX BFT512 cooled charge-coupled device camera (Roper Scientific, Trenton, NJ)
2. Xenon lamp: XBO 75 W (Zeiss). Neutral density filters 0.6 and 0.3 (Chroma Technology, Bellows Falls, VT)
3. Filter sets for individual channels
   a. CFP—420DF20 excitation filter, 450DRLP dichroic mirror, 475DF40 emission filter
   b. YFP-FRET—420DF20 excitation filter, 450DRLP dichroic mirror, 535DF25 emission filter
   c. YFP—495DF10 excitation filter, 515DRLP dichroic mirror, 535DF25 emission filter
4. Lambda 10-2 filter changer (Sutter Instruments)
5. Immersol® 518F fluorescence free immersion oil (Zeiss)
6. Imaging medium: 1× Hanks balanced salt solution (Gibco) with 2.0 g/L D-glucose; pH adjusted to 7.4 using NaOH and sterilized using a.22-μm filter (see Note 4). Store at 4°C and bring to room temperature before imaging (see Note 5)
7. 1000× stocks of forskolin (Fsk, Calbiochem), 3-isobutyl-1-methylxanthine (IBMX, Sigma), and sodium nitroprusside (SNP, Sigma)

5.2.3 Imaging and Analysis Software

1. Metafluor 7.7 (Molecular Devices) or other imaging software for image acquisition
2. Microsoft Excel for converting imaging software files into spreadsheet format and calculating background subtracted and normalized FRET ratios
3. Image-J (NIH) or other software for image processing and compilation
4. GraphPad Prism version 5 for creating graphing and statistical analyses

5.2.4 Step-by-Step Protocol

5.2.4.1 Cell Culture and Transfection

1. Propagate Cos-7 cells in T-25 flask with the appropriate medium. When they are 70% to 80% confluent, aspirate the growth media and wash with 2 mL of 1× DPBS. To dissociate cells, add 300 μL of Trypsin/EDTA, tilt the flask gently from side to side, and incubate for 2 to 5 min (see Note 6). Add 4.7 mL of fresh medium and resuspend the cells gently (see Note 7). Split the cells 1:10 into 35 mm glass-bottomed imaging dishes. In 24 h, the cells should be 50% to 70% confluent (see Note 8). At this confluence, transfect the cells with the desired construct and image after 24 h.
2. To transfect Cos-7 cells, aliquot 50 μL of OPTI-MEM into two 1.5-mL Eppendorf tubes. One tube is for the plasmid, the other is for the Lipofectamine 2000. Add 2.4 μL of Lipofectamine 2000 per 1.0 μg of DNA to one tube and incubate for 5 min. Add 0.5 μg of DNA to the other tube. After 5 min of incubation, add the DNA into the Lipofectamine tube drop-wise, tap lightly to mix, and incubate for 30 min at room temperature (see Note 9).
3. After 30 min, add the DNA and Lipofectamine complex onto the cells drop-wise. There is no need to change the medium after transfection. Many cells should be transfected using this protocol with very little cell death. Grow cells for another 24 h to allow for good plasmid expression.

5.2.4.2 Microscope Setup and Image Acquisition

1. Prepare 1.5-mL Eppendorf tubes with 2 μL of 1000× stock of drug. Keep on ice until needed.
2. Turn on the lamp, microscope, filter changer, camera controller, and computer sequentially.
3. Configure imaging software to prepare to acquire images for the CFP, FRET, and YFP channels. Typical exposure times for the CFP and FRET channels is 500 ms. YFP images should be acquired using shorter exposures such as 50 ms (see Note 10).
4. Before imaging, wash transfected cells with 1 mL of HBSS imaging buffer ×2. Add 2 mL of HBSS imaging buffer to the dish and place onto the stage (see Note 11). Identify transfected cells with normal morphology and biosensor expression (see Note 12).
5. Acquire images of each channel for 5 min to establish a baseline FRET ratio.
6. To treat cells, aspirate 500 μL of HBSS imaging buffer from the dish and add to the 1.5 mL Eppendorf tube containing the drug. Mix drug with HBSS and add back to the cells using the side of the dish. Mix the solution several times in the dish. Mark the addition of the first drug treatment as an event and resume imaging. For each subsequent drug treatment, repeat step 6, marking the drug additions as you go (see Note 13).
7. Allow any changes in FRET ratio to plateau for at least 5 min.

5.2.4.3 Image Analysis

1. Export the logged channel intensity files into Microsoft Excel.
2. Use the following equations for calculating the background subtracted FRET ratio for each region of interest (ROI; see Note 15).

$$\text{ICUE3 emission ratio} = \frac{(\text{ROI CFP intensity} - \text{Background CFP intensity})}{(\text{ROI FRET intensity} - \text{Background FRET intensity})}$$

$$\text{cGES-DE5 emission ratio} = \frac{(\text{ROI FRET intensity} - \text{Background FRET intensity})}{(\text{ROI CFP intensity} - \text{Background CFP intensity})}$$

3. Set time of drug addition to 0 and normalize the emission ratio using the following equation:

$$\text{Normalized emission ratio at a time t} = \frac{\text{Emission ratio at a time t}}{\text{Emission ratio at a time 0}}$$

4. Graph the emission ratio over time using GraphPad Prism or other software.
5. Open the channel images in ImageJ if you wish to generate a time-lapse of the experiment.
   a. Open each file you wish to include in the movie by dragging the image file into ImageJ
   b. Adjust the brightness and contrast if desired and propagate changes to each image automatically
   c. Stack images by clicking Image>Stacks>Images to stack. Name the stack and save as an .avi file at an appropriate number of frames per second. The more frames per second, the faster the movie will play
6. Use an image processing software such as ImageJ to generate pseudo-colored images for each acquisition. The pseudo color image represents the FRET/CFP ratio (cGES-DE5) or CFP/FRET ratio (ICUE3). These images can be stacked into a movie or turned into a panel for illustrating the real-time changes in FRET. Figure 5.2b and d are examples of these ratiometric images (see Note 14).

5.2.4.4 Calibrating FRET to Cyclic Nucleotide Concentration

1. Use a nondenaturing mammalian cell lysis buffer to obtain cell lysates from HEK293 or other heterologous cells expressing ICUE3.
2. Add 200 μM of IBMX to block degradation of cAMP by PDEs.
3. Measure fluorescence spectra with a fluorimeter, such as Fluoro-Max-3 (Jobin Yvon, Inc.), using a 434 nm excitation light before and after the addition of cAMP (Sigma). The CFP donor emission peak is at 477 nm and the YFP acceptor emission peak is at 525 nm.

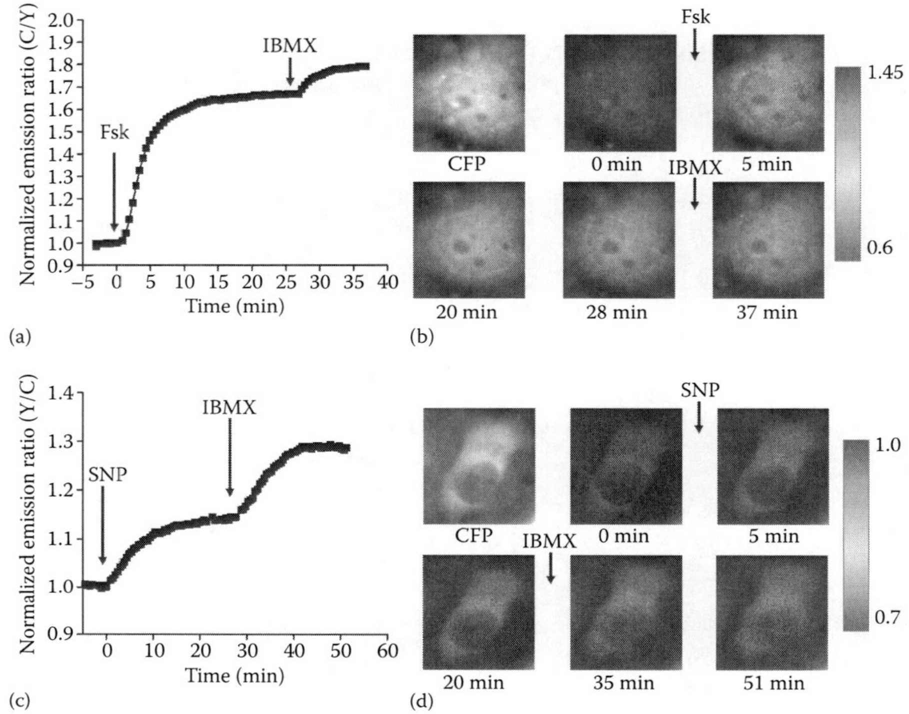

FIGURE 5.2 **(See color insert.)** ICUE3 and cGES-DE5 response in live cells. Cos-7 cells were transiently transfected with ICUE3 and cGES-DE5 plasmid DNA for 24 h before imaging at 40×. The cells were treated at time 0 with (a, b) 50 μM Fsk, an adenylyl cyclase agonist, or (c, d) 80 μM SNP, a soluble guanylyl cyclase agonist. After the response plateued, the cells were further treated with 100 μM IBMX, a general PDE inhibitor. (a, c) Plot of the background-subtracted and normalized emission ratio over time for (a) ICUE3 (CFP/FRET) and (c) cGES-DE5 (FRET/CFP) shows the change in cyclic nucleotide levels. (b, d) Pseudocolor ratiometric images of the cell ROI for (b) ICUE3 and (d) cGES-DE5 responses.

4. Calculate the EC_{50} values using a sigmoidal dose–response equation with variable slope. X represents the logarithmic concentration, Ratio is the FRET ratio response (525 nm/477 nm), and $Ratio_{min}$ and $Ratio_{max}$ are the minimum and maximum FRET ratios.[28]

$$\text{FRET ratio} = \text{Ratio}_{\text{min}} + \frac{\text{Ratio}_{\text{max}} - \text{Ratio}_{\text{min}}}{1 + 10^{\log(\text{EC}_{50} - X) \times \text{Hillslope}}}$$

5. Use the following equation to calculate the intracellular free cAMP from measured FRET ratios. Ratio represents the instantaneous FRET ratio, $Ratio_{max}$ is the maximum change in FRET ratio determined by saturating ICUE3 with cAMP elicited from 50 μM Fsk in the presence of 200 μM IBMX.[28]

$$[cAMP](\mu M) = \frac{EC_{50} \cdot Ratio}{Ratio_{max} - Ratio}$$

5.2.5 Notes

1. When choosing a cell line for FRET imaging, it is important to understand the variability not only between cells but also between cell lines. Some cell types may lack a certain enzyme that is involved in the signaling pathway you are trying to investigate. In some cases, a FRET reporter may work in one cell line, but not in another. Choosing an appropriate cell system is important, so choose one cell line for each line of investigation.
2. All solutions used for cell culture and imaging should be 0.2 μm sterile-filtered to prevent contamination.
3. Because of the similar structures of cyclic nucleotides, the CNBD of different proteins may have some cross-affinity for other cyclic nucleotides. Before performing critical experiments, ensure that the biosensor is specific for the cyclic nucleotide under investigation.
4. All solutions must be made with 18.2 MΩ resistivity unless otherwise specified.
5. Certain cell lines are more sensitive to temperature change. An optional Heatable Insert P for Scanning Stage and Mechanical Stage (Zeiss) may be used. Preheat the imaging HBSS to 37°C prior to imaging.
6. Ensure that the cells are fully detached from the flask by tilting the dish from side to side.
7. Triturate cells by pipetting up and down gently. This will break up clumps and create a single cell suspension. Overly vigorous pipetting may damage and kill the cells.
8. This protocol applies to different cell lines commonly used in laboratories. Make sure to verify the doubling time of the cells, as well as the specific cell culture and transfection guidelines particular to each cell line.
9. Tap the end of the tube lightly to mix the DNA and Lipofectamine. Do not vortex the solution to mix.
10. The FRET channel logs sensitized YFP emission intensity upon CFP excitation, the CFP channel logs direct CFP emission intensity upon CFP excitation, and the YFP channel logs direct YFP emission intensity upon YFP excitation. The YFP channel serves to control for YFP photobleaching and does not factor into emission ratio calculations. Long exposure may cause phototoxicity to the cells and photobleaching of the FPs. Calibrate the exposure times appropriately.
11. The dish must be secured in the stage holder to prevent movement during image acquisition. Movement on the stage while imaging will shift the focus and the regions of interest.
12. Normal Cos-7 adherent cells are flat and fairly uniform in cell morphology. Rounded cells are not good to use. The intensity value should not be beyond

the dynamic range of the imaging system. Do not use overly bright or dim cells. Low intensities will produce low signal-to-noise ratios.
13. The addition of drugs or other perturbations may move the dish. Adjust the regions manually, or use an ROI tracker like Imaris Track.
14. If the software you are using does not have the option to record the ratiometric image, this can be done in ImageJ or another program using the raw emission intensity images from individual channels.
15. In these equations, FRET is used for clarity and refers to the YFP emission upon donor excitation. This can also be referred to as YFP-FRET.

5.3 EXAMPLES OF MONITORING cAMP AND cGMP USING FRET BIOSENSORS

5.3.1 Descriptions and Applications

5.3.1.1 ICUE3 Specifics

Many cAMP and cGMP probes have been developed and further improved over time. One such line of biosensors for cAMP called Indicator of cAMP Using Exchange protein directly activated by cAMP (ICUE) was developed in our laboratory by DiPilato and colleagues.[29] ICUE1 is based on the full-length Epac1, a protein involved in guanine nucleotide exchange for Rap GTPases.[30] Epac1 acts as the sensing unit for ICUE1 by detecting and binding cAMP. The reporting unit is composed of enhanced CFP (ECFP) as the FRET donor and citrine, an enhanced variant of YFP, as the FRET acceptor. These FPs flank the Epac1 protein with ECFP at the N-terminus and citrine at the C-terminus. In its unbound form, ICUE1 is in a high FRET conformation in which ECFP and citrine undergo FRET with ECFP excitation. Upon binding cAMP, Epac undergoes a conformational change, which reorients ECFP and citrine, leading to a decrease in FRET (Figure 5.1b). Next in the lineage is ICUE2, which is generated by replacing full-length Epac1 in ICUE1 with the N-terminally truncated Epac1$^{149-881}$.[28] Finally, ICUE3 is the most recent version of ICUE, using Epac1$^{149-881}$ flanked by ECFP and cpVenus-L194 (Figure 5.1a).[23] It exhibited a 67% improvement in its maximum response as compared with ICUE2. The ICUE probes express diffusely throughout the cytosol and are nuclear excluded when expressed in HEK-293 cells. An example of an ICUE3 FRET imaging experiment in Cos-7 cells is shown in Figure 5.2a and b.

The ICUE line of cAMP sensors has been used to monitor cAMP dynamics in many different biological settings. One such application is to monitor cAMP in different cellular states such as migration. The spatiotemporal dynamics of cAMP are tightly controlled, and depend on the activities of cAMP modulators such as cAMP PDEs and ACs. A study by Lim et al. used the PKA biosensor AKAR3 targeted to the plasma membrane and cAMP sensor ICUE2 to understand cellular signaling dynamics during cell migration. By scratching a monolayer of CHO cells to induce migration, they observed that the leading edge of migrating cells exhibited an integrin-mediated PKA activity. Furthermore, the PKA activity gradient was due to a front-to-back cAMP gradient as reported by ICUE2. This study highlights the use of cAMP and PKA biosensors to elucidate the spatiotemporal dynamics involved in specific cellular functions.[31]

5.3.1.2 cGES-DE5 Specifics

The cGMP probe cGES-DE5 was developed by Nikolaev et al. to monitor cGMP levels inside living cells.[27] The acronym stands for cGMP energy transfer sensor derived from PDE5A. The sensing unit of cGES-DE5 is the regulatory GAF-A domain from PDE5A and the reporting unit is composed of EYFP at the N-terminus and ECFP at the C-terminus (Figure 5.1a). The regulatory GAF-A domain from PDE5A binds cGMP and undergoes a conformational change bringing the two FPs close together and facilitates FRET (Figure 5.1c).[32,33] In conjunction with this probe, the authors developed a similar sensor based on PDE2 and PKGIB, but found cGES-DE5 to be the best with high selectivity for cGMP/cAMP and a large FRET response. This single-chain cGMP probe generates a 40% increase in emission ratio, representing a twofold improvement over the older Cygnet construct (Table 5.1), making it an attractive alternative to previous cGMP biosensors. An example of a FRET imaging experiment in Cos-7 cells is shown in Figure 5.2c and d.

As mentioned previously, cAMP and cGMP are involved in many biological processes. A study by Shelly et al. provides an example of utilizing ICUE3 and cGES-DE5 to study neuronal development.[4] cAMP is a well-known determinant of axonal polarization, whereas cGMP has been implicated in dendrite differentiation.[34,35] The authors conducted several experiments using primary rat cortical and hippocampal neurons transfected with ICUE3 or cGES-DE5. Exposing a neurite tip to cAMP agonists using glass beads caused a cAMP increase in the exposed neurite and a decrease in the other neurites. The same stimulation caused a decrease in cGMP in the stimulated neurite, and an increase in the distal neurites. This phenomenon was dubbed long-range inhibition of cAMP. The authors argue for a mechanism of axon/dendrite polarization that involves localized increases of cAMP and cGMP and activation of their downstream effectors. Thus, ICUE3 and cGES-DE5 were used together to study the effect of reciprocal regulation of cyclic nucleotides in the developing nervous system. The ability of these genetically encoded biosensors to report free cyclic nucleotide dynamics in living cells with high spatial and temporal resolution makes them a very powerful tool for understanding the molecular mechanisms underlying critical biological processes.

5.4 ADVANTAGES AND LIMITATIONS OF GENETICALLY ENCODED FLUORESCENT BIOSENSORS

Enzyme-linked immunosorbent assay and radioimmunoassays are sensitive, specific, and provide a snapshot of the biochemical state of a population of cells. However, these assays require many cells, lack spatial information, and measure total levels of cyclic nucleotides. Early studies of cyclic nucleotide compartmentation comes from the use of cyclic nucleotide–gated ion channels, but these were limited by single compartment localization and lack of generalizability.[36]

FP-based biosensors have many advantages over traditional biochemical techniques including their ability to (1) measure free cyclic nucleotides in live cells, (2) provide information on temporal dynamics, (3) provide information on spatially compartmentalized nucleotides, (4) report on the cyclic nucleotide status of single cells, and (5) provide a quantitative ratiometric readout. Because these biosensors are genetically encodable, they are produced by the cell and can be continuously monitored in live cells to track cyclic nucleotide dynamics. By attaching subcellular targeting sequences to the N-terminus or C-terminus of the biosensor, it can report on dynamics within specific compartments or microdomains. Furthermore, genetically encoded biosensors help uncover the complexities of subcellular signaling events at the level of individual cells. This information generates insights into single cell behaviors and reveals variations between cells.

The ability to provide a quantitative measurement of intracellular cyclic nucleotides is an attractive feature of ratiometric biosensors such as ICUE3 and cGES-DE5. By using the ratio of acceptor emission over donor emission upon donor excitation, some experimental variations such as light source and cell thickness are minimized, making it possible to use calibration curves to help determine nucleotide concentration. This can be accomplished by obtaining lysate from a cell expressing a biosensor and adding known concentrations of cAMP or cGMP while monitoring the FRET ratio using a fluorimeter (see Section 5.2.4.4). The FRET ratio can then be correlated to the concentration of free cyclic nucleotide in live cells.[37] In addition, single-chain FRET biosensors do not have the problem of unequal donor–acceptor expression. Taken together, the many advantages offered by genetically encoded biosensors for cyclic nucleotides make them a great option for researchers wishing to study real-time cyclic nucleotide signaling.

Every technology has its limitations, which need to be kept in mind when using that technology to answer biological questions. In the case of genetically encoded cyclic nucleotide biosensors, endogenous signaling pathways may be affected by the introduction of a cyclic nucleotide biosensor. Because of the high affinity of CNBDs for their cyclic nucleotides, overexpressing them may buffer cAMP or cGMP, leaving the cell without enough cyclic nucleotides to complete the signaling process. Therefore, one needs to assess any perturbations to the cell by comparing cyclic nucleotide–dependent signal transduction responses between cells transfected with a biosensor versus a control GFP. Because every FRET reporter has a set dynamic range and cyclic nucleotide sensitivity, the level of cyclic nucleotide may not be faithfully reported. For example, the FRET change might plateau while the concentration of cyclic nucleotides is still increasing. This limitation can be addressed by designing sensors with varying sensitivities to cAMP and cGMP. If a particular signaling phenomenon induces small changes in cAMP, for example, a biosensor with a high sensitivity would be able to detect it. This same biosensor may not be suitable, however, for larger increases that saturate it early on during the generation of cAMP. In this case a biosensor with a less sensitive FRET response would be more appropriate. To this extent, Russwurm et al.[38] have developed a series of FRET sensors for cGMP based on the tandem GAF domains of PKG called cGi-500, cGi-3000, and cGi-6000, in which the numbers refer to the EC_{50} values for cGMP. Lastly, cyclic nucleotide specificity must be high to accurately report on cyclic nucleotide levels.

Some domains bind to other cyclic nucleotides in addition to the target, making it difficult to study signaling in complex interacting systems.

There are several advantages to using BRET-based reporters for live-cell imaging. Because there is no external excitation light, background autofluorescence is eliminated and sensitivity is increased. The lack of exogenous excitation also makes BRET more suitable for studying photosensitive systems such as photoreceptors or most plant tissues. Additionally, because there is no direct donor excitation, BRET eliminates two of the problems associated with fluorescence imaging, direct excitation of the acceptor and donor photobleaching.[13] In contrast to FRET, BRET requires the availability of an exogenous substrate to generate energy. Bioluminescence is also dimmer than fluorescence and therefore requires sensitive imaging equipment; making high-resolution imaging of cells and tissues more challenging.[39] Because of its useful properties, BRET technology is becoming increasingly popular for high-throughput screening assays.[14,40] Depending on the application and sensitivity requirements, either fluorescence imaging or BRET may be the better choice; nevertheless, both have been used in developing genetically encodable fluorescent reporters to shed light on the complexities of cellular signaling.

5.5 OUTLOOK

The future is bright for FP-based cyclic nucleotide biosensors, which have already provided many great insights into cell signaling. Many groups are working on improved versions of currently available sensors using new and enhanced FPs. Improvements in sensitivity, selectivity, kinetics, and the robustness of the response upon binding will be valuable contributions to monitoring cAMP and cGMP. In addition to cyclic mononucleotides, several groups are pursuing reporters for cyclic dinucleotides, which are important with regard to bacterial pathogenesis. By utilizing these types of reporters in high-throughput screens, researchers may find compounds that act to modulate this unique bacterial signaling network. Developing more advanced far-red and near-infrared cyclic nucleotide sensors for deep tissue imaging will be a remarkable development in the field of biology, especially for neuroscience and cancer research, in which noninvasive imaging techniques are essential.

ACKNOWLEDGMENTS

This work is supported by funding from the NSF GRFP 1232825 (to K.G.) and NIH grant R01DK073368 (to J.Z.). We also thank the members of the Zhang Lab and the Johns Hopkins School of Medicine, Department of Pharmacology and Molecular Sciences for support.

REFERENCES

1. Tengholm, A. Cyclic AMP dynamics in the pancreatic beta-cell. *Ups J Med Sci* 117, 4 (2012): 355–369.
2. Garelick, M.G., Chan, G.C.K., DiRocco, D.P. et al. Overexpression of type I adenylyl cyclase in the forebrain impairs spatial memory in aged but not young mice. *J Neurosci* 29, 35 (2009): 10835–10842.

3. Hofmann, F. The biology of cyclic GMP-dependent protein kinases. *J Biol Chem* 280, 1 (2005): 1–4.
4. Shelly, M., Lim, B.K., Cancedda, L. et al. Local and long-range reciprocal regulation of cAMP and cGMP in axon/dendrite formation. *Science* 327, 5965 (2010): 547–552.
5. Diviani, D., Dodge-Kafka, K.L., Li, J. et al. A-kinase anchoring proteins: Scaffolding proteins in the heart. *Am J Physiol Heart Circ Physiol* 301, 5 (2011): H1742–H1753.
6. Castro, L.R.V., Verde, I., Cooper, D.M.F. et al. Cyclic guanosine monophosphate compartmentation in rat cardiac myocytes. *Circulation* 113, 18 (2006): 2221–2228.
7. Mehta, S., and Zhang, J. Reporting from the field: Genetically encoded fluorescent reporters uncover signaling dynamics in living biological systems. *Annu Rev Biochem* 80, 1 (2011): 375–401.
8. Zhang, J., Ma, Y., Taylor, S.S. et al. Genetically encoded reporters of protein kinase A activity reveal impact of substrate tethering. *Proc Natl Acad Sci U S A* 98, 26 (2001): 14997–15002.
9. Tsien, R.Y. The green fluorescent protein. *Annu Rev Biochem* 67 (1998): 509–544.
10. Sample, V., Newman, R.H., and Zhang, J. The structure and function of fluorescent proteins. *Chem Soc Rev* 38, 10 (2009): 2852–2864.
11. Zhang, J., Campbell, R.E., Ting, A.Y. et al. Creating new fluorescent probes for cell biology. *Nat Rev Mol Cell Biol* 3, 12 (2002): 906–918.
12. Förster, T. Zwischenmolekulare energiewanderung und fluoreszenz. *Annalen der Physik* 437, 1–2 (1948): 55–75.
13. Newman, R.H., Fosbrink, M.D., and Zhang, J. Genetically encodable fluorescent biosensors for tracking signaling dynamics in living cells. *Chem Rev* 111, 5 (2011): 3614–3666.
14. Robinson, K., Yang, J., and Zhang, J. FRET and BRET-based biosensors in live cell compound screens. In: *Fluorescent Protein-Based Biosensors*, eds. J. Zhang, Q. Ni, and R.H. Newman. Humana Press, New York, 2014.
15. Nakai, J., Ohkura, M., and Imoto, K. A high signal-to-noise Ca(2+) probe composed of a single green fluorescent protein. *Nat Biotechnol* 19, 2 (2001): 137–141.
16. Nausch, L.W., Ledoux, J., Bonev, A.D. et al. Differential patterning of cGMP in vascular smooth muscle cells revealed by single GFP-linked biosensors. *Proc Natl Acad Sci U S A* 105, 1 (2008): 365–370.
17. Prinz, A., Diskar, M., Erlbruch, A. et al. Novel, isotype-specific sensors for protein kinase A subunit interaction based on bioluminescence resonance energy transfer (BRET). *Cell Signal* 18, 10 (2006): 1616–1625.
18. Jiang, L.I., Collins, J., Davis, R. et al. Use of a cAMP BRET sensor to characterize a novel regulation of cAMP by the sphingosine 1-phosphate/G13 pathway. *J Biol Chem* 282, 14 (2007): 10576–10584.
19. Biswas, K.H., Sopory, S., and Visweswariah, S.S. The GAF domain of the cGMP-binding, cGMP-specific phosphodiesterase (PDE5) is a sensor and a sink for cGMP. *Biochemistry* 47, 11 (2008): 3534–3543.
20. Zhang, J., Qiang, N., and Newman, R. Fluorescent protein-based biosensors. In: *Methods in Molecular Biology*, eds. J. Zhang, Q. Ni, and R.H. Newman. Humana Press, New York, 2014.
21. Sprenger, J., and Nikolaev, V. Biophysical techniques for detection of cAMP and cGMP in living cells. *Int J Mol Sci* 14, 4 (2013): 8025–8046.
22. Zaccolo, M., and Pozzan, T. Discrete microdomains with high concentration of cAMP in stimulated rat neonatal cardiac myocytes. *Science* 295, 5560 (2002): 1711–1715.
23. DiPilato, L.M., and Zhang, J. The role of membrane microdomains in shaping [small beta]2-adrenergic receptor-mediated cAMP dynamics. *Mol Biosyst* 5, 8 (2009): 832–837.

24. Nikolaev, V.O., Bünemann, M., Hein, L. et al. Novel single chain cAMP sensors for receptor-induced signal propagation. *J Biol Chem* 279, 36 (2004): 37215–37218.
25. Nikolaev, V.O., Bünemann, M., Schmitteckert, E. et al. Cyclic AMP imaging in adult cardiac myocytes reveals far-reaching β1-adrenergic but locally confined β2-adrenergic receptor–mediated signaling. *Circ Res* 99, 10 (2006): 1084–1091.
26. Honda, A., Adams, S.R., Sawyer, C. L. et al. Spatiotemporal dynamics of guanosine 3′,5′-cyclic monophosphate revealed by a genetically encoded, fluorescent indicator. *Proc Natl Acad Sci U S A* 98, 5 (2001): 2437–2442.
27. Nikolaev, V.O., Gambaryan, S., and Lohse, M.J. Fluorescent sensors for rapid monitoring of intracellular cGMP. *Nat Methods* 3, 1 (2006): 23–25.
28. Violin, J.D., DiPilato, L.M., Yildirim, N. et al. β2-Adrenergic receptor signaling and desensitization elucidated by quantitative modeling of real time cAMP dynamics. *J Biol Chem* 283, 5 (2008): 2949–2961.
29. DiPilato, L.M., Cheng, X., and Zhang, J. Fluorescent indicators of cAMP and Epac activation reveal differential dynamics of cAMP signaling within discrete subcellular compartments. *Proc Natl Acad Sci U S A* 101, 47 (2004): 16513–16518.
30. de Rooij, J., Zwartkruis, F.J., Verheijen, M.H. et al. Epac is a Rap1 guanine-nucleotide-exchange factor directly activated by cyclic AMP. *Nature* 396, 6710 (1998): 474–477.
31. Lim, C.J., Kain, K.H., Tkachenko, E. et al. Integrin-mediated protein kinase A activation at the leading edge of migrating cells. *Mol Biol Cell* 19, 11 (2008): 4930–4941.
32. Martinez, S.E., Heikaus, C.C., Klevit, R.E. et al. The structure of the GAF A domain from phosphodiesterase 6C reveals determinants of cGMP binding, a conserved binding surface, and a large cGMP-dependent conformational change. *J Biol Chem* 283, 38 (2008): 25913–25919.
33. Martinez, S.E., Wu, A.Y., Glavas, N.A. et al. The two GAF domains in phosphodiesterase 2A have distinct roles in dimerization and in cGMP binding. *Proc Natl Acad Sci U S A* 99, 20 (2002): 13260–13265.
34. Zheng, J.Q., Zheng, Z., and Poo, M. Long-range signaling in growing neurons after local elevation of cyclic AMP-dependent activity. *J Cell Biol* 127, 6 (Pt 1) (1994): 1693–1701.
35. Cheng, P.-L., and Poo, M.-M. Early events in axon/dendrite polarization. *Annu Rev Neurosci* 35, 1 (2012): 181–201.
36. Rochais, F., Vandecasteele, G., Lefebvre, F. et al. Negative feedback exerted by cAMP-dependent protein kinase and cAMP phosphodiesterase on subsarcolemmal cAMP signals in intact cardiac myocytes: An in vivo study using adenovirus-mediated expression of CNG channels. *J Biol Chem* 279, 50 (2004): 52095–52105.
37. Thunemann, M., Fomin, N., Krawutschke, C. et al. Visualization of cGMP with cGi biosensors. *Methods Mol Biol* 1020 (2013): 89–120.
38. Russwurm, M., Mullershausen, F., Friebe, A. et al. Design of fluorescence resonance energy transfer (FRET)-based cGMP indicators: A systematic approach. *Biochem J* 407, 1 (2007): 69–77.
39. Xu, Y., Kanauchi, A., von Arnim, A.G. et al. Bioluminescence resonance energy transfer: Monitoring protein–protein interactions in living cells. *Methods Enzymol* 360 (2003): 289–301.
40. Fan, F., and Wood, K.V. Bioluminescent assays for high-throughput screening. *Assay Drug Dev Technol* 5, 1 (2007): 127–136.

6 Structural Characterization of Epac by X-Ray Crystallography

Holger Rehmann

CONTENTS

6.1 INTRODUCTION

Epac1 and Epac2, the most recently discovered cAMP receptors,[1,2] are guanine nucleotide exchange factors (GEFs) for the small G protein Rap. Small G proteins are GTPases that cycle between an inactive GDP and an active GTP-bound state,[3] and GEFs such as Epac bind cAMP so that they may catalyze the release of GDP from Rap, after which GTP may then bind to Rap.[4] The ability of cAMP to signal through Epac allows the GTP-bound form of Rap to interact with its downstream effector proteins and thereby to transduce the cAMP signal so that cellular functions will be altered.

Epac consists of an N-terminal regulatory region and a C-terminal catalytic region (Figure 6.1). The catalytic region is responsible for GEF activity and it mediates the interaction of Epac with its substrate protein Rap. The regulatory region of Epac inhibits the catalytic region of Epac in the absence of cAMP. The regulatory region of Epacl and Epac2 contains one and two cyclic nucleotide binding (CNB) domains, respectively. The first CNB domain in Epac2 is not required to maintain the auto-inhibited state, and neither is binding of cAMP to the first CNB domain required to induce activation.[5] Because the first CNB domain of Epac2 is not involved in direct regulation of the catalytic region, it is likely and in agreement with the available experimental data that the molecular basis of regulation is the same in both Epac proteins. To date, crystallographic studies have been successfully applied for Epac2 only. The three constructs of Epac2 that were crystallized are listed here in historical order: (1) the regulatory region containing both CNB domains and the DEP domain,[6] (2) the full-length protein in the autoinhibited state,[5] and (3) an N-terminal truncated version lacking the first CNB domain and the DEP domain, and in complex with a cyclic nucleotide and the substrate protein Rap.[7] Thus, the latest structure reflects the active state whereas the first two structures are determined in the absence of cyclic nucleotide and provide insights concerning the inactive state of Epac2.

The CNB domain of Epacl was extensively studied by nuclear magnetic resonance (NMR).[8–13] These studies gave detailed insights into the dynamics of the cyclic nucleotide binding process. This chapter will discuss the main findings obtained from the characterization of Epac by x-ray crystallography. Detailed protocols concerning the crystallization of Epac are also provided. Characterization of Epac CNB domains by NMR-based techniques is the subject of Chapter 10.

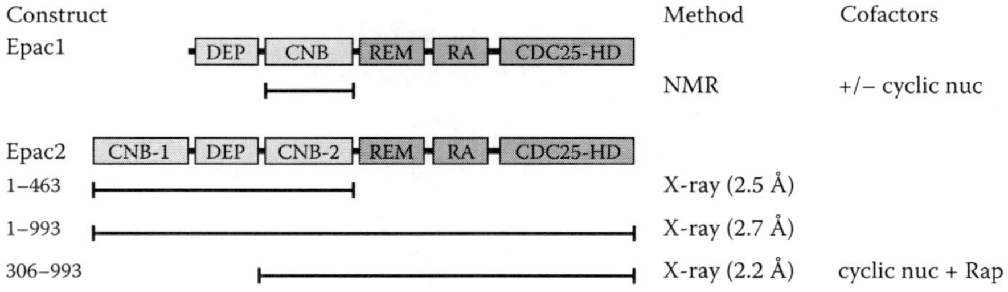

FIGURE 6.1 Constructs of Epac used for structural investigations. The domain arrangement in the primary structure of Epacl and Epac2 is shown. The regulatory region is depicted in light gray and the regulatory region in dark gray. The boundaries of the constructs used for structural investigation are indicated underneath. On the right, the method used for structural investigation is indicated including the resolution obtained in crystallographic studies as well as its cofactors. CNB, cyclic nucleotide binding domain; DEP, Dishevelled, Egl-10, Pleckstrin domain; REM, Ras exchange motif; RA, Ras association domain; CDC25-HD, CDC25-homology domain; cyclic nuc, cyclic nucleotide.

6.2 THE CNB DOMAIN AS A SENSOR OF CYCLIC NUCLEOTIDES

In eukaryotes, the CNB domains are found in cAMP-dependent kinase (PKA), cGMP-dependent kinase (PKG), cyclic nucleotide–regulated ion channels, and in Epacl and Epac2. In prokaryotes, a CNB domain is found in the transcription factor catabolite activated protein (CAP).[14] In all cases, the activities of these proteins are dependent on the binding of cyclic nucleotides. The CNB domain is thus controlling very distinct activities, which raised the question of how cyclic nucleotide binding is translated into biological activity at the molecular level. For a long time, structural information was available on cyclic nucleotide–bound CNB domains only, namely, that of CAP[15] and the regulatory subunit of PKA.[16,17] These structures provided great insights into the binding mode of cyclic nucleotides but left many questions unanswered regarding the activation mechanism. The structure of the regulatory region of Epac2 was the first structure of a CNB domain determined in the absence of a cyclic nucleotide. Based on the comparison of Epac2 with the known cAMP-bound structures of PKA and CAP, a universal model of ligand-induced conformational changes within CNB domains was proposed.[6] However, it remained unknown how these changes control the activities of these cAMP-binding proteins. The answers to these questions required comparisons of the cyclic nucleotide–bound and free structures for the individual proteins. For Epac proteins, the answer was obtained by the determination of the structure of full-length Epac2 in the absence of a cyclic nucleotide,[5] and by the determination of the structure of a truncated $Epac2^{306-993}$ in complex with cyclic nucleotide and Rap.[7] In the active complex of Epac2, Rap is bound to the catalytic site within the CDC-25-homology domain, where it can be trapped in the absence of GDP or GTP bound to Rap (see Section 6.3). The catalytic site is unperturbed in the inactive state of Epac but the access of Rap to the catalytic site is blocked trough steric hindrance by the CNB domain. The CNB domain hangs over the catalytic site, but direct interactions between the CNB domain and the catalytic site are very limited.[5] Upon activation of Epac, cAMP binds first through its phosphate sugar moiety to the CNB domain that is still in the inactive state. These initial interactions induce small rearrangements in a loop that is in direct proximity to cAMP. This allows the core of the CNB domain together with the bound cAMP to swing away, which has two consequences. First, the cAMP-binding site becomes fully established, meaning that after initial movement of the CNB, cAMP can establish a tighter interaction with Epac and thereby traps the CNB domain at the new position. Second, the catalytic site becomes accessible for Rap.[6,7]

6.3 CATALYTIC MECHANISM OF GUANINE NUCLEOTIDE EXCHANGE

The catalytic region of Epac contains a Ras exchange motif (REM) domain, a Ras association (RA) domain, and a CDC25-homology domain (Figure 6.1). CDC25-homology domains are found in all known GEFs for small G proteins of the Ras family.[4,18] Although the CDC25-homology domain is sufficient for catalytic activity, the adjacent REM domain stabilizes the CDC25-homology domain by shielding hydrophobic residues. Fundamental insight into the catalytic mechanism of CDC25-homology domains was originally obtained from the structure of the GEF Son of Sevenless (SOS) in complex with the small G protein Ras.[19] In SOS, the REM domain lies immediately N-terminal to the CDC25-homology domain, whereas in Epac proteins, an RA domain is inserted between the two domains. In the Epac2 tertiary structure, the REM domain and the CDC25-homology domain have the same relative orientation as in SOS.

Even though the CDC25-homology domain is similar in all known exchange factors for small G proteins of the Ras family, each individual exchange factor displays a unique selectivity profile to only a subset of the Ras family members. Thus, Ras family G proteins activated by GEFs can be subdivided as Ras proteins, Rap proteins, and Ral proteins. The comparison of SOS in complex with Ras[19] and Epac in complex with Rap[7] allowed the identification of amino acid residues that determine the selectivity with which GEFs interact with small G proteins.[20] Because mutations of these residues switch the selectivity of GEFs for Ras and Rap, structural studies have revealed the molecular basis for such selectivity, thereby providing an explanation for the specificity with which individual GEFs activate Ras or Rap signal transduction pathways.

6.4 METHODS FOR CHARACTERIZATION OF EPAC2 BY X-RAY CRYSTALLOGRAPHY

6.4.1 Protein Expression and Purification

6.4.1.1 Epac2

The following protocol describes protein expression and purification as one continuous process. It is possible to interrupt the process where indicated.

I. Protein expression

The plasmids pGEX4T2:Epac2_1–463,[6] pGEX4T2:Epac2_1–993,[5] or pGEX4T1:Epac2_306–993[7] were transformed into the bacterial strain CK600K and kept as a glycerol stock at −80°C.

Day 1:

1. Preparation of preculture: Inoculate 500 mL LB medium supplemented with 100 mg/L ampicillin and 50 mg/L kanamycin from glycerol stock and incubate overnight while shaking at 37°C in a 2 L Erlenmeyer flask.

Day 2:

2. Inoculate 10 L Standard I medium (15 g/L peptones, 3 g/L yeast extract, 6 g/L NaCl, 1 g/L glucose) supplemented with 100 mg/L ampicillin and 50 mg/L kanamycin by adding (parts of) the preculture such that an initial OD_{600} of approximately 0.06 is reached. Distribute the culture over six 5-L Erlenmeyer flaks (1.7 L per flask) and incubate while shaking at 25°C. Induce protein expression with 100 μM IPTG by adding 170 μL of 1 M IPTG stock solution per Erlenmeyer flask when an OD_{600} of 0.8 is reached. Continue culturing overnight.

Day 3:

3. Collect bacteria by centrifugation and wash the bacteria once with 0.9% NaCl. The bacterial sediment can be frozen and stored at −20°C for several months prior to protein purification.

II. Protein purification

All steps are performed at 4°C. Make sure the buffers are temperature-equilibrated before use.

Day 3:

4. Equilibrate a 20 mL glutathione sepharose column with 100 mL of buffer A containing 50 mM Tris–HCl (pH 7.5), 50 mM NaCl, 5 mM EDTA, 5 mM β-mercapthoethanol, and 5% glycerol.
5. Resuspend the bacterial sediment in 200 mL buffer A containing 500 μM PMSF. Lyse the bacteria by 5 min sonication using a 1 cm horn in a beaker glass cooled in an ice-water bath.
6. Remove cell debris by centrifugation at 60,000*g* or higher for 1 h. Carefully decant the supernatant into a glass flask.
7. Load the supernatant onto the equilibrate glutathione sepharose column at a flow rate of 1 mL/min.
8. Wash the column with at least 200 mL of buffer B containing 50 mM Tris–HCl (pH 7.5), 400 mM NaCl, 5 mM β-mercapthoethanol, and 5% glycerol.
9. Wash the column at a flow rate of 0.4 mL/min with 350 mL of buffer C containing 50 mM Tris–HCl (pH 7.5), 100 mM KCl, 10 mM $MgCl_2$, 5 mM β-mercapthoethanol, 250 μM ATP, and 5% glycerol. Steps 5 and 6 are typically continued overnight.

Day 4:

10. Wash the column at a flow rate of 1 mL/min with 60 mL of buffer D containing 50 mM Tris–HCl (pH 7.5), 50 mM NaCl, 10 mM $CaCl_2$, 5 mM β-mercapthoethanol, and 5% glycerol.
11. Dissolve 60 units of thrombin (Serva Electrophoresis) in 18 mL of buffer D and load the solution onto the column. Leave the column without flow overnight.

    Note: One unit of thrombin is defined as 1 NIH unit, which clots a standard fibrinogen solution in 15 s at 37°C.[21] Different suppliers of thrombin may use different unit definitions.
12. Equilibrate a gel filtration column with buffer E containing 40 mM Tris–HCl (pH 7.5), 50 mM NaCl, 5 mM β-mercapthoethanol, and 2% glycerol according to the following scheme:

    $Epac2^{1-463}$: Equilibrate Superdex 70 26/60 with 1000 mL buffer at a flow rate of 1 mL/min

    $Epac2^{1-993}$: Equilibrate Superdex 200 16/60 with 300 mL buffer at a flow rate of 0.5 mL/min

    $Epac2^{306-993}$: Equilibrate Superdex 200 16/60 with 300 mL buffer at a flow rate of 0.5 mL/min

    (this is not required on day 4 if the process will be interrupted at step 14 on day 5)

Day 5:

13. Elute the cleaved Epac protein from the column by applying a flow of 1 mL/min with buffer D in 3 mL fractions. Elution of the protein will start immediately.

14. Pool the Epac-containing fractions and concentrate using a spin filter column. Determine the protein concentration. If desired, the concentrated protein can be flash-frozen in liquid nitrogen and the protein stored at −80°C.
15. Regenerate the glutathione sepharose column by washing with 80 mL buffer containing 50 mM Tris–HCl (pH 7.5), 50 mM NaCl, and 20 mM glutathione followed by 100 mL 6 M guanidine hydrochloride and 300 mL water.
16. Load 100 mg protein to the equilibrated gel filtration column. The volume to be loaded should not exceed 3 mL in case of $Epac2^{1-463}$ or 1 mL in case of $Epac2^{1-993}$ and $Epac2^{306-993}$. Apply a flow rate of 1 mL/min (Superdex 75 26/60) or of 0.5 mL/min (Superdex 200 16/60) with buffer E.
    Superdex 75 26/60: Use a total of 330 mL buffer. Collect 3 mL fractions between 110 and 330 mL.
    Superdex 200 16/60: Use a total of 120 mL buffer. Collect 2 mL fractions between 20 and 120 mL.
17. Analyze fractions by SDS-PAGE. Pool Epac-containing fractions and concentrate using a spin filter column. Determine the protein concentration and flash freeze the protein in liquid nitrogen.

6.4.1.2 Rap

The protocol is based on the original description of the purification of Rap by Herrmann and coworkers.[22]

I. Protein expression:
The plasmid ptac:Rap1b_1–167 was transformed into the bacterial strain CK600K and kept as a glycerol stock at −80°C.

Day 1:

1. Preparation of preculture: Inoculate 500 mL LB medium supplemented with 100 mg/L ampicillin and 50 mg/L kanamycin from glycerol stock and incubate overnight while shaking at 37°C in a 2-L Erlenmeyer flask.

Day 2:

2. Inoculate 10 L standard I medium (15 g/L peptones, 3 g/L yeast extract, 6 g/L NaCl, and 1 g/L glucose) supplemented with 100 mg/L ampicillin and 50 mg/L kanamycin by adding (parts of) the preculture such that an initial OD_{600} of approximately 0.06 is reached. Distribute the culture over six 5-L Erlenmeyer flasks (1.7 L per flask) and incubate while shaking at 25°C. Induce protein expression with 100 μM of IPTG by adding 170 μL of 1 M IPTG stock solution per Erlenmeyer flask when an OD_{600} of 0.8 is reached. Continue culturing overnight.
   Start equilibration of column for protein purification (see protein purification; step 7).

Day 3:

3. Collect bacteria by centrifugation and wash the bacteria once with 0.9% NaCl. The bacterial sediment can be frozen and stored at −20°C for several months prior to protein purification.

II. Protein purification:

All steps are performed at 4°C. Make sure the buffers are temperature-equilibrated before use.

Day 2:

4. Equilibrate a 700 mL Q-sepharose column with at least 1.5 L of buffer F containing 32 mM Tris–HCl (pH 7.6), 5 mM $MgCl_2$ and 5 mM β-mercapthoethanol.

Day 3:

5. Resuspend the bacterial sediment in 200 mL buffer F containing 500 μM PMSF. Lyse the bacteria by 5 min sonication using a 1 cm horn in a beaker glass cooled in an ice-water bath.
6. Remove cell debris by centrifugation at 60,000*g* or higher for 1 h. Carefully decant the supernatant into a glass flask.
7. Load the supernatant onto the equilibrated Q-sepharose column at a flow rate of 4 mL/min.
8. Wash the column with at least 1 L of buffer F. The column will appear white again.
9. Elute the column in a linear 4 L gradient (0–300 mM NaCl) in buffer F. Start collecting after 1 L in 12 mL fractions. Steps 5 and 6 are typically performed overnight.

Day 4:

10. Analyze every other fraction by 15% SDS-PAGE. Rap appears as a band of about 20 kDa protein. Pool Rap-containing fractions in a beaker glass and measure the volume.
11. Precipitate Rap with ammonium sulfate: Add a magnetic stirring bar to the protein solution, mix gently, and add the ammonium sulfate (156 g $(NH_4)_2SO_4$ per liter protein solution) in small portions during 3 h to the protein solution. Continue stirring overnight.
12. Regenerate Q-sepharose column: wash with 1 L of 5 M NaCl, 1 L water, 300 mL of 6 M guanidine hydrochloride, and 2 L of 1 mM NaN_3.
13. Equilibrate gel filtration column (superdex 26/60) with 1 L buffer G containing 50 mM Tris–HCl (pH 7.6), 5 mM $MgCl_2$, 50 mM NaCl, and 5 mM β-mercapthoethanol (this is not required on day 4 if the process will be interrupted at step 15 on day 5).

Day 5:

14. Spin down the precipitated protein. Carefully decant the supernatant. Spin the centrifuge tube containing the sediment briefly to collect remaining liquid and remove the liquid carefully with a pipette.
15. Dissolve the protein pellet in 10 mL of buffer G by gentle *up and down* pipetting. The protein solution can be flash-frozen in liquid nitrogen and stored. It is recommended to determine the protein concentration and aliquot the protein at portions of 100 mg.
16. Load 100 mg protein to the equilibrated gel filtration column. The volume to be loaded should not exceed 3 mL. Elute the column with 330 mL buffer G at a flow rate of 1 mL/min. Collect 3 mL fractions between 110 and 330 mL.

17. Analyze fractions by 15% SDS-PAGE. Typically, Rap-containing fractions start at about fraction 30.
18. Pool the Rap-containing fractions and concentrate the protein by using the spin filter column. Determine the protein concentration and flash freeze the protein in liquid nitrogen. Store the protein at −80°C. If desired, the gel filtration can be done overnight such that steps 13 and 14 are performed at day 6.
19. Repeat steps 12 to 14 with additional 100 mg of aliquots obtained at step 11. The gel filtration is highly reproducible and it is possible to skip the analysis of the fractions by SDS-PAGE in the consecutive runs.

6.4.2 Crystallization

6.4.2.1 Crystallization of Epac2$^{1-463}$

Epac2$^{1-463}$ is prepared at a concentration of 40 g/L in buffer E. This buffer is identical to the buffer used in the last purification step of the protein.

Reservoir solution containing 25 mM HEPES (pH 7.5), 250 mM NaH_2PO_4, 750 mM K_2HPO_4, and 10 mM DTT is prepared from the following stock solutions and distilled water:

- 1 M HEPES in distilled water, pH adjusted to 7.5 with 1 N NaOH
- 0.5 M NaH_2PO_4 and 1.5 M K_2HPO_4 in distilled water (referred to a 2 M phosphate)
- 0.5 M DTT in distilled water

All stock solutions are filtered through a 0.2 μm filter.

Crystals are grown either in the sitting drop setup or the hanging drop setup at 293 K. Drops are prepared by mixing 1.5 μL of protein solution with 1.5 μL of reservoir solution. The drops are equilibrated against 500 μL of the reservoir solution. Crystals are typically obtained after 2 to 3 days. To obtain optimally sized crystals, it is recommended that the concentration of phosphate should vary from 0.8 to 1.2 M in 0.1-M steps.

6.4.2.2 Cryoprotection of Epac2$^{1-463}$ Crystals

The cryosolution used is the same as the reservoir solution plus 23% glycerol. Crystals are fished out of the mother drop with a crystal mounting loop of the appropriate size, transferred into a drop of cryosolution of 5 μL volume, and soaked for 5 to 10 min. Soaked crystals are then flash-frozen and stored in ice-free liquid nitrogen.

6.4.2.3 Crystal Properties of Epac2$^{1-463}$

Crystals diffract up to a 2.5 Å resolution at a synchrotron radiation source. Epac2$^{1-463}$ crystals belong to the space group $P2_12_12_1$ with the following unit cell parameters: a = 67 Å, b = 96 Å, c = 103 Å, α = 90°, β = 90°, γ = 90°. The unit cell contains one molecule per asymmetric unit. Residues 1 to 12, 168 to 179, and 446 to 463 are not visible in the electron density.

6.4.2.4 Crystallization of Epac2$^{1-993}$

Epac2$^{1-993}$ is prepared at a concentration of 10 g/L in buffer E. This buffer is identical to the buffer used in the last purification step of the protein.

Reservoir solution containing 100 mM Bis-Tris propane (pH 7.5), 200 mM $NaNO_3$, and 12% PEG3350 is prepared from the following stock solutions and distilled water:

- 1 M Bis-Tris propane in distilled water, pH adjusted to 7.5 with hydrochloric acid
- 1 M $NaNO_3$ dissolved in distilled water
- 50% weight per volume PEG3350 in distilled water

All stock solutions are filtered through a 0.2 μm filter.

Crystals are grown in the sitting drop setup at 293 K. Drops are prepared by mixing 1.5 μL of protein solution with 1.5 μL of reservoir solution. The drops are equilibrated against 500 μL of the reservoir solution. Crystals are typically obtained after 2 to 3 days. To obtain optimally sized crystals, it is recommended that the concentration of PEG3350 be varied from 11% to 13% in 0.25% steps.

6.4.2.5 Cryoprotection of Epac2$^{1-993}$ Crystals

The cryosolution used is the same as the reservoir solution plus 16% glycerol. Crystals are fished out of the mother drop with a crystal mounting loop of the appropriate size, transferred into a drop of cryosolution of 5 μL volume, and soaked in cryosolution briefly (about 1 min) and fished out again with a mounting loop. Crystals are flash-frozen and stored in ice-free liquid nitrogen.

6.4.2.6 Crystal Properties of Epac2$^{1-993}$

Crystals diffract up to a 2.7 Å resolution at a synchrotron radiation source. Crystals belong to the space group $P2_12_12_1$ with the following unit cell parameters: a = 76 Å, b = 97 Å, c = 172 Å, α = 90°, β = 90°, γ = 90°. The unit cell contains one molecule per asymmetric unit. Residues 1 to 14, 171 to 178, 465 to 476, 614 to 641, 725 to 731, and 992 to 993 are not visible in the electron density.

6.4.2.7 Crystallization of Epac2$^{306-993}$ in Complex with Rap1b and Cyclic Nucleotide

The complex is prepared directly prior to crystallization by mixing the proteins and the cyclic nucleotides in buffer containing 40 mM Tris–HCl (pH 7.5), 50 mM NaCl, 2.5% glycerol, and 1 mM EDTA to the following final concentrations:

- 8.7 g/L Epac2$^{306-993}$
- 3.4 g/L Rap1b$^{1-167}$
- 3 mM cyclic nucleotide

If desired, the mixture can be flash-frozen in liquid nitrogen and stored at −80°C. This allows the preparation of follow-up crystallization trials from exactly the same protein solution. It is recommended to keep freezing and thawing cycles to a minimum.

Reservoir solution containing 100 mM sodium citrate (pH 5.6), 0.4 M $(NH_4)_2SO_4$, and 1.2 M $LiSO_4$ is prepared from the following stock solutions and distilled water:

- 1 M citrate (pH 5.6) prepared in distilled water from sodium citrate and citric acid
- 4 M $(NH_4)_2SO_4$ in distilled water
- 2.7 M $LiSO_4$ in distilled water

All stock solutions are filtered through a 0.2 μm filter.

Crystals are grown in the sitting drop setup at 277 K. Drops are prepared by mixing 1.5 μL protein solution with 1.5 μL reservoir solution. The drops are equilibrated against 500 μL of reservoir solution. Crystals typically appear after 4 to 5 days and continue growing for 2 or 3 weeks. Crystal growth may be initiated by seeding. Seeding is performed by mechanical destruction of a crystal with a sharp needle and by wiping the needle successively trough several drops of freshly mixed crystallization drops.

6.4.2.8 Cryoprotection of Crystals of Epac2$^{306-993}$ in Complex with Rap1b and Cyclic Nucleotide

Cryosolution containing 100 mM sodium citrate (pH 5.6), 0.4 M $(NH_4)_2SO_4$, and 2.1 M $LiSO_4$ is used. Crystals are fished out of the mother drop with a crystal mounting loop of the appropriate size, transferred into a drop of 5 μL cryosolution and briefly soaked in cryosolution (about 1 min). Crystals are flash-frozen and in ice-free liquid nitrogen.

6.4.2.9 Crystal Properties of Epac2$^{306-993}$ in Complex with Rap1b and Cyclic Nucleotide

Crystals diffract up to a 2.2 Å resolution at a synchrotron radiation source. Crystals belong to the space group $I2_12_12_1$ with the following unit cell parameters: a = 125 Å, b = 149 Å, c = 225 Å, $\alpha = 90°$, $\beta = 90°$, $\gamma = 90°$. The unit cell contains one molecule per asymmetric unit. Residues 306 to 309, 463 to 477, 613 to 642, 953 to 961, and 991 to 993 of Epac2$^{306-993}$ and 1 to 2, 45 to 49, 135 to 142, and 165 to 167 of Rap1b are not visible in the electron density.

The complex contains three non–water ligand molecules:

A sulfate ion is bound to Rap. The Rap molecule itself is nucleotide-free, meaning that it is neither bound to GDP or GTP. This is typical for crystal structures of small G proteins in complex with their exchange factors. The exchange factor and the nucleotide compete for binding to the G protein by partially overlapping interactions areas. The trimeric complexes of G protein, nucleotide, and exchange factor are thus transient in nature, whereas the G protein–nucleotide complex and the G protein–exchange factor complex are of high affinity.[4] However, the sulfate ion is found at the position in Rap, which would, in the GTP bound state, be occupied by the γ-phosphate of GTP. The sulfate ion thus stabilizes the conformation of the residue normally involved in γ-phosphate binding. The sulfate ion originates from the crystallization solution and its presence in the nucleotide binding site of Rap may partially rationalize the success of crystallization in these conditions. Interestingly, a sulfate ion is found at the equivalent position in the small G protein Ran in the structure of Ran in complex with its exchange factor RCC1.[23]

Two molecules of the cyclic nucleotide are found in the crystal. The first cyclic nucleotide is found in the cAMP binding site within the cyclic nucleotide binding domain. The core of this binding site is defined by the interactions of the phosphate sugar moiety of the cyclic nucleotide with the protein. The residues involved in this interactions define typical consensus motifs found in cyclic nucleotide binding domains.[24] The basic binding mode is thus identical to that found in the cyclic nucleotide binding domains of cAMP-dependent kinase,[16] cyclic nucleotide–regulated ion channels,[25] and the bacterial catabolic activated protein.[15] Mutations of residues constituting this binding site in Epac interfere with cyclic nucleotide binding to Epac and with the activation of Epac by cyclic nucleotides.[6,7,26,27] This cyclic nucleotide is thus reflecting the physiological relevant binding mode.

The binding site of the second cyclic nucleotide is constituted by residues from two symmetry-related Epac2$^{306-993}$ molecules (Figure 6.2). The binding surface contributed by the individual Epac molecules is relatively small. It is therefore very

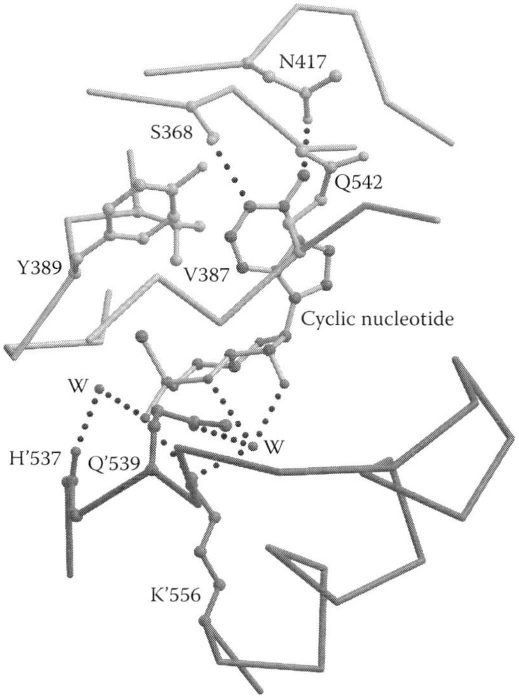

FIGURE 6.2 The second cyclic nucleotide establishes a crystal contact. The cyclic nucleotide is shown in ball-and-stick with bonds in light gray and atoms in dark gray. The cyclic nucleotide forms interactions with two symmetry-related Epac molecules shown in light and dark gray. Amino acids are labeled by single-letter codes. W, water molecule. The figure was generated using the programs Molscript[28] and Raster3D.[29]

unlikely that the surface of just one molecule would be sufficient for binding with reasonable affinity. The binding of the second cyclic nucleotide is thus considered as a crystallization artifact without physiological relevance.

6.4.3 Structure Solution and Model Building

The choices of the strategies to solve the phase problems need to be seen in the historical context. To solve the structure of Epac2$^{1-463}$, protein labeled with selenomethionine was prepared for experimental phase determination by MAD. Trials to solve the structure by molecular replacement with CNB domains of PKA had failed, probably because the C-terminal borders of the search models were not chosen optimally. The structure of Epac2$^{1-993}$ was solved by molecular replacement with the structure of Epac2$^{1-463}$ and the CDC25 homology domain and the REM domain of SOS as search models. The structure of Epac2$^{306-993}$ in complex with Rap1b and cyclic nucleotide could be solved by using the catalytic region from the structure of Epac2$^{1-993}$ as a search model. Model building and refinement were done by common crystallographic procedures, the details of which are provided in the original literature and in the entries of the protein data bank.[5–7]

It is expected that future structures of different Epac isoforms or of novel complexes can be solved by molecular replacement protocols.

6.5 SUMMARY AND OPEN QUESTIONS

The structural characterization of Epac2 has identified the molecular mechanism by which cAMP controls the exchange activity of Epac proteins and has contributed to a general understanding of how CNB domains couple the cellular cAMP concentration to cellular signaling.[14] However, several questions still remain open.

A high-resolution structure of cyclic nucleotide bound to full-length Epac2 is still missing. Such a structure may give insights into the function of the first CNB domain. Structural information of Epac1 is limited to the isolated CNB domain. A full-length structure of Epac1 may point out differences in regulation mechanisms between Epac1 and Epac2. In the cellular context, Epac signaling occurs in protein complexes. For example, Epac1 is known to interact with Ezrin[30] and the nuclear pore protein RanBP2[31] and Epac2 is known to interact with Rim,[32] a protein involved in exocytosis. The structural basis for this interaction is unknown and the structures of these protein complexes remain to be solved. Finally, further development of small molecule Epac1 and Epac2 agonists or antagonists will profit from cocrystal structures of lead compounds.

REFERENCES

1. Kawasaki, H. et al. A family of cAMP-binding proteins that directly activate Rap1. *Science* 282, 2275–2279 (1998).
2. de Rooij, J. et al. Epac is a Rap1 guanine-nucleotide-exchange factor directly activated by cyclic AMP. *Nature* 396, 474–477 (1998).
3. Vetter, I. R., and Wittinghofer, A. The guanine nucleotide-binding switch in three dimensions. *Science* 294, 1299–1304 (2001).

4. Bos, J. L., Rehmann, H., and Wittinghofer, A. GEFs and GAPs: Critical elements in the control of small G proteins. *Cell* 129, 865–877 (2007).
5. Rehmann, H., Das, J., Knipscheer, P., Wittinghofer, A., and Bos, J. L. Structure of the cyclic-AMP-responsive exchange factor Epac2 in its auto-inhibited state. *Nature* 439, 625–628 (2006).
6. Rehmann, H. et al. Structure and regulation of the cAMP-binding domains of Epac2. *Nat Struct Biol* 10, 26–32 (2003).
7. Rehmann, H. et al. Structure of Epac2 in complex with a cyclic AMP analogue and RAP1B. *Nature* 455, 124–127 (2008).
8. Mazhab-Jafari, M. T. et al. Understanding cAMP-dependent allostery by NMR spectroscopy: Comparative analysis of the EPAC1 cAMP-binding domain in its apo and cAMP-bound states. *J Am Chem Soc* 129, 14482–14492 (2007).
9. Das, R. et al. Entropy-driven cAMP-dependent allosteric control of inhibitory interactions in exchange proteins directly activated by cAMP. *J Biol Chem* 283, 19691–19703 (2008).
10. Das, R. et al. Dynamically driven ligand selectivity in cyclic nucleotide binding domains. *J Biol Chem* 284, 23682–23696 (2009).
11. Selvaratnam, R., Chowdhury, S., Van Schouwen, B., and Melacini, G. Mapping allostery through the covariance analysis of NMR chemical shifts. *Proc Natl Acad Sci U S A* 108, 6133–6138 (2011).
12. Van Schouwen, B., Selvaratnam, R., Fogolari, F., and Melacini, G. Role of dynamics in the autoinhibition and activation of the exchange protein directly activated by cyclic AMP (EPAC). *J Biol Chem* 286, 42655–42669 (2011).
13. Selvaratnam, R. et al. The projection analysis of NMR chemical shifts reveals extended EPAC autoinhibition determinants. *Biophys J* 102, 630–639 (2012).
14. Rehmann, H., Wittinghofer, A., and Bos, J. L. Capturing cyclic nucleotides in action: Snapshots from crystallographic studies. *Nat Rev Mol Cell Biol* 8, 63–73 (2007).
15. Weber, I. T., and Steitz, T. A. Structure of a complex of catabolite gene activator protein and cyclic AMP refined at 2.5 A resolution. *J Mol Biol* 198, 311–326 (1987).
16. Su, Y. et al. Regulatory subunit of protein kinase A: Structure of deletion mutant with cAMP binding domains. *Science* 269, 807–813 (1995).
17. Diller, T. C., Madhusudan, Xuong, N. H., and Taylor, S. S. Molecular basis for regulatory subunit diversity in cAMP-dependent protein kinase: Crystal structure of the type II beta regulatory subunit. *Structure (Camb)* 9, 73–82 (2001).
18. Quilliam, L. A., Rebhun, J. F., and Castro, A. F. A growing family of guanine nucleotide exchange factors is responsible for activation of Ras-family GTPases. *Prog Nucleic Acid Res Mol Biol* 71, 391–444 (2002).
19. Boriack-Sjodin, P. A., Margarit, S. M., Bar-Sagi, D., and Kuriyan, J. The structural basis of the activation of Ras by Sos. *Nature* 394, 337–343 (1998).
20. Popovic, M., Rensen-de, L. M., and Rehmann, H. Selectivity of CDC25 homology domain-containing guanine nucleotide exchange factors. *J Mol Biol* 425, 2782–2794 (2013).
21. Baughman, D. J. Thrombin assay. *Methods Enzymol* 19, 145–157 (1970).
22. Herrmann, C., Horn, G., Spaargaren, M., and Wittinghofer, A. Differential interaction of the ras family GTP-binding proteins H-Ras, Rap1A, and R-Ras with the putative effector molecules Raf kinase and Ral-guanine nucleotide exchange factor. *J Biol Chem* 271, 6794–6800 (1996).
23. Renault, L., Kuhlmann, J., Henkel, A., and Wittinghofer, A. Structural basis for guanine nucleotide exchange on Ran by the regulator of chromosome condensation (RCC1). *Cell* 105, 245–255 (2001).
24. Canaves, J. M., and Taylor, S. S. Classification and phylogenetic analysis of the cAMP-dependent protein kinase regulatory subunit family. *J Mol Evol* 54, 17–19 (2002).

25. Clayton, G. M., Silverman, W. R., Heginbotham, L., and Morais-Cabral, J. H. Structural basis of ligand activation in a cyclic nucleotide regulated potassium channel. *Cell* 119, 615–627 (2004).
26. Rehmann, H., Schwede, F., Doskeland, S. O., Wittinghofer, A., and Bos, J. L. Ligand-mediated activation of the cAMP-responsive guanine nucleotide exchange factor Epac. *J Biol Chem* 278, 38548–38556 (2003).
27. Rehmann, H., Rueppel, A., Bos, J. L., and Wittinghofer, A. Communication between the regulatory and the catalytic region of the cAMP-responsive guanine nucleotide exchange factor Epac. *J Biol Chem* 278, 23508–23514 (2003).
28. Kraulis, P. J. Molscript—A program to produce both detailed and schematic plots of protein structures. *J Appl Crystallogr* 24, 946–950 (1991).
29. Merritt, E. A., and Murphy, M. E. P. Raster3D version-2.0—A program for photorealistic molecular graphics. *Acta Crystallogr D Biol Crystallogr* 50, 869–873 (1994).
30. Gloerich, M. et al. Spatial regulation of cyclic AMP-Epac1 signaling in cell adhesion by ERM proteins. *Mol Cell Biol* 30, 5421–5431 (2010).
31. Gloerich, M. et al. The nucleoporin RanBP2 tethers the cAMP effector Epac1 and inhibits its catalytic activity. *J Cell Biol* 193, 1009–1020 (2011).
32. Kashima, Y. et al. Critical role of cAMP-GEFII—Rim2 complex in incretin-potentiated insulin secretion. *J Biol Chem* 276, 46046–46053 (2001).

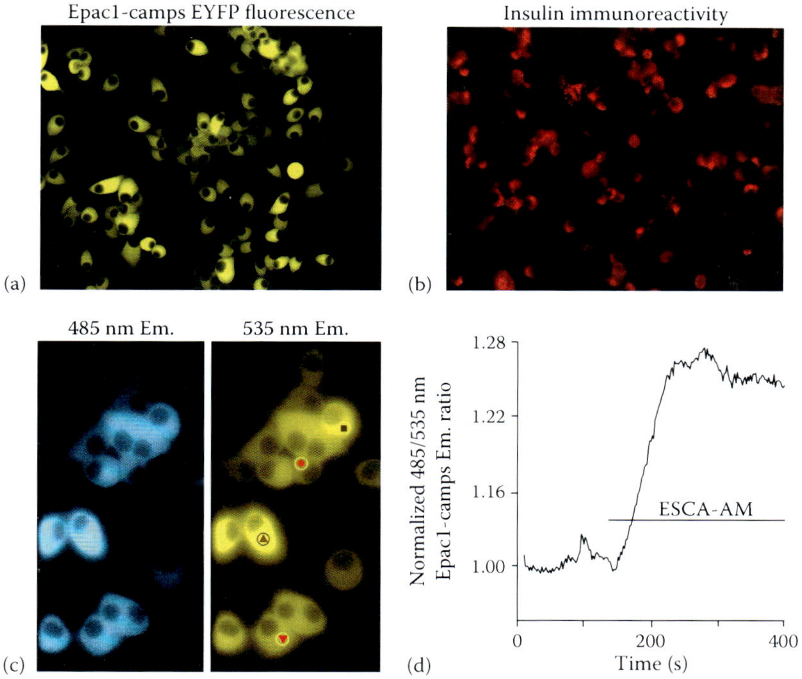

FIGURE 3.2 INS-1 cell clone C-10 stably expresses Epac1-camps, as detected using an EYFP filter set (a). These cells also express insulin immunoreactivity that is detectable by fluorescence immunocytochemistry using an anti-insulin primary antiserum in combination with an Alexa Fluor 594 conjugated secondary antiserum (b). Live-cell imaging of C-10 cells using 440 nm excitation light allows detection of Epac1-camps in which the donor ECFP emission fluorescence is measured at 485 nm, whereas the acceptor EYFP emission fluorescence is measured at 535 nm (c). Activation of Epac1-camps by the Epac-selective cAMP analog 8-pCPT-2′-*O*-Me-cAMP-AM (ESCA-AM, bath application indicated by horizontal line) results in a decrease of FRET that is measured within defined regions of interest (circles over individual cells) as an increase of the 485/535 nm emission ratio (d).

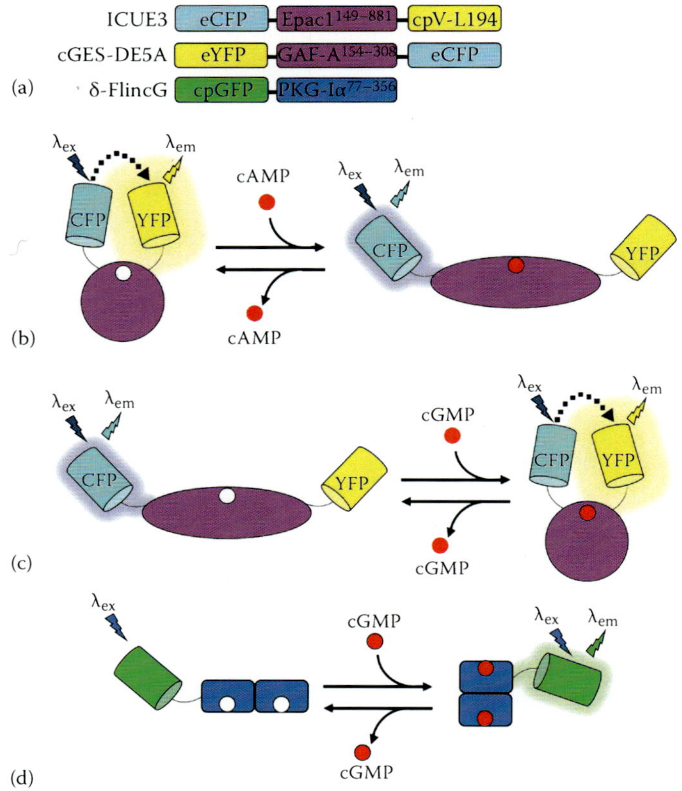

FIGURE 5.1 The design of ICUE3, cGES-DE5A, and δ-FlincG. (a) ICUE3 incorporates truncated Epac1$^{149-881}$ positioned between eCFP (FRET donor) and cpVenus L194 (FRET acceptor). cGES-DE5A incorporates truncated GAF-A$^{154-308}$ domain of PDE5A. δ-FlincG uses cpGFP as the FP and PKGIα$^{77-356}$. (b) General schematic of ICUE3. A cAMP-induced conformational change decreases the proximity of the donor and acceptor FPs. Donor excitation causes nonradiative energy transfer to the acceptor FP and emits light. (c) General schematic of cGES-DE5A. A cGMP-induced conformational change increases the proximity of the donor and acceptor FPs and increases FRET. (d) General schematic for δ-FlincG. δ-FlincG is a single FP sensor that uses tandem cGMP binding domains from PKG-I$^{77-356}$. When cGMP binds, the domains undergo a conformational change resulting in fluorescence from cpGFP.

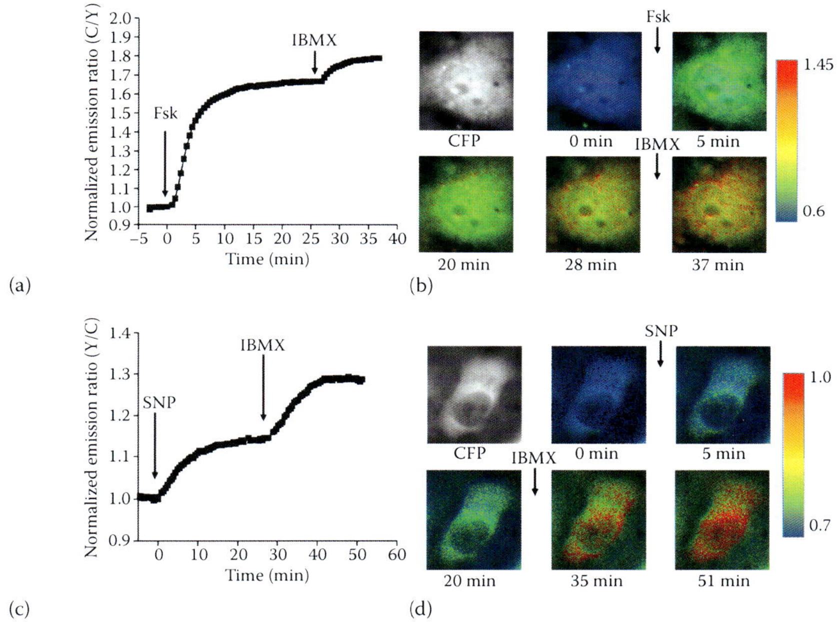

FIGURE 5.2 ICUE3 and cGES-DE5A response in live cells. Cos-7 cells were transiently transfected with ICUE3 and cGES-DE5A plasmid DNA for 24 h before imaging at 40×. The cells were treated at time 0 with (a, b) 50 μM Fsk, an adenylyl cyclase agonist, or (c, d) 80 μM SNP, a soluble guanylyl cyclase agonist. After the response plateued, the cells were further treated with 100 μM IBMX, a general PDE inhibitor. (a, c) Plot of the background-subtracted and normalized emission ratio over time for (a) ICUE3 (CFP/FRET) and (c) cGES-DE5A (FRET/CFP) shows the change in cyclic nucleotide levels. (b, d) Pseudo-color ratiometric images of the cell ROI for (b) ICUE3 and (d) cGES-DE5A responses.

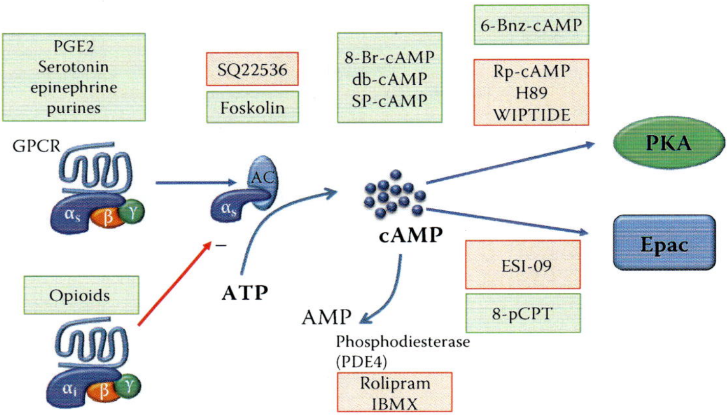

FIGURE 7.1 Pharmacological approaches to studying the role of cAMP pathways in pain. A schematic drawing of the cAMP signaling cascade and the tools used to inhibit (red) and activate (green) specific parts of the cAMP signaling cascades to elucidate cAMP pathways in chronic pain. Inflammatory mediators that are ligands of G_s-coupled GPCRs promote the generation of cAMP through activation of adenylate cyclase (AC), whereas agonists of G_i-coupled GPCR inhibit adenylate cyclase and cAMP generation. Activators or inhibitors of adenylate cyclase respectively promote or inhibit the generation of cAMP from ATP. Inhibitors of phosphodiesterases (rolipram or IBMX), such as PDE4, prevent the degradation of cAMP thereby increasing cAMP levels. cAMP analogues mimic increased cAMP levels and cause the activation of cAMP sensors, PKA and Epac. Modifications of cAMP analogues have led to the generation of specific PKA and Epac activators and inhibitors to study the role of these two cAMP sensors in chronic pain.

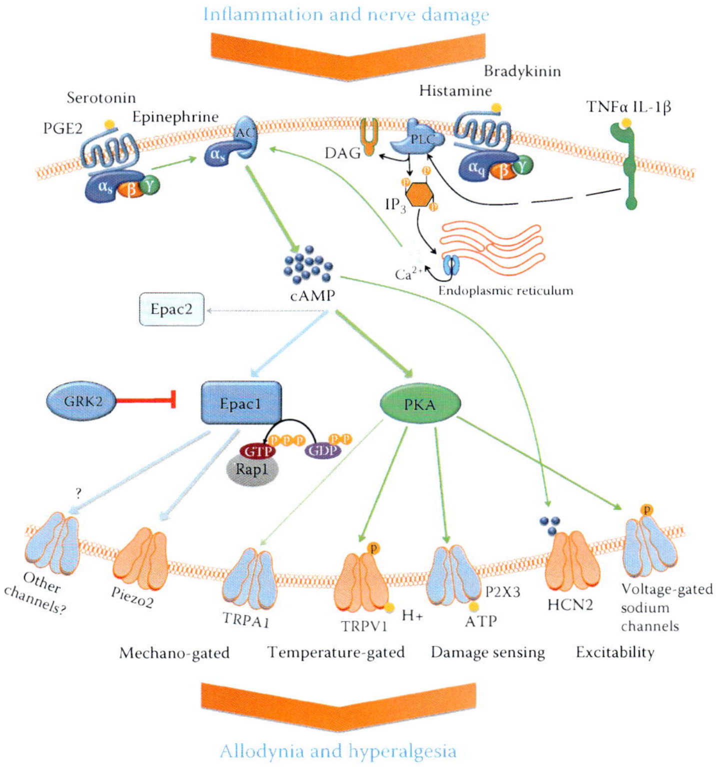

FIGURE 7.2 Downstream target of cAMP signaling in sensory neurons. A schematic overview of the signaling cascades in sensory neurons that can lead to the generation of cAMP and activation of downstream effector molecules causing allodynia and hyperalgesia. Inflammatory mediators produced during inflammation or damage act directly on sensory neurons and cause, through a diversity of signaling cascades, an increase in intracellular cAMP. cAMP can directly act on cyclic nucleotide–gated ion channels or activate the cAMP sensors, PKA and Epac. PKA and Epac each modulate specific ion channels, thereby affecting sensory neuron properties such as mechanotransduction, thermosensation and excitability thereby leading to increased sensitivity for sensory stimuli.

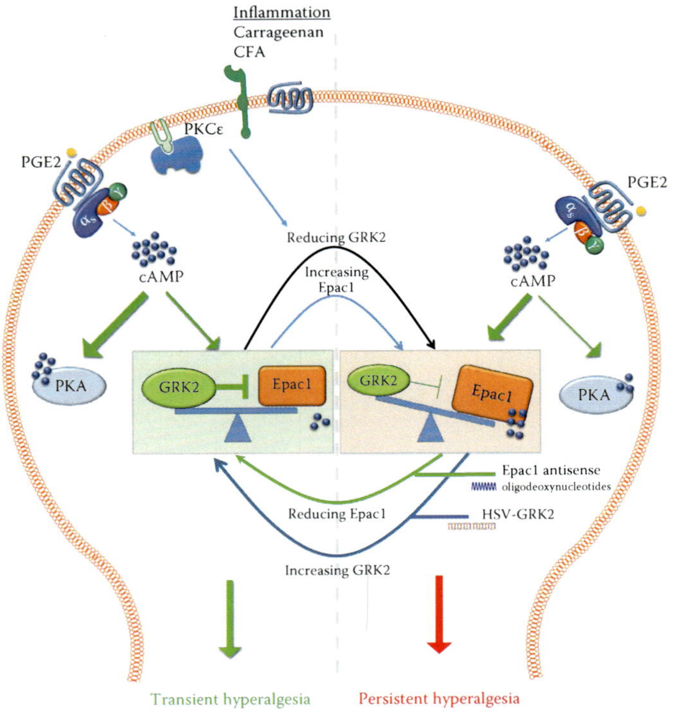

FIGURE 7.3 Role of GRK2 in regulating cAMP signaling in sensory neurons. In healthy mice, PGE2 induces transient inflammatory hyperalgesia that is completely dependent on PKA. GRK2 binds Epac1 and prevents the activation of Epac1 by cAMP. Peripheral inflammation induced by the seaweed extract carrageenan or CFA reduce GRK2 levels in primary sensory neurons innervating the inflamed tissue. CFA-induced inflammation or direct activation of PKCε increase Epac1 expression in sensory neurons. These changes in either GRK2 or Epac1 expression change the GRK2/Epac balance, thereby favoring Epac1 activation by cAMP after intraplantar PGE2 injection causing long-lasting persistent hyperalgesia. Restoring the GRK2/Epac1 balance either through HSV-mediated sensory neuron–specific overexpression of GRK2 or through reducing Epac1 expression using intrathecal Epac1 antisense oligodeoxynucleotides diminishes chronic hyperalgesia.

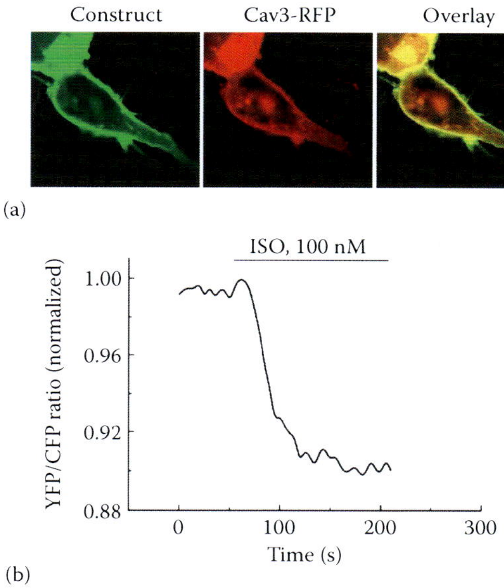

FIGURE 8.3 Testing the targeted FRET construct in 293A cells. (a) pmEpacl-camps was cotransfected together with cav3-RFP plasmid and imaged by confocal microscopy. The biosensor shows good colocalization with cav3-RFP at the plasma membrane, suggesting the successful targeting to the desired compartment. (b) FRET analysis of local cAMP levels measured in transfected cells using the pmEpacl-camps sensor. Cells were stimulated with 100 nM of the β-adrenergic agonist isoproterenol, and the changes of the YFP/CFP (acceptor/donor) ratio were monitored online. A decrease of the ratio indicated an increase in cAMP measured with the sensor at the membrane.

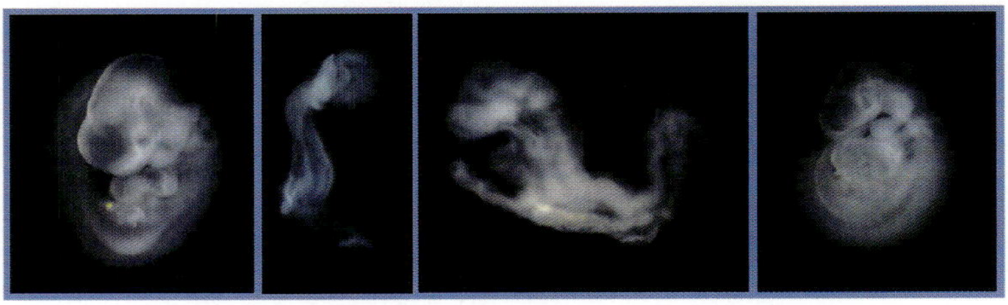

FIGURE 13.1 E9.5 embryos of R1α or Cα mutant mice. Loss of RIα results in early embryonic lethality that is partially rescued if Cα is also reduced or deleted.

7 Sensory Neuron cAMP Signaling in Chronic Pain

Niels Eijkelkamp, Pooja Singhmar, Cobi J. Heijnen, and Annemieke Kavelaars

CONTENTS

7.1 INTRODUCTION

Chronic pain is a major debilitating disorder of the nervous system that affects more than one-fifth of the adult population. Chronic pain is the result of a mixture of pathogenic mechanisms and there is a lack of effective, side effect–free therapeutic options. Many researchers have tried to elucidate specific signaling pathways that are responsible for aberrant pain states. The first pathway identified in regulating pain sensitivity is the cAMP signaling pathway. Major advances have been made and have helped to elucidate the role of cyclic nucleotide signaling in the sensory system. A myriad of effectors of the cAMP signaling cascade have been attributed to the development of chronic pain. In this chapter, we will provide an overview of the important cAMP-mediated cellular processes in chronic pain and the methods used to advance this exciting field. We will specifically focus on cAMP in sensory neurons.

7.2 PAIN

Pain can exist in different forms. In general, it serves to protect the body. It is an early warning system helping to protect us against harmful noxious stimuli. Pressure, something cold, hot, or sharp will induce protective pain. Another form of adaptive pain is when the sensory system heightens its sensitivity to sensory stimuli after tissue damage, such as inflammation. This pain will alert the body to avoid further contact with the affected tissue to promote healing.[1,2] However, when pain persists, or when pain is sensed without noxious stimuli, or when no, or minimal, peripheral inflammatory activity is present, this pain can be considered pathological. Such pathological pain can persist for months, is characterized by abnormal sensitivity to sensory stimuli, and is often difficult to treat. An important feature of pain is sensitization of sensory neurons leading to increased pain sensitivity (hyperalgesia) or pain in response to innocuous stimuli (allodynia). The term *chronic pain*, defined as pain that lasts longer than 3 to 6 months, encompasses a wide variety of chronic pathological pain states that are each caused by specific (yet unknown) mechanisms but also share similarities in their neurobiological underpinnings.

7.3 CHRONIC PAIN

Chronic pain of neuropathic origin is caused by damage or disease affecting the somatosensory system, for example, resulting from trauma to peripheral nerves, toxicity of cancer therapy, or occurring in diabetics and HIV-infected patients. Neuropathic pain is divided into peripheral neuropathic, central neuropathic, or mixed neuropathic pain depending on the anatomic site of the sensory system that is affected, respectively, by the peripheral sensory nerves, the central nervous system (brain and spinal cord), or both. Neuropathic pain is often associated with dysesthesia (abnormal sensations) and allodynia.[3]

Many chronic inflammatory diseases, including rheumatoid arthritis, osteoarthritis, and inflammatory bowel disease, are associated with persistent pain. In these inflammatory disorders, local peripheral inflammation sensitizes a subset of sensory neurons that normally detect noxious stimuli, leading to hyperalgesia. In most cases, hyperalgesia recovers when inflammation resolves. However, in a significant subgroup of patients, pain persists even after the resolution of inflammation. Clinically, this type of chronic pain is observed in, for example, arthritis patients with minimal disease activity due to successful anti-inflammatory therapy, in patients with sustained remission of arthritis, or in individuals with postherpetic neuralgia following a shingles infection.[1]

In addition, earlier episode(s) of acute pain induced by inflammation or injury can also trigger maladaptive adaptations in the sensory nervous system that cause long-lasting (chronic) pain responses after a subsequent inflammatory stimulus. This phenomenon, known as hyperalgesic priming, is clinically observed in conditions such as fibromyalgia, repetitive strain injury, complex regional pain syndrome type I, repeat surgery, and occupational repetitive stress disorders in which earlier episodes of pain, inflammation, or injury likely contribute to the chronic pain states.[4]

Visceral pain, in contrast to somatic pain, is typically diffuse, vague, and poorly defined and involves sensitization processes similar to those in somatic pain, but also involves the activation of previously silent nociceptors. Chronic visceral pain is observed in functional disorders such as irritable bowel syndrome (IBS), and in conditions like dysmenorrhea and chronic pelvic pain, in which interstitial cystitis and painful bladder syndrome are the most common forms. In addition, a significant number of patients with IBS continue to experience pain while clinical remission is achieved or inflammation has resolved.[5] In all these different types of chronic pain, the involvement of peripheral as well as central components have been suggested to contribute to the development and maintenance of chronic pain.

Alterations in the sensory system can occur at different levels. At the level of the primary sensory neurons, which detect and relay sensory information from the periphery to the spinal cord, or at the level of the central nervous system, where sensory information is processed and integrated into emotional and cognitive aspects. Specific sets of primary sensory neurons detect innocuous and noxious stimuli and play a key role in transducing sensory stimuli from the periphery to the central nervous system.[6] Several processes at these different levels, including enhanced transduction, increased excitability, decreased inhibition, and structural reorganization, have been shown to occur and may contribute to a wide range of chronic inflammatory and neuropathic pain disorders.

In vertebrates, sensory neurons connect peripheral tissues with the central nervous system. Different sets of sensory neurons are specialized for the detection of different modalities of sensory information. For example, Aβ fibers are mainly involved in the detection of innocuous stimuli such as light touch and are involved in proprioception. In contrast, thinly myelinated Aδ and nonmyelinated c-fibers are associated with the detection of painful stimuli and are therefore also known as nociceptors.[2] These nociceptors specialize in the detection of noxious stimuli such as noxious mechanical pressure, chemicals, and heat or cold. The functional properties of sensory neurons is controlled by a myriad of factors present in the extracellular space, which could be the result of tissue trauma or chronic inflammatory processes that affect the processing of specific sensory modalities. The factors range from molecules such as potassium, sodium, nitric oxide to lipids or lipid-derived messengers and peptide and proteins. Importantly, the proinflammatory mediators produced after tissue damage or inflammation act through receptors expressed on sensory neurons and activation of these receptors changes sensory neuron functioning. These receptors are:

1. G protein–coupled receptors (GPCRs) activated by molecules produced during inflammation, which can act as algogenic factors, such as bradykinin, histamine, prostanoids, purines, and serotonin.
2. Tyrosine kinase receptors, such as TrkA and TrkB, activated by the respective neurotrophins nerve growth factor (NGF) and brain-derived neurotrophic factor (BDNF) or
3. Cytokine receptors activated, for example, by interleukin-1 (IL-1) and tumor necrosis factor α (TNFα).[7]

Changes in the excitability and transduction capacity of these primary sensory neurons severely affect pain perception. Not only are different sets of sensory neurons involved in detecting different sensory modalities, specific subsets of neurons also contribute to either inflammatory or neuropathic pain. For example, a subset of sensory neurons, the so-called nonpeptidergic small diameter sensory neurons that are characterized by expression of isolectin B4 (IB4), are involved in rodent models of spinal nerve injury and cancer chemotherapy–induced chronic neuropathic pain.[8,9] Acute inflammatory pain depends on IB4$^+$ as well as IB4$^-$ neurons.[10,11] In contrast, nociceptors expressing the voltage-gated sodium channel Nav1.8, which included IB4$^+$ and IB4$^-$ neurons, are required for inflammatory pain but not neuropathic pain.[12]

Increasing efforts are being used to advance our understanding of signaling cascades in sensory neurons that cause maladaptive changes in sensory neurons leading to the development or maintenance of chronic pain. The identification of these signaling cascades that specifically contribute to the generation of chronic pain states might open new and highly needed therapeutic options to prevent debilitating pain states without affecting the ability to detect noxious stimuli. This is important because complete inhibition of pain sensation is harmful because it deprives us of a warning signal that serves an important protective function. One of the second messengers to be implicated in sensory neuron sensitization is cAMP. Although cAMP contributes to alterations in sensory neurons at different anatomic and cellular levels. Here, we will primarily focus on the role of cAMP signaling in primary sensory neurons in relation to chronic pain. Pharmacological tools to specifically activate cAMP pathways (Figure 7.1) or the use of

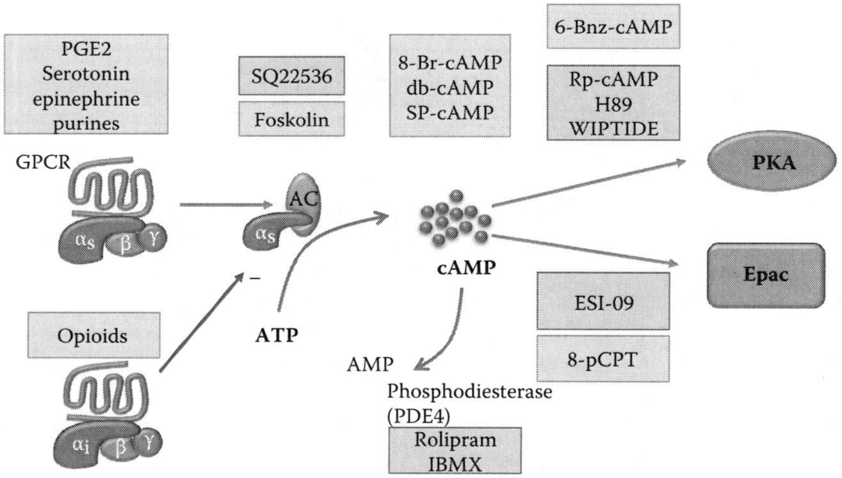

FIGURE 7.1 **(See color insert.)** Pharmacological approaches to studying the role of cAMP pathways in pain. A schematic drawing of the cAMP signaling cascade and the tools used to inhibit (red) and activate (green) specific parts of the cAMP signaling cascades to elucidate cAMP pathways in chronic pain. Inflammatory mediators that are ligands of G_s-coupled GPCRs promote the generation of cAMP through activation of adenylate cyclase (AC), whereas agonists of G_i-coupled GPCR inhibit adenylate cyclase and cAMP generation. Activators or inhibitors of adenylate cyclase respectively promote or inhibit the generation of cAMP from ATP. Inhibitors of phosphodiesterases (rolipram or IBMX), such as PDE4, prevent the degradation of cAMP thereby increasing cAMP levels. cAMP analogues mimic increased cAMP levels and cause the activation of cAMP sensors, PKA and Epac. Modifications of cAMP analogues have led to the generation of specific PKA and Epac activators and inhibitors to study the role of these two cAMP sensors in chronic pain.

(conditional) knockout mice have furthered our understanding of the role of cAMP signaling in sensory neurons in chronic pain. Here, we will review the current perspectives on cAMP signaling in chronic pain and the methods used to unravel cAMP-mediated pathways in chronic pain.

7.4 PHARMACOLOGICAL AND GENETIC APPROACHES TO UNRAVEL THE ROLE OF cAMP IN CHRONIC PAIN

Although a myriad of proinflammatory mediators and their receptors can cause the activation of a range of signaling pathways in sensory neurons, the second messenger cAMP was the first intracellular signaling molecule implicated in the sensitization of sensory neurons. Because of the technical difficulties of studying individual nerve fibers, most of the knowledge about signal transduction cascades in sensory neurons has been gathered by combining in vitro cultures of the cell bodies of sensory neurons that have been obtained from dorsal root ganglia of neonatal or adult mice and rats from in vivo studies. In Figure 7.1, we have depicted the pharmacological tools used to unravel cAMP signaling in pain.

The prostanoid, prostaglandin E2 (PGE2), is probably the most well-known inflammatory agent that increases pain sensitivity in both animals and humans.[13,14] The lipid mediator PGE2 is produced downstream of activation of cyclooxygenases. The central role of cAMP-inducing prostanoids in pain is highlighted by the fact that nonsteroidal anti-inflammatory drugs (NSAIDs), which mostly target the prostaglandin-generating COX enzymes, are currently one of the most widely prescribed classes of pharmaceutical agents for treating pain. For example, indomethacin (COX inhibitor) treatment blocks inflammatory pain induced by lipopolysaccharides (LPS) or carrageenan.[15] PGE2 causes peripheral sensitization via binding to specific prostaglandin receptors expressed on primary nociceptors. The four subtypes of PGE2 receptors (EP1–4) belong to the subclass of seven-transmembrane GPCR. Binding of PGE2 to EP4 receptors activates adenylate cyclase via a heterotrimeric G protein of the G_s class and thus increases intracellular levels of cAMP. Using an in vitro approach, several groups have shown that the addition of inflammatory mediators such as PGE2 to cultures of primary sensory neurons causes an increase in intracellular cAMP.[16] Intraplantar injection of PGE2 in rodents induces local transient mechanical as well as thermal hyperalgesia that is thought to be predominantly due to EP4-mediated increases in cAMP,[17–19] Similarly, other agonists of G_s-coupled GPCRs such as serotonin, epinephrine, and purines produce hyperalgesia when injected into the hind paw.[20–23] Importantly, hyperalgesia induced by PGE2 can be blocked by coinjection of the inactive cAMP analogue Rp-cAMP, which keeps the regulatory subunit of protein kinase A (PKA) in a locked conformation, preventing the release the catalytic subunit that is normally a prerequisite for the activation of PKA.[24,25] Conversely, increasing cAMP levels through inhibiting PDE4, the phosphodiesterase that metabolizes cAMP with isobutylmethylxanthine (IBMX), or rolipram, enhances PGE2 hyperalgesia.[24,26] Intradermal injections of membrane-permeable cAMP analogues, such as dibutyryl(db)-cAMP and 8-Bromo(br)-cAMP into the hind paw of rats induced strong reduction in the detection threshold for mechanical stimuli.[24,27] In addition, activation of adenylate cyclase in the hind paw by intraplantar injection of forskolin induces profound hyperalgesia.[24]

Overall, these studies clearly indicate that cAMP and mediators that induce cAMP promote sensitization of sensory neurons leading to hyperalgesia.

The generation of cAMP does not only affect acute hyperalgesia. The competitive inhibitor of PKA, Rp-cAMP, also reduces hyperalgesia even if hyperalgesia is already established; these findings indicate that ongoing cAMP signaling is involved in maintaining already established hyperalgesia.[24,28] In various models of chronic neuropathic pain, COX-2 and PGE2 are chronically upregulated and are thought to contribute to neuropathic pain conditions.[29] In a rodent model of neuropathic pain, nerve damage leads to IL-6 synthesis in medium to large diameter sensory neurons, and this IL-6 synthesis is dependent on EP4, COX2, and ultimately, PKA.[30] In line with increased cAMP levels in sensory neurons promoting hyperalgesia, the peripheral antinociceptive actions of morphine and other opioids can be explained by the fact that these opioids act through G_i-coupled μ opioid receptors expressed in sensory neurons that inhibit adenylate cyclases and thereby prevent increased levels of cAMP.[31,32]

Sensory neurons express ion channels that can be directly modulated by cAMP. For example, sensory neurons express members of the hyperpolarization-activated cyclic nucleotide–modulated family of ion channels (HCN1–4) that carry an inward current that is activated by membrane hyperpolarization. HCN2 is highly expressed in sensory neurons and is sensitive to cAMP, which promotes the inward current that could drive repetitive firing in sensory neurons under chronic pain conditions. The addition of adenylate cyclase activator forskolin or PGE2 to sensory neurons accelerated action potential firing upon current injection to depolarize neurons, which were ablated in sensory neurons lacking HCN2. Importantly, inflammatory and neuropathic pain was attenuated in mice with specific deletion of HCN2 in nociceptors.[33,34] These data indicate that cAMP-gated HCN2 channels are central in chronic pain development through regulating sensory neuron excitability.

Although cAMP can directly alter sensory neuron function through regulation of ion channels, most of the effects of cAMP in regulating sensory neuron function have been attributed to the cAMP sensor PKA.

7.5 PAIN AND cAMP SENSORS

Although an increase in intracellular cAMP level has long been held to be synonymous with the activation of PKA, in 1998, the Bos and Graybiel groups independently discovered the cAMP sensor exchange protein activated by cAMP (Epac), a.k.a. cAMP-regulated guanine nucleotide exchange factor (cAMP-GEF).[35,36] It is now known that Epac1 and Epac2 are guanine GEFs that activate Rap, which is a small GTP-binding protein in the Ras family of GTPases. Upon binding of cAMP, Epac stimulates the exchange of GDP bound to Rap for GTP, resulting in the activation of Rap that is upstream of various other effector proteins, including adaptor proteins that affect the cytoskeleton, phospholipases, and mitogen-activated protein kinases.[37,38] Most research in the pain field has focused on the role of PKA and resulted in the discovery of many sensory targets regulated by PKA. Since the identification of Epacs as an alternative cAMP sensor, increasing evidence has indicated an additional role of the cAMP/Epac signaling axis in chronic pain. Although specific PKA inhibitors that can also be used in vivo are readily available and have helped to further our understanding of the role of cAMP-PKA signaling in pain regulation, the lack of specific Epac inhibitors has, until recently, prevented us from specifically investigating the effects of Epac signaling in chronic pain conditions. As such, most studies have focused on the involvement of PKA in nociceptor sensitization and chronic pain as outlined in later sections. More recently, the use of specific Epac activators, Epacl and Epac2 knockout mice, and antisense oligodeoxynucleotides strategies to knock down Epac in sensory neurons in vivo have helped us take the first steps to better understanding the role of cAMP to Epac signaling in chronic pain.[39–43] A novel Epac inhibitor, ESI-09, was recently described and might prove valuable in further elucidating the role of Epac in chronic pain conditions.[44]

7.5.1 Protein Kinase A

The first evidence of the involvement of PKA in controlling inflammatory hyperalgesia was shown by Ferreira and Levine.[26,28] Intraplantar injection of PGE2 induced profound hyperalgesia that was blocked by the PKA inhibitors, H89 and Rp-cAMP, showing that PGE2 hyperalgesia is mediated through PKA activation.[17,24] Similarly, evidence shows that intradermal injections of serotonin and purines, such as adenosine, which signal through G_s-coupled GPCRs and promote the activation of adenylate cyclase and the generation of cAMP, cause the lowering of thresholds for mechanical and thermal stimuli.[20,45] Interestingly, intradermal injection of epinephrine, a hormone and neurotransmitter that is released upon stressful events but is also thought to be released as a neurotransmitter following tissue trauma and peripheral neuropathies,[46] induces hyperalgesia that in part depends on PKA signaling.[22,47]

PKA can be differentially involved in the development of pain sensitization and acute pain caused by algogenic mediators. For example, bee venom–induced hyperalgesia depends on PKA, whereas acute nocifensive behaviors induced by the same bee venom are independent of PKA.[48] These data indicate that PKA is involved in the generation of hypersensitivity to painful stimuli but not directly in the generation of pain. Interestingly, inflammatory mediators whose receptors are not coupled directly to adenylate cyclase, such as TNFα or IL-1β, induced hyperalgesia via a route that at least in part depends on the activation of PKA. For example, the application of TNFα to dorsal root ganglia containing the soma of sensory neurons in vivo elicits mechanical hypersensitivity. By using excised L4 and L5 dorsal root ganglia (DRG) and the connecting sciatic nerve, the authors showed that TNFα promoted neuronal discharges in most c-fibers and in some Aβ fiber, and enhanced electrical excitability in these same neuronal fibers. Most importantly, coapplication of the PKA inhibitors, H89 or Rp-cAMP, completely blocked the TNFα-induced enhancements in sensory neuron excitability.[49] Subcutaneous injection of the prototypic inflammatory cytokine IL-1 induces transient thermal hyperalgesia and mechanical allodynia.[50] The signaling cascades involved in the generation of IL-1–induced thermal hypersensitivity are different from that of mechanical hypersensitivity. Mechanical hypersensitivity is mediated through PKA activity in large-diameter primary afferent nerve fibers normally associated with the detection of innocuous stimuli. IL-1β–induced thermal hyperalgesia, in contrast, is mediated by PKC-mediated signals in the small-diameter primary afferent nerves.[51] A question that remains is, how does IL-1β or TNFα signaling trigger elevated levels of cAMP in these neurons? Possibly, the cytokine increase intracellular calcium levels that activate adenylate cyclases such as AC1 and AC8. Alternatively, an indirect pathway depending on the generation of inflammatory mediators, such as PGE2 though induction of COX2, could contribute.

The use of pharmacological tools to promote the generation of cAMP or inhibit PKA signaling (Figure 7.1) has clearly identified the important role of PKA in the sensitization of sensory neurons leading to hyperalgesia under inflammatory conditions. Although PKA has a central role in inflammatory hyperalgesia, the role of PKA in neuropathic pain is less clear. Several groups have used pharmacological and genetic tools to identify whether cAMP or its sensor PKA plays a role in mechanical allodynia in neuropathic pain conditions. In mice deficient for adenylate cyclase 5, the development of mechanical allodynia in models of neuropathic pain is severely impaired.[52] Chronic compression of the DRG is a model of neuropathic pain and leads to neuronal hyperexcitability and thermal hyperalgesia. Inhibition of adenylate cyclase by SQ22536 or PKA by Rp-cAMP attenuated the hyperexcitability induced by chronic compression of rat DRG in vitro and blocked hyperalgesia in vivo.[53] After partial sciatic nerve ligation in rats, spinal activation of PKA and the cellular transcription factor cAMP response element-binding protein, which is downstream of cAMP and PKA signaling, was observed. Spinal administration of a PKA inhibitor or adenylate cyclase inhibitor prevented partial sciatic nerve ligation–induced mechanical allodynia.[54] In contrast, deletion of the neuronal-specific isoform of the type I regulatory subunit (RIβ) of PKA attenuated inflammatory pain but did not affect neuropathic pain.[55] In line with these findings, Isensee et al.[56] showed that the RIIβ regulatory subunit of PKA specifically regulates nociceptive processing in the terminals of small-diameter primary afferent fibers that are shown to be involved in inflammatory pain. Finally, there is some evidence that PKA mediates inflammatory pain, whereas neuropathic pain is dependent on PKC signaling.[57]

7.5.2 Exchange Factor Activated by cAMP

The discovery of the novel cAMP signal transduction mechanism utilizing Epac as cAMP sensors has also generated interest in the pain field. In 2010, Suzuki and colleagues[39] generated the first Epacl knockout mice. Mice lacking Epacl showed profound attenuation of the development of nerve damage–induced mechanical allodynia. Interestingly, the 50% reduction in Epacl protein that is observed in Epacl heterozygous knockout mice is already sufficient to attenuate mechanical allodynia. In line with the role of Epacl in chronic mechanical allodynia, intraplantar injection of a specific Epac activator 8-(4-chlorophenylthio)-2′-*O*-methyladenosine 3′,5′-cAMP (8-pCPT) induces profound mechanical hypersensitivity that is longer in duration compared with mechanical hypersensitivity induced by intraplantar injection of a selective PKA activator.[41] In chronic inflammatory pain states, Epacl also seems to have a dominant role. In Epacl$^{-/-}$ mice, *transient inflammatory hyperalgesia* induced by an intraplantar injection of a small dose of carrageenan or prostaglandin did not produce any effect, whereas knockdown of Epacl during already *established persistent inflammatory hyperalgesia* induced by intraplantar injection of complete Freund's adjuvant (CFA) attenuated mechanical hyperalgesia.[40] Overall, these results indicate that Epacl signaling contributes to chronic inflammatory and neuropathic pain states.

Evidence indicates that Epac and PKA signaling have distinct roles in regulating pain sensitivity, in particular with respect to the contribution of specific sensory neuron subsets involved in PKA versus Epac-mediated mechanical allodynia.

Small-diameter nociceptors have been divided into two major classes according to their dependence on the growth factor NGF or GDNF. The nonpeptidergic nociceptor population expressing the canonical GDNF receptor, GDNF family receptor α1 (GFRα1), also binds IB4[58] and primarily targets interneurons in lamina II. The peptidergic, IB4-negative population targets dorsal horn projection neurons in lamina I and interneurons in superficial lamina II.[1] In nonpeptidergic IB4-positive sensory neurons, Epac signaling mediates cAMP to PKCε signaling, leading to inflammatory hyperalgesia.[59] We recently showed that Epac-mediated mechanical allodynia is independent of Nav1.8 nociceptors, a subset of sensory neurons both consisting of $IB4^+$ and $IB4^-$ sensory neurons that are normally associated with inflammatory hyperalgesia.[12,41] Mice in which Nav1.8-expressing nociceptors were ablated by cell-specific expression of diphtheria toxin developed mechanical allodynia after an intraplantar injection of the specific Epac agonist 8-pCPT to the same extent as WT mice, indicating that mechanical hypersensitivity is independent of Nav1.8 nociceptors. Thermal hypersensitivity induced by intraplantar injection of 8-pCPT is, however, completely abrogated in mice without Nav1.8 nociceptors. Intraplantar injection of the specific PKA activator, 6-Bnz-cAMP, also induces both thermal and mechanical hypersensitivity; however, both thermal and mechanical hyperalgesia are completely abrogated in mice without Nav1.8 nociceptors.[41] Thus, these data indicate that PKA and Epac signaling induce mechanical hypersensitivity that depend on different sensory neurons subsets, whereas thermal hypersensitivity is dependent on the same $Nav1.8^+$ subset. The difference in sensory neuron subset dependence of PKA- and Epac-mediated hypersensitivity was further substantiated by in vivo electrophysiological recordings in which the responses of spinal cord–wide dynamic range neurons that receive and integrate input from Aβ, Aδ, and c fibers are measured. Epac activation with the specific agonist 8-pCPT resulted in increased responses of spinal-wide dynamic range neurons to light mechanical stimuli whereas electrical excitability of Aβ, Aδ, and c fibers was unaffected. In contrast, PKA activation using the specific agonist 6-Bnz-cAMP primarily enhanced mechanical stimuli to the noxious range, which was associated with increased electrical excitability in c fibers.[41] Thus, cAMP signaling to PKA likely promotes noxious mechanical hypersensitivity through regulating c fiber excitability, whereas cAMP signaling to Epac promotes mechanical hypersensitivity to innocuous mechanical stimuli through direct regulation of mechanotransduction, likely in fibers normally associated with the detection of touch such as Aβ fibers.

7.6 cAMP REGULATION OF SENSORY SPECIFIC FUNCTIONS

A great interest in the downstream targets of cAMP in sensory neurons has yielded a wealth of studies showing the role of cAMP, PKA, and Epac signaling in the regulation of specific sensory neuron functions. As such, we will highlight key cAMP-mediated regulation of the ion channels involved in the transduction of sensory stimuli or sensory neuron excitability that is summarized in Figure 7.2.

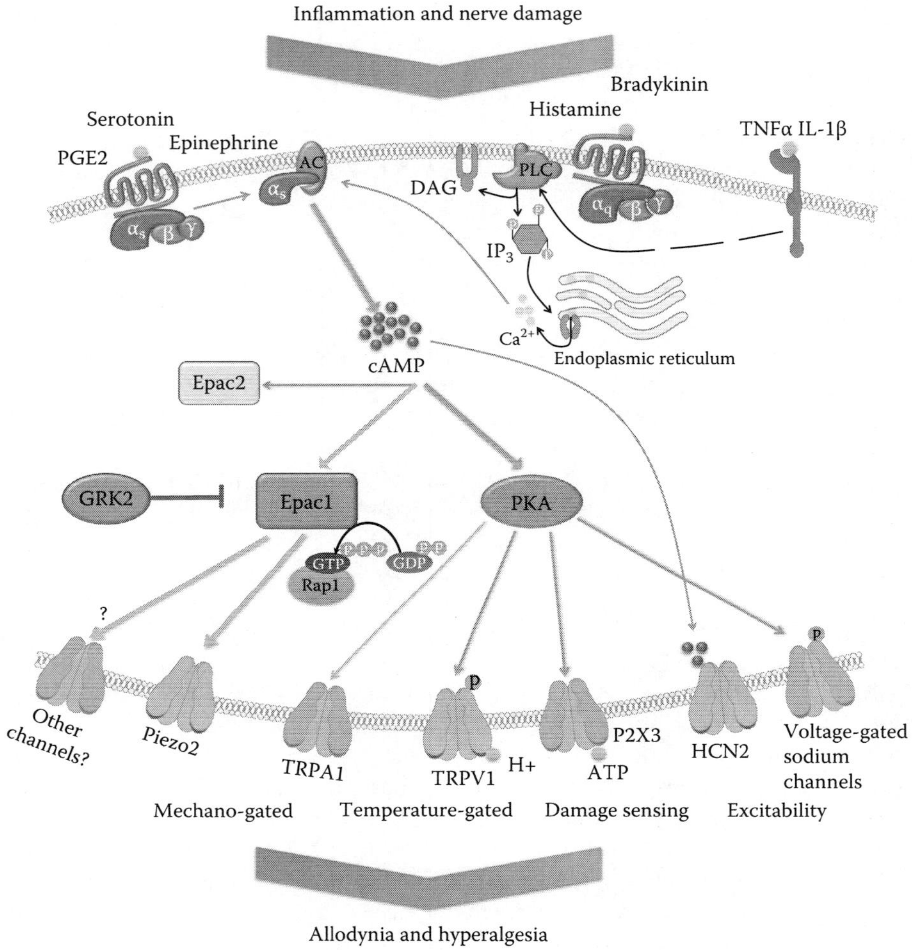

FIGURE 7.2 **(See color insert.)** Downstream target of cAMP signaling in sensory neurons. A schematic overview of the signaling cascades in sensory neurons that can lead to the generation of cAMP and activation of downstream effector molecules causing allodynia and hyperalgesia. Inflammatory mediators produced during inflammation or damage act directly on sensory neurons and cause, through a diversity of signaling cascades, an increase in intracellular cAMP. cAMP can directly act on cyclic nucleotide–gated ion channels or activate the cAMP sensors, PKA and Epac. PKA and Epac each modulate specific ion channels, thereby affecting sensory neuron properties such as mechanotransduction, thermosensation and excitability thereby leading to increased sensitivity for sensory stimuli.

7.6.1 Thermosensing

Sensory neurons express a range of ion channels that mediate the transduction of physicochemical stimuli by eliciting a generator current after activation. First experiments showed that heat-activated ionic currents were potentiated after exposure to the cAMP activator forskolin in rat nociceptive neurons.[60] Temperature sensitivity (hot and cold) is detected by different classes of ion channels. An important class of such channels is the thermosensitive transient receptor potential channel family, of which TRPV1 good example, and is activated by heat, protons, and capsaicin, the pungent ingredient of hot peppers.[61,62] In addition, the calcium-activated chloride channel anoctamine 1 is also important for the detection noxious heat.[63] TRPV1 is essential for the development of chronic inflammatory thermal hyperalgesia. TRPV1 is normally activated by temperatures higher than 42°C, but inflammatory conditions sensitize TRPV1 so that TRPV1 is activated by temperatures as low 37°C and thus may contribute to spontaneous pain at body temperature.[61] TRPV1, like many other ion channels and receptors, is desensitized upon continuous activation. Phosphorylation of serine/threonine residues within the N- and C-termini of TRPV1 are implicated in receptor sensitization and activation. In DRG neurons, the activation of PKA leads to the phosphorylation of TRPV1 on Ser-116, preventing TRPV1 desensitization and thus promoting activation of TRPV1 by lower temperatures.[60,64] The scaffolding protein AKAP79/150 promotes the formation of a complex that involves TPRV1, PKA, and adenylyl cyclase (AC). Dissociation of AC from this complex abrogates TRPV1 sensitization by PGE2 or forskolin[65] and the direct anchoring of both PKA and AC to TRPV1 by AKAP79/150 is therefore thought to be critical in the development of thermal hyperalgesia. In line with the role of TRPV1 sensitization and thermal hyperalgesia, genetic or pharmacological inhibition of TRPV1 prevents thermal hypersensitivity induced by cAMP-inducing agents.[62,66,67]

7.6.2 Mechanosensing

Although specific populations of sensory neurons can respond to increased temperatures, sensory neurons have been characterized electrophysiologically when triggered with mechanical cues using a technique originally developed by Dr. Jon Levine.[68] Mechanically sensitive sensory neurons can be divided into subpopulations according to the properties of the mechanically activated currents generated after membrane deflections.[6] Addition of the Epac-specific activator 8-pCPT to sensory neurons promotes mechanically activated, rapidly adapting currents in these cells that represent a population associated with the detection of light touch. In a heterologous expression system, activation of Epac1 but not Epac2 promotes the sensitization of a recently identified family of mechanically activated channels, Piezo.[41] The Piezo family, consists of Piezo1 and Piezo2 that are large transmembrane proteins conserved among various species. Piezo2 is expressed in sensory neurons and produces rapidly adapting currents upon mechanical stimulation of sensory neurons. Moreover, expression of Piezo2 in HEK293 cells renders these cells sensitive to mechanical stimuli; mechanical stimulation produces large inward currents at negative holding potentials. Thus, Piezo2 is likely part of the mechanically activated cation channels that transduce mechanical stimuli.[69,70] Addition of 8-Br-cAMP to HEK293 cells expressing Piezo2 promotes mechano-current amplitudes and slows the inactivation of Piezo2 currents.[71] In contrast, in a similar heterologous expression system, specific activation of PKA with 6-Bnz-cAMP failed to enhance Piezo2 currents.[41] These in vitro changes in mechanotransduction also have implications for in vivo chronic pain conditions. In vivo Piezo2 antisense applications reduce mechanical hypersensitivity induced by intraplantar injection of an Epac activator or allodynia in two different models of neuropathic pain.[41] Regulation of mechanotransduction in sensory neurons by cAMP is not restricted to the regulation of the Piezo family. For example, TRPA1, a member of the transient receptor potential channel family, is thought to contribute to noxious mechanosensation.[72] Hyperalgesia induced by PGE2 or PKA activation is abridged when TRPA1 expression is reduced or TRPA1 function is blocked by a specific channel blocker,[73] indicating that cAMP could potentially affect mechanotransduction through other mechano-gated channels.

7.6.3 Neuronal Excitability

An important aspect of altered neuronal function in chronic pain states is the changes in the excitability of sensory neurons. Voltage-gated sodium channels are key in action potential generation in sensory neurons. Different populations of sensory neurons express specific sets of voltage-gated sodium channels important in the generation of action potentials. For example, Nav1.8 and Nav1.9 expression is predominantly found in small-diameter nociceptors.[74] Importantly, sodium channels have multiple sites for phosphorylation by PKA and PKC. Phosphorylation of sodium channels is an important mechanism of modulation of peripheral neurons.[75] In isolated primary afferent sensory neurons, PGE2 increases excitability that is associated with a hyperpolarizing shift in the activation curve of the tetrodotoxin-resistant (TTX-R) Na^+ currents, which include Nav1.8.[76] Moreover forskolin and 8-br-cAMP enhance TTX-resistant sodium channels consistent with PKA phosphorylation.[23,76,77] In addition, some reports show that PGE2 enhances TTX-sensitive sodium currents through an AC/PKA-dependent signaling pathway,[75] whereas PKA activation has been shown to reduce the TTX-sensitive sensory neurons specific sodium channel Nav1.7.[78] Interestingly, alternative splicing of human Nav1.7 modulates cAMP-mediated regulation of this channel.[79] Continuous activation of PKA and its effect on the TTX-R sodium channel Nav1.8 is associated with persistent inflammatory hypernociception.[80] PKA activation also promotes the trafficking of Nav1.8 channels to the cell membrane, thereby enhancing current density and excitability of sensory neurons.[81] PKA-mediated regulation of sensory neuron excitability has mostly focused on voltage-gated sodium channels. However, some reports show that PGE2 inhibits an outward potassium current in sensory neurons via activation of PKA thereby, in part, inhibiting delayed rectifier-like potassium current.[82] Likely, cAMP signaling also promotes other channels such as voltage-gated potassium and calcium channels critically involved in neuronal excitability.

In contrast with PKA, the role of Epac1 in sensory neuron excitability has not yet been described. However, some reports show that cAMP-Epac signaling can enhance neuronal excitability through the regulation of calcium-dependent potassium channels in cerebellar granule cells,[83] opening the possibility that cAMP-EPAC signaling could regulate the excitability of sensory neurons.

7.7 cAMP AND SENSORY NEURON PLASTICITY

Activity-dependent neuronal plasticity is important for the development of many aspects of adaptive and pathological behavior including chronic pain. Several signaling cascades have been implicated in sensory neuron plasticity in chronic pain, including the cAMP signaling pathway.

7.7.1 Hyperalgesic Priming

Short-lived episodes of acute pain induced by inflammation can cause chronic pain. In 2000, Jon Levine developed a model in which the mechanisms that maintained chronic pain could be discriminated from those that caused acute pain in rats. He coined the term *hyperalgesic priming*,[4,84] and others have adapted this model in other species and further extended this model.[40,85,86] In this model of hyperalgesic priming, rodents are challenged with a transient inflammatory stimulus, such as carrageen (a seaweed extract), IL-6, or growth factors such as GDNF to induce a short-lasting hyperalgesia. After recovery from this transient inflammatory hyperalgesia, when the signs of inflammation have resolved (even weeks later), animals are challenged with a second inflammatory stimulus, for example, PGE2, serotonin, or adenosine. Injection of a low dose of PGE2 normally induces transient inflammatory hyperalgesia, but in mice that were *primed* with an earlier episode of inflammation, PGE2, serotonin, or adenosine evoke hyperalgesia that lasts much longer.

Importantly, this transition from acute to chronic pain is associated with nociceptor plasticity that involves changes in cAMP signaling. In naive healthy rats, PGE2 induces short-lasting hyperalgesia dependent of the stimulatory G protein (G_s) and PKA. By using specific inhibitors of PKA, MEK/ERK, and PKCε, Levine and coworkers originally described that in primed rats, PGE2 hyperalgesia shifts from this PKA-dependent pathway to one that is PKA-independent but dependent on PKCε or MEK/ERK (or both).[87] In multiple follow-up studies, the Levine group has further shown that novel signaling pathways are recruited to PGE2 signaling in the primed state, which includes inhibitory G-protein (G_i), phospholipase C β3 (PLCβ3), and PKCε.[88–90] An interesting question the authors tried to tackle is, which signaling cascades lead to the activation of PKCε after PGE2 signaling in primed rats? Some evidence already indicated that Epac can signal to PLC, and that PLC activation can lead to PKCε translocation, suggesting that Epac is a likely candidate to mediate cAMP to PKCε signaling.[37,38] Indeed, in sensory neurons, Epac was upstream of phospholipase C and phospholipase D signaling; both required for plasma membrane translocation and thereby activation of PKCε. Intriguingly, this cAMP-to-PKCε signaling cascade is functional in a specific subset of sensory neurons, the nonpeptidergic IB4$^+$ sensory neuron population.[59] Overall, these studies indicate that sensory neurons show plasticity in cAMP signaling after a transient inflammation that has major implications regarding how sensory neurons respond to a subsequent challenge.

7.7.2 GRK2 and Nociceptor cAMP Signaling

We have more recently followed up on mechanisms of neuronal plasticity in sensory neuron signaling that contribute to chronic pain development in the context of cAMP signaling. As such, we discovered a novel mechanism through which nociceptor GPCR kinase 2 (GRK2) contributes to regulation of inflammatory hyperalgesia through cAMP-mediated pathways (Figure 7.3). GRK2, also known as β-adrenergic

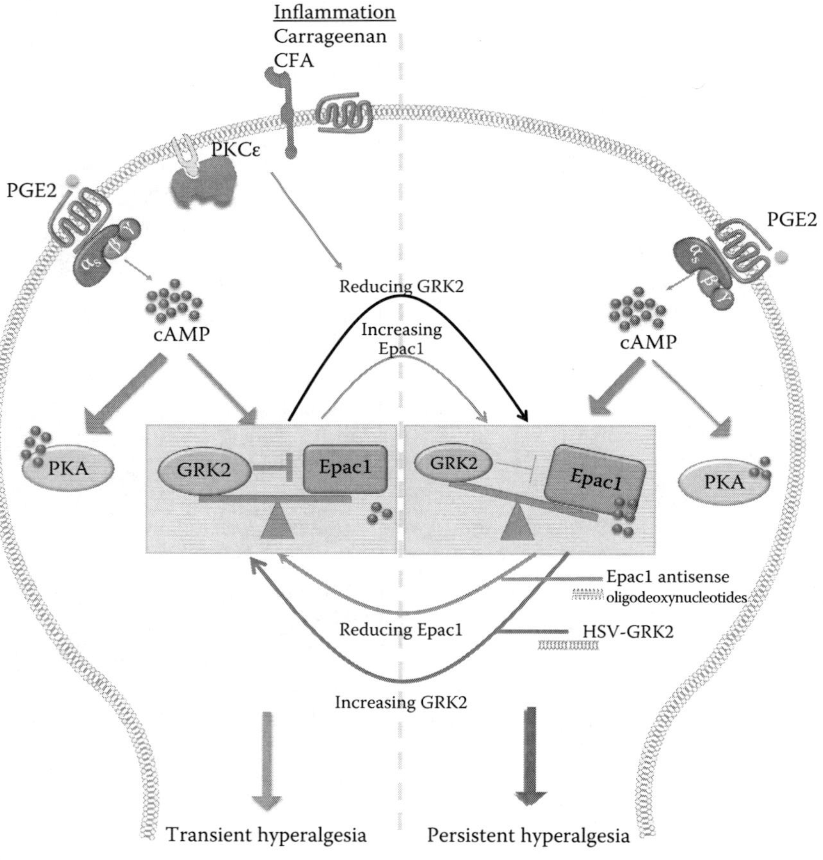

FIGURE 7.3 **(See color insert.)** Role of GRK2 in regulating cAMP signaling in sensory neurons. In healthy mice, PGE2 induces transient inflammatory hyperalgesia that is completely dependent on PKA. GRK2 binds Epac1 and prevents the activation of Epac1 by cAMP. Peripheral inflammation induced by the seaweed extract carrageenan or CFA reduce GRK2 levels in primary sensory neurons innervating the inflamed tissue. CFA-induced inflammation or direct activation of PKCε increase Epac1 expression in sensory neurons. These changes in either GRK2 or Epac1 expression change the GRK2/Epac balance, thereby favoring Epac1 activation by cAMP after intraplantar PGE2 injection causing long-lasting persistent hyperalgesia. Restoring the GRK2/Epac1 balance either through HSV-mediated sensory neuron–specific overexpression of GRK2 or through reducing Epac1 expression using intrathecal Epac1 antisense oligodeoxynucleotides diminishes chronic hyperalgesia.

receptor kinase 1, was originally described as a kinase that phosphorylates GPCR, and in particular the β-adrenergic receptor, on serine and threonine residues leading to the uncoupling of G proteins and subsequent recruitment of β-arrestin resulting in the internalization of these receptors.[91] It is now clear that GRK2 phosphorylates a wide range of GPCRs to regulate signaling. Importantly, increasing evidence shows that GRK2 interacts with a variety of signaling molecules other than GPCRs, including those involved in cAMP signaling.[18,92–95]

The reduction of GRK2 specifically in Nav1.8-expressing nociceptors by using CRE-LOX technology (giving rise to SNS-GRK2$^{+/-}$ mice), severely prolongs the duration of hyperalgesia induced by PGE2 or epinephrine.[18,22] In addition, injection of 8-Br-cAMp into the hind paw of mice induces transient hyperalgesia that is severely prolonged in these SNS-GRK2$^{+/-}$ mice with reduced GRK2 in nociceptors, indicating that GRK2 affects PGE2 signaling in sensory neurons downstream of cAMP, thereby severely prolonging pain duration. In SNS-GRK2$^{+/-}$ mice, PGE2 or 8-Br-cAMP hyperalgesia was PKA independent, but was blocked by PKCε and MEK/ERK inhibitors. These data suggested a possible link of between cAMP and Epac1 signaling in nociceptors with reduced GRK2 expression. Immunoprecipitation studies indicated that Epac1 and GRK2 are found in the same protein complex. Intraplantar injection of a selective Epac agonist induced hyperalgesia that was severely prolonged in mice with reduced GRK2 expression, whereas mechanical hypersensitivity induced by a selective PKA agonist was unaffected by GRK2.[18] Thus, low nociceptor GRK2 protein levels switch PGE2-cAMP mediated hyperalgesia from a PKA-dependent pathway to a PKCε-MEK-ERK–dependent pathway, likely through facilitating Epac signaling. A similar switch in cAMP signaling is also observed in cultured sensory neurons after persistent peripheral inflammation. In DRG neurons isolated from normal rats, PGE2 enhances fast inactivating ATP currents in a PKA-dependent fashion. In DRG neurons from sensory neurons isolated from rats with CFA-induced paw inflammation, PGE2 produces a larger increase in ATP currents that also involve PKC possibly downstream of Epac1.[96] The functional consequences of this switch from cAMP-PKA to a cAMP-Epac pathway in sensory neurons also has therapeutic implications because the prevention of the signaling switch or cAMP to Epac signaling in sensory neurons could diminish chronic pain. Indeed, our preclinical studies showed that in models of chronic pain, including hyperalgesic priming as well as chronic inflammatory pain, the balance in GRK2 and Epac1 expression levels shifted toward low GRK2/Epac1 balance. Restoring the reduced GRK2/Epac1 balance by increasing GRK2 expression levels in sensory neurons using Herpes simplex virus (HSV) amplicons encoding GRK2 prevented development of chronic hyperalgesia in two *hyperalgesic priming* models or inhibited chronic inflammatory pain. In addition, normalizing the GRK2/Epac1 balance through reduction of Epac1 expression was also sufficient to inhibit chronic pain in a hyperalgesic priming and inflammatory pain model. Thus, these studies highlight the importance of the cAMP-GRK2/EPAC pathway in chronic pain development.[40] New studies have to show whether similar mechanisms are operative in other chronic pain states such as painful neuropathies. In line with this hypothesis, knockdown of Epac1 attenuates neuropathic pain.[41] However, it remains to be determined whether GRK2 expression promotes recovery from neuropathic pain, or whether pharmacological inhibition of Epacl with the recently discovered Epac inhibitor would block neuropathic pain.

7.8 FINAL REMARKS

In this chapter, we show that the use of specific adenylate cyclase activators, PKA and Epac agonists, and PKA antagonists have helped to further our understanding of cAMP signaling in chronic pain. In addition, the development of knockout mice has helped to better understand the roles of cAMP sensors in pathological pain states. The role of cAMP signaling in different sets of sensory neurons has specific implications for the development of chronic pain states. In particular the cAMP sensory Epac has an intriguing role in chronic pain states, whereas PKA activity is more associated with adaptive transient changes in pain sensitivity. Cell-specific inducible Epac and PKA subunit knockout mice will be needed to further explore the cell-specific roles of these cAMP sensors in chronic pain.

ACKNOWLEDGMENT

This work was supported in part by the National Institutes of Health, United States grants R01NS073939 and R01NS074999.

REFERENCES

1. Basbaum, A.I., Bautista, D.M., Scherrer, G., and Julius, D. Cellular and molecular mechanisms of pain. *Cell* 2009 (139), 267–284.
2. Woolf, C.J., and Ma, Q. Nociceptors—Noxious stimulus detectors. *Neuron* 2007 (55), 353–364.
3. von Hehn, C.A., Baron, R., and Woolf, C.J. Deconstructing the neuropathic pain phenotype to reveal neural mechanisms. *Neuron* 2012 (73), 638–652.
4. Reichling, D.B., and Levine, J.D. Critical role of nociceptor plasticity in chronic pain. *Trends Neurosci* 2009 (32), 611–618.
5. Bielefeldt, K., Davis, B., and Binion, D.G. Pain and inflammatory bowel disease. *Inflamm Bowel Dis* 2009 (15), 778–788.
6. Wood, J.N., and Eijkelkamp, N. Noxious mechanosensation—Molecules and circuits. *Curr Opin Pharmacol* 2012 (12), 4–8.
7. McMahon, S.B., and Bevan, S. Inflammatory mediators and modulators of pain. In *Wall and Melzack's Textbook of Pain Edition*. Edited by S. McMahon and M. Koltzenburg, Churchill Livingstone (Elsevier Health Sciences), Amsterdam, 2005, pp. 49–72.
8. Tarpley, J.W., Kohler, M.G., and Martin, W.J. The behavioral and neuroanatomical effects of IB4-saporin treatment in rat models of nociceptive and neuropathic pain. *Brain Res* 2004 (1029), 65–76.
9. Joseph, E.K., Chen, X., Bogen, O., and Levine, J.D. Oxaliplatin acts on IB4-positive nociceptors to induce an oxidative stress-dependent acute painful peripheral neuropathy. *J Pain* 2008 (9), 463–472.
10. Joseph, E.K., and Levine, J.D. Hyperalgesic priming is restricted to isolectin B4-positive nociceptors. *Neuroscience* 2010 (169), 431–435.
11. Ferrari, L.F., Bogen, O., and Levine, J.D. Nociceptor subpopulations involved in hyperalgesic priming. *Neuroscience* 2010 (165), 896–901.

12. Abrahamsen, B., Zhao, J., Asante, C.O. et al. The cell and molecular basis of mechanical, cold, and inflammatory pain. *Science* 2008 (321), 702–705.
13. Collier, H.O., and Schneider, C. Nociceptive response to prostaglandins and analgesic actions of aspirin and morphine. *Nat New Biol* 1972 (236), 141–143.
14. Ferreira, S.H., Nakamura, M., and de Abreu Castro, M.S. The hyperalgesic effects of prostacyclin and prostaglandin E2. *Prostaglandins* 1978 (16), 31–37.
15. Cunha, F.Q., Poole, S., Lorenzetti, B.B., and Ferreira, S.H. The pivotal role of tumour necrosis factor alpha in the development of inflammatory hyperalgesia. *Br J Pharmacol* 1992 (107), 660–664.
16. Wise, H. Lack of interaction between prostaglandin E2 receptor subtypes in regulating adenylyl cyclase activity in cultured rat dorsal root ganglion cells. *Eur J Pharmacol* 2006 (535), 69–77.
17. Taiwo, Y.O., and Levine, J.D. Prostaglandin effects after elimination of indirect hyperalgesic mechanisms in the skin of the rat. *Brain Res* 1989 (492), 397–399.
18. Eijkelkamp, N., Wang, H., Garza-Carbajal, A. et al. Low nociceptor GRK2 prolongs prostaglandin E2 hyperalgesia via biased cAMP signaling to Epac/Rap1, protein kinase Cepsilon, and MEK/ERK. *J Neurosci* 2010 (30), 12806–12815.
19. Lin, C.R., Amaya, F., Barrett, L. et al. Prostaglandin E2 receptor EP4 contributes to inflammatory pain hypersensitivity. *J Pharmacol Exp Ther* 2006 (319), 1096–1103.
20. Taiwo, Y.O., Heller, P.H., and Levine, J.D. Mediation of serotonin hyperalgesia by the cAMP second messenger system. *Neuroscience* 1992 (48), 479–483.
21. Dina, O.A., Hucho, T., Yeh, J., Malik-Hall, M., Reichling, D.B., and Levine, J.D. Primary afferent second messenger cascades interact with specific integrin subunits in producing inflammatory hyperalgesia. *Pain* 2005 (115), 191–203.
22. Wang, H., Heijnen, C.J., Eijkelkamp, N. et al. GRK2 in sensory neurons regulates epinephrine-induced signalling and duration of mechanical hyperalgesia. *Pain* 2011 (152), 1649–1658.
23. Gold, M.S., Reichling, D.B., Shuster, M.J., and Levine, J.D. Hyperalgesic agents increase a tetrodotoxin-resistant Na^+ current in nociceptors. *Proc Natl Acad Sci U S A* 1996 (93), 1108–1112.
24. Taiwo, Y.O., and Levine, J.D. Further confirmation of the role of adenyl cyclase and of cAMP-dependent protein kinase in primary afferent hyperalgesia. *Neuroscience* 1991 (44), 131–135.
25. Dostmann, W.R. (RP)-cAMPS inhibits the cAMP-dependent protein kinase by blocking the cAMP-induced conformational transition. *FEBS Lett* 1995 (375), 231–234.
26. Cunha, F.Q., Teixeira, M.M., and Ferreira, S.H. Pharmacological modulation of secondary mediator systems—Cyclic AMP and cyclic GMP—On inflammatory hyperalgesia. *Br J Pharmacol* 1999 (127), 671–678.
27. Ferreira, S.H., Lorenzetti, B.B., and De Campos, D.I. Induction, blockade and restoration of a persistent hypersensitive state. *Pain* 1990 (42), 365–371.
28. Aley, K.O., and Levine, J.D. Role of protein kinase A in the maintenance of inflammatory pain. *J Neurosci* 1999 (19), 2181–2186.
29. Ma, W., and Quirion, R. Does COX2-dependent PGE2 play a role in neuropathic pain? *Neurosci Lett* 2008 (437), 165–169.
30. St-Jacques, B., and Ma, W. Role of prostaglandin E2 in the synthesis of the pro-inflammatory cytokine interleukin-6 in primary sensory neurons: An in vivo and in vitro study. *J Neurochem* 2011 (118), 841–854.
31. Ingram, S.L., and Williams, J.T. Opioid inhibition of Ih via adenylyl cyclase. *Neuron* 1994 (13), 179–186.
32. Ferreira, S.H., and Nakamura, M. III—Prostaglandin hyperalgesia: Relevance of the peripheral effect for the analgesic action of opioid-antagonists. *Prostaglandins* 1979 (18), 201–208.

33. Schnorr, S., Eberhardt, M., Kistner, K. et al. HCN2 channels account for mechanical (but not heat) hyperalgesia during long-standing inflammation. *Pain* 2014 (155), 1079–1090.
34. Emery, E.C., Young, G.T., Berrocoso, E.M., Chen, L., and McNaughton, P.A. HCN2 ion channels play a central role in inflammatory and neuropathic pain. *Science* 2011 (333), 1462–1466.
35. Kawasaki, H., Springett, G.M., Mochizuki, N. et al. A family of cAMP-binding proteins that directly activate Rap1. *Science* 1998 (282), 2275–2279.
36. de Rooij, J., Zwartkruis, F.J., Verheijen, M.H. et al. Epac is a Rap1 guanine-nucleotide-exchange factor directly activated by cyclic AMP. *Nature* 1998 (396), 474–477.
37. Grandoch, M., Roscioni, S.S., and Schmidt, M. The role of Epac proteins, novel cAMP mediators, in the regulation of immune, lung and neuronal function. *Br J Pharmacol* 2010 (159), 265–284.
38. Holz, G.G., Kang, G., Harbeck, M., Roe, M.W., and Chepurny, O.G. Cell physiology of cAMP sensor Epac. *J Physiol* 2006 (577), 5–15.
39. Suzuki, S., Yokoyama, U., Abe, T. et al. Differential roles of Epac in regulating cell death in neuronal and myocardial cells. *J Biol Chem* 2010 (285), 24248–24259.
40. Wang, H., Heijnen, C.J., van Velthoven, C.T. et al. Balancing GRK2 and EPAC1 levels prevents and relieves chronic pain. *J Clin Invest* 2013 (123), 5023–5034.
41. Eijkelkamp, N., Linley, J.E., Torres, J.M. et al. A role for Piezo2 in EPAC1-dependent mechanical allodynia. *Nat Commun* 2013 (4), 1682.
42. Shibasaki, T., Takahashi, H., Miki, T. et al. Essential role of Epac2/Rap1 signaling in regulation of insulin granule dynamics by cAMP. *Proc Natl Acad Sci U S A* 2007 (104), 19333–19338.
43. Christensen, A.E., Selheim, F., de Rooji, J. et al. cAMP analog mapping of Epac1 and cAMP kinase. Discriminating analogs demonstrate that Epac and cAMP kinase act synergistically to promote PC-12 cell neurite extension. *J Biol Chem* 2003 (278), 35394–35402.
44. Almahariq, M., Tsalkova, T., Mei, F.C. et al. A novel EPAC-specific inhibitor suppresses pancreatic cancer cell migration and invasion. *Mol Pharmacol* 2013 (83), 122–128.
45. Taiwo, Y.O., and Levine, J.D. Direct cutaneous hyperalgesia induced by adenosine. *Neuroscience* 1990 (38), 757–762.
46. Janig, W., Levine, J.D., and Michaelis, M. Interactions of sympathetic and primary afferent neurons following nerve injury and tissue trauma. *Prog Brain Res* 1996 (113), 161–184.
47. Chen, X., and Levine, J.D. Epinephrine-induced excitation and sensitization of rat C-fiber nociceptors. *J Pain* 2005 (6), 439–446.
48. Chen, H.S., Lei, J., He, X. et al. Peripheral involvement of PKA and PKC in subcutaneous bee venom-induced persistent nociception, mechanical hyperalgesia, and inflammation in rats. *Pain* 2008 (135), 31–36.
49. Zhang, J.M., Li, H., Liu, B., and Brull, S.J. Acute topical application of tumor necrosis factor alpha evokes protein kinase A-dependent responses in rat sensory neurons. *J Neurophysiol* 2002 (88), 1387–1392.
50. Eijkelkamp, N., Heijnen, C.J., Carbajal, A.G. et al. G protein-coupled receptor kinase 6 acts as a critical regulator of cytokine-induced hyperalgesia by promoting phosphatidylinositol 3-kinase and inhibiting p38 signaling. *Mol Med* 2012 (18), 556–564.
51. Kim, M.J., Lee, S.Y., Yang, K.Y. et al. Differential regulation of peripheral IL-1beta–induced mechanical allodynia and thermal hyperalgesia in rats. *Pain* 2014 (155), 723–732.
52. Kim, K.S., Kim, J., Back, S.K., Im, J.Y., Na, H.S., and Han, P.L. Markedly attenuated acute and chronic pain responses in mice lacking adenylyl cyclase-5. *Genes Brain Behav* 2007 (6), 120–127.
53. Song, X.J., Wang, Z.B., Gan, Q., and Walters, E.T. cAMP and cGMP contribute to sensory neuron hyperexcitability and hyperalgesia in rats with dorsal root ganglia compression. *J Neurophysiol* 2006 (95), 479–492.

54. Liou, J.T., Liu, F.C., Hsin, S.T., Yang, C.Y., and Lui, P.W. Inhibition of the cyclic adenosine monophosphate pathway attenuates neuropathic pain and reduces phosphorylation of cyclic adenosine monophosphate response element-binding in the spinal cord after partial sciatic nerve ligation in rats. *Anesth Analg* 2007 (105), 1830–1837 (table).
55. Malmberg, A.B., Brandon, E.P., Idzerda, R.L., Liu, H., McKnight, G.S., and Basbaum, A.I. Diminished inflammation and nociceptive pain with preservation of neuropathic pain in mice with a targeted mutation of the type I regulatory subunit of cAMP-dependent protein kinase. *J Neurosci* 1997 (17), 7462–7470.
56. Isensee, J., Diskar, M., Waldherr, S. et al. Pain modulators regulate the dynamics of PKA-RII phosphorylation in subgroups of sensory neurons. *J Cell Sci* 2014 (127), 216–229.
57. Yajima, Y., Narita, M., Shimamura, M., Narita, M., Kubota, C., and Suzuki, T. Differential involvement of spinal protein kinase C and protein kinase A in neuropathic and inflammatory pain in mice. *Brain Res* 2003 (992), 288–293.
58. Molliver, D.C., Wright, D.E., Leitner, M.L. et al. IB4-binding DRG neurons switch from NGF to GDNF dependence in early postnatal life. *Neuron* 1997 (19), 849–861.
59. Hucho, T.B., Dina, O.A., and Levine, J.D. Epac mediates a cAMP-to-PKC signaling in inflammatory pain: An isolectin B4(+) neuron-specific mechanism. *J Neurosci* 2005 (25), 6119–6126.
60. Rathee, P.K., Distler, C., Obreja, O. et al. PKA/AKAP/VR-1 module: A common link of Gs-mediated signaling to thermal hyperalgesia. *J Neurosci* 2002 (22), 4740–4745.
61. Caterina, M.J., Schumacher, M.A., Tominaga, M., Rosen, T.A., Levine, J.D., and Julius, D. The capsaicin receptor: A heat-activated ion channel in the pain pathway. *Nature* 1997 (389), 816–824.
62. Julius, D. TRP channels and pain. *Annu Rev Cell Dev Biol* 2013 (29), 355–384.
63. Cho, H., Yang, Y.D., Lee, J. et al. The calcium-activated chloride channel anoctamin 1 acts as a heat sensor in nociceptive neurons. *Nat Neurosci* 2012 (15), 1015–1021.
64. Bhave, G., Zhu, W., Wang, H., Brasier, D.J., Oxford, G.S., and Gereau, R.W. cAMP-dependent protein kinase regulates desensitization of the capsaicin receptor (VR1) by direct phosphorylation. *Neuron* 2002 (35), 721–731.
65. Efendiev, R., Bavencoffe, A., Hu, H., Zhu, M.X., and Dessauer, C.W. Scaffolding by A-kinase anchoring protein enhances functional coupling between adenylyl cyclase and TRPV1 channel. *J Biol Chem* 2013 (288), 3929–3937.
66. Bolcskei, K., Helyes, Z., Szabo, A. et al. Investigation of the role of TRPV1 receptors in acute and chronic nociceptive processes using gene-deficient mice. *Pain* 2005 (117), 368–376.
67. Moriyama, T., Higashi, T., Togashi, K. et al. Sensitization of TRPV1 by EP1 and IP reveals peripheral nociceptive mechanism of prostaglandins. *Mol Pain* 2005 (1), 3.
68. McCarter, G.C., and Levine, J.D. Ionic basis of a mechanotransduction current in adult rat dorsal root ganglion neurons. *Mol Pain* 2006 (2), 28.
69. Coste, B., Mathur, J., Schmidt, M. et al. Piezo1 and Piezo2 are essential components of distinct mechanically activated cation channels. *Science* 2010 (330), 55–60.
70. Woo, S.H., Ranade, S., Weyer, A.D. et al. Piezo2 is required for Merkel-cell mechanotransduction. *Nature* 2014 (509), 622–626.
71. Dubin, A.E., Schmidt, M., Mathur, J. et al. Inflammatory signals enhance piezo2-mediated mechanosensitive currents. *Cell Rep* 2012 (2), 511–517.
72. Eijkelkamp, N., Quick, K., and Wood, J.N. Transient receptor potential channels and mechanosensation. *Annu Rev Neurosci* 2013 (36), 519–546.
73. Dall'acqua, M.C., Bonet, I.J., Zampronio, A.R., Tambeli, C.H., Parada, C.A., and Fischer, L. The contribution of transient receptor potential ankyrin 1 (TRPA1) to the in vivo nociceptive effects of prostaglandin E. *Life Sci* 2014 (105), 7–13.
74. Eijkelkamp, N., Linley, J.E., Baker, M.D. et al. Neurological perspectives on voltage-gated sodium channels. *Brain* 2012 (135), 2585–2612.

75. Bevan, S., and Storey, N. Modulation of sodium channels in primary afferent neurons. *Novartis Found Symp* 2002 (241), 144–153.
76. England, S., Bevan, S., and Docherty, R.J. PGE2 modulates the tetrodotoxin-resistant sodium current in neonatal rat dorsal root ganglion neurones via the cyclic AMP-protein kinase A cascade. *J Physiol* 1996 (495, Pt 2), 429–440.
77. Fitzgerald, E.M., Okuse, K., Wood, J.N., Dolphin, A.C., and Moss, S.J. cAMP-dependent phosphorylation of the tetrodotoxin-resistant voltage-dependent sodium channel SNS. *J Physiol* 1999 (516, Pt 2), 433–446.
78. Vijayaragavan, K., Boutjdir, M., and Chahine, M. Modulation of Nav1.7 and Nav1.8 peripheral nerve sodium channels by protein kinase A and protein kinase C. *J Neurophysiol* 2004 (91), 1556–1569.
79. Chatelier, A., Dahllund, L., Eriksson, A., Krupp, J., and Chahine, M. Biophysical properties of human Na v1.7 splice variants and their regulation by protein kinase A. *J Neurophysiol* 2008 (99), 2241–2250.
80. Villarreal, C.F., Sachs, D., Funez, M.I., Parada, C.A., de Queiroz, C.F., and Ferreira, S.H. The peripheral pro-nociceptive state induced by repetitive inflammatory stimuli involves continuous activation of protein kinase A and protein kinase C epsilon and its Na(V)1.8 sodium channel functional regulation in the primary sensory neuron. *Biochem Pharmacol* 2009 (77), 867–877.
81. Liu, C., Li, Q., Su, Y., and Bao, L. Prostaglandin E2 promotes Na1.8 trafficking via its intracellular RRR motif through the protein kinase A pathway. *Traffic* 2010 (11), 405–417.
82. Evans, A.R., Vasko, M.R., and Nicol, G.D. The cAMP transduction cascade mediates the PGE2-induced inhibition of potassium currents in rat sensory neurones. *J Physiol* 1999 (516, Pt 1), 163–178.
83. Ster, J., De Bock, F., Guerineau, N.C. et al. Exchange protein activated by cAMP (Epac) mediates cAMP activation of p38 MAPK and modulation of Ca^{2+}-dependent K^{+} channels in cerebellar neurons. *Proc Natl Acad Sci U S A* 2007 (104), 2519–2524.
84. Aley, K.O., Messing, R.O., Mochly-Rosen, D., and Levine, J.D. Chronic hypersensitivity for inflammatory nociceptor sensitization mediated by the epsilon isozyme of protein kinase C. *J Neurosci* 2000 (20), 4680–4685.
85. Tillu, D.V., Melemedjian, O.K., Asiedu, M.N. et al. Resveratrol engages AMPK to attenuate ERK and mTOR signaling in sensory neurons and inhibits incision-induced acute and chronic pain. *Mol Pain* 2012 (8), 5.
86. Asiedu, M.N., Tillu, D.V., Melemedjian, O.K. et al. Spinal protein kinase M zeta underlies the maintenance mechanism of persistent nociceptive sensitization. *J Neurosci* 2011 (31), 6646–6653.
87. Dina, O.A., McCarter, G.C., de Coupade, C., and Levine, J.D. Role of the sensory neuron cytoskeleton in second messenger signaling for inflammatory pain. *Neuron* 2003 (39), 613–624.
88. Dina, O.A., Khasar, S.G., Gear, R.W., and Levine, J.D. Activation of Gi induces mechanical hyperalgesia poststress or inflammation. *Neuroscience* 2009 (160), 501–507.
89. Joseph, E.K., Bogen, O., Essandri-Haber, N., and Levine, J.D. PLC-beta 3 signals upstream of PKC epsilon in acute and chronic inflammatory hyperalgesia. *Pain* 2007 (132), 67–73.
90. Hucho, T., and Levine, J.D. Signaling pathways in sensitization: Toward a nociceptor cell biology. *Neuron* 2007 (55), 365–376.
91. Pitcher, J.A., Freedman, N.J., and Lefkowitz, R.J. G protein-coupled receptor kinases. *Annu Rev Biochem* 1998 (67), 653–692.
92. Evron, T., Daigle, T.L., and Caron, M.G. GRK2: Multiple roles beyond G protein-coupled receptor desensitization. *Trends Pharmacol Sci* 2012 (33), 154–164.

93. Kleibeuker, W., Jurado-Pueyo, M., Murga, C. et al. Physiological changes in GRK2 regulate CCL2-induced signaling to ERK1/2 and Akt but not to MEK1/2 and calcium. *J Neurochem* 2008 (104), 979–992.
94. Peregrin, S., Jurado-Pueyo, M., Campos, P.M. et al. Phosphorylation of p38 by GRK2 at the docking groove unveils a novel mechanism for inactivating p38MAPK. *Curr Biol* 2006 (16), 2042–2047.
95. Jimenez-Sainz, M.C., Murga, C., Kavelaars, A. et al. G protein-coupled receptor kinase 2 negatively regulates chemokine signaling at a level downstream from G protein subunits. *Mol Biol Cell* 2006 (17), 25–31.
96. Wang, C., Gu, Y., Li, G.W., and Huang, L.Y. A critical role of the cAMP sensor Epac in switching protein kinase signalling in prostaglandin E2-induced potentiation of P2X3 receptor currents in inflamed rats. *J Physiol* 2007 (584), 191–203.

8 Monitoring Real-Time Cyclic Nucleotide Dynamics in Subcellular Microdomains

Zeynep Bastug and Viacheslav O. Nikolaev

CONTENTS

8.1 INTRODUCTION

Cyclic nucleotides 3′,5′-cyclic adenosine monophosphate (cAMP) and 3′,5′-cyclic guanosine monophosphate (cGMP) are ubiquitous second messengers that regulate myriads of physiological functions ranging from memory formation to vasorelaxation and cardiac contractility (Beavo and Brunton 2002; Hofmann and Wegener 2013).

The specificity of multiple intracellular responses engaged by these two cyclic nucleotide second messengers is believed to be achieved by compartmentation of their signaling in numerous differentially regulated subcellular microdomains where cAMP and cGMP activate local pools of the downstream effector proteins (Fischmeister et al. 2006; Lefkimmiatis and Zaccolo 2014; Perera and Nikolaev 2013). Such microdomains contain, for example, local pools of cyclic nucleotide–dependent protein kinases, anchoring proteins and phosphodiesterases (PDEs), and enzymes responsible for local cAMP and cGMP degradation and therefore for spatial segregation of the microdomains (Dodge-Kafka et al. 2006; Diviani et al. 2011; Conti and Beavo 2007; Baillie 2009).

Local cAMP or cGMP dynamics are believed to underpin the specificity of multiple functional effects. However, their direct visualization, especially using classic biochemical techniques, is extremely challenging. In this chapter, we present a short review of the available techniques to study local cyclic nucleotide signaling and a protocol for design and testing of targeted cAMP biosensors.

8.2 CLASSIC METHODS TO STUDY CYCLIC NUCLEOTIDE SIGNALING AND THEIR LIMITATIONS

8.2.1 Enzyme-Linked Immunosorbent Assays and Radioimmunoassays

Radioimmunoassays can be used for quantitative measurements of total (i.e., the sum of free and protein-bound) cAMP or cGMP concentrations in various cells and tissues (Brooker et al. 1979; Harper and Brooker 1975; Williams 2004). Radioimmunoassays are based on anti-cAMP or anti-cGMP antibodies and radioactively labeled tracer molecules such as ^{125}I-cAMP/cGMP. The tracer binds to the specific antibodies and can be displaced by cyclic nucleotides present in cell or tissue samples, thereby decreasing antibody-bound radioactivity. Using calibration curves, radioactivity measurements can be easily converted into the net amount of cAMP and cGMP in the sample. The nonradioactive enzyme-linked immunosorbent assays (ELISA) can be alternatively used to measure cAMP and cGMP in cell or tissue lysates. They are based on primary polyclonal cAMP/cGMP antibodies and secondary antibodies, which are coated on a 96-well plate. In these assays, cyclic nucleotides from the sample usually compete with cAMP or cGMP molecules covalently attached to alkaline phosphatase. During incubation, this enzyme binds via antibodies to the plate and then converts its substrate into a colored product that can be measured and calibrated to the cyclic nucleotide content of the sample (Williams 2004).

These biochemical methods are sensitive and measure cyclic nucleotides in a linear fashion to cover a broad concentration range. However, they require the disruption of large amounts of cells or tissues and only provide a multicell average measure of cAMP or cGMP content. Therefore, they are not useful for the analysis of real-time cyclic nucleotides with high spatial resolutions. Moreover, some ELISA assays are prone to artifacts caused by cell lysis using hydrochloric acid, which might lead to false-positive responses.

8.2.2 Biochemical Techniques Based on Cell Fractionation and Coimmunoprecipitation

cAMP levels and cAMP-dependent protein kinase (PKA) activities can be also analyzed in isolated cellular fractions by the classic methods described previously. Using isolated membrane and cytosolic fractions from adult rabbit cardiomyocytes, Brunton and colleagues showed that while isoproterenol could stimulate the accumulation of cAMP in both fractions, prostaglandin increased cAMP only in the cytosolic fraction (Brunton et al. 1981; Buxton and Brunton 1983). These seminal studies were the first to demonstrate cAMP compartmentation. However, this classic approach is limited to only a small number of potential subcellular compartments that can be fractionated using a density gradient, and provides only an approximate snapshot of the signaling occurring in such compartments.

More refined biochemical approaches have been developed based on the coimmunoprecipitation of individual protein complexes formed around A-kinase anchoring proteins, receptors, or distinct PDE isoforms (Richter et al. 2008; Lehnart et al. 2005; Lygren et al. 2007). Using this approach, it is possible to measure PDE and PKA activities in individual biochemically characterized microdomains to better understand the mechanisms of cyclic nucleotide compartmentation. These methods require good antibodies for coimmunoprecipitation and elaborate back-phosphorylation assays. These assays involve careful extraction of particular proteins from cells or tissues followed by in vitro phosphorylation using radioactively labeled ATP. The amount of incorporated labeled phosphate is inversely proportional to the amount of phosphorylated protein in the starting cell or tissue material. They can usually report only the downstream kinase and PDE activity and not the direct cyclic nucleotide levels in each individual microdomain.

8.2.3 Analysis of Substrate Phosphorylation

In addition to measuring protein kinase activity in a back-phosphorylation assay, it is possible to analyze the phosphorylation of the downstream kinase substrate using phosphospecific antibodies. Such substrates may include the L-type calcium channels, the ryanodine receptor, phospholamban, several contractile proteins, and other proteins (Di Benedetto et al. 2008; Mika et al. 2014). This approach can potentially provide a functional readout of the microdomain-specific cAMP signaling but has a limitation that these substrates cannot always be assigned to the defined individual microdomains and may be phosphorylated by cAMP, which originates from neighboring compartments.

8.3 ELECTROPHYSIOLOGICAL AND BIOPHYSICAL TECHNIQUES

8.3.1 Cyclic Nucleotide–Gated Channels as Sensors

An interesting approach to measuring real-time cAMP and cGMP dynamics has been developed based on the electrophysiological and biochemical properties of the so-called cyclic nucleotide–gated (CNG) channels (Rich et al. 2000, 2001). These are homotetrameric nonselective cation channels that show an increased opening probability upon binding of cAMP or cGMP to the cyclic nucleotide binding domain in the intracellular C-terminus. When ectopically expressed in intact cells, they allow an indirect monitoring of cAMP levels by measuring CNG channel currents or by calcium imaging.

Native CNG channels and their mutants with increased affinity for cAMP have been used in several studies to monitor cAMP in the subsarcolemmal compartment where these channels are localized. These studies have further supported the concept of cAMP compartmentation. For example, the existence of submembrane cAMP pools with transient kinetic properties after adenylyl cyclase (AC) stimulation has been shown in C6-2B glioma cells and human embryonic kidney 293 (HEK293) cells (Rich et al. 2000, 2001). Stimulation of these cells by either natriuretic peptides or nitric oxide (NO) also led to cGMP synthesis in two distinct and differentially regulated subcellular microdomains (Piggott et al. 2006). In adult rat cardiomyocytes, various G_s-coupled receptors were able to induce various patterns of cAMP signals and involve distinct subsets of PDEs (Rochais et al. 2004, 2006). Wild-type CNG channels have also been used to monitor subsarcolemmal pools of cGMP in adult rat ventricular myocytes. This approach could reveal PDE-dependent differential regulation of cGMP pools associated with natriuretic peptide receptors (stimulated with atrial natriuretic peptide and controlled by both PDE2 and PDE5) and with soluble guanylyl cyclase (stimulated with NO donors and controlled by PDE5; Castro et al. 2006). Interestingly, the cGMP-dependent protein kinase affects each pool differently. In the former, it increased the ANP-mediated increase of cGMP, whereas in the latter pool, it exerted a negative feedback on local cGMP levels by activating PDE5 (Castro et al. 2010).

8.3.1.1 Advantages

CNG channel–based sensors for cAMP and cGMP have several advantages. They show fast kinetics of gating and no catalytic activity. They do not desensitize and have a broad dynamic range.

8.3.1.2 Requirements and Limitations

However, CNG measurements require a high degree of electrophysiological expertise and transfection of the channel constructs into the cells. Moreover, the selectivity of these channels for cAMP or cGMP is rather low, which does not always allow us to distinguish between either cyclic nucleotide, especially in such cases when they cross-regulate each other (Zaccolo and Movsesian 2007). The main limitation remains the exclusive targeting of the channel to poorly defined plasma membrane fractions, presumably associated with noncaveolar domains (Rich et al. 2000), and the inability to use these channels for cAMP/cGMP measurements in other subcellular microdomains.

8.3.2 Targeted FRET Biosensors

Probably the most specific way of measuring cyclic nucleotide levels in a clearly defined microdomain is the use of targeted Förster resonance energy transfer (FRET)–based biosensors specially tailored to localize in given subcellular loci. To date, several targeted biosensors for cAMP have been developed, which are summarized in a recent review (Sprenger and Nikolaev 2013). For example, fusion of a 10 amino acid peptide sequence from the Lyn kinase to the N-terminus of the sensor would lead to its localization in caveolin-rich membrane compartments (Wachten et al. 2010), whereas the use of a nuclear localization signal or a mitochondrial matrix protein motif results in biosensors that target the respective organelles (DiPilato et al. 2004).

8.3.2.1 Advantages

Targeted biosensors are especially attractive because they provide a possibility to freely choose the microdomain of interest and to develop a respective targeted construct that is exclusively localized to this microdomain. Through fusion to a large number of proteins, this approach offers a high degree of freedom and flexibility. It provides reporters that can directly monitor cyclic nucleotide dynamics directly into the desired subcellular location.

8.3.2.2 Requirements and Limitations

To guarantee the specificity and the local nature of the measured response, it is advisable to confirm the exclusive localization of a targeted sensor in the microdomain under study and its absence from the bulk cytosol or other microdomains, which might *contaminate* the local response. It is also crucial to confirm that the presence of the sensor in this subcellular location does not interfere with normal microdomain function and composition that would preclude from reliable conclusions about the true local cyclic nucleotide dynamics and its regulation by local pools of PKA and PDEs presented in these microdomains.

8.4 DEVELOPMENT OF TARGETED FRET BIOSENSORS

8.4.1 Design and Cloning

In this chapter, we would like to provide two examples of targeted cAMP biosensors, which are designed to report cyclic nucleotide dynamics at the plasma membrane or at the sarcoplasmic reticulum.

To monitor cAMP directly at the sarcoplasmic reticulum, the ICUE3 biosensor sequence without the stop codon can be fused in-frame to the transmembrane domain of the sarcoplasmic reticulum protein phospholamban, as shown in Figure 8.1a. ICUE3 is a cytosolic cAMP biosensor comprised of the full-length Epac1 protein sandwiched between the donor, enhanced cyan fluorescent protein (eCFP), and the acceptor,

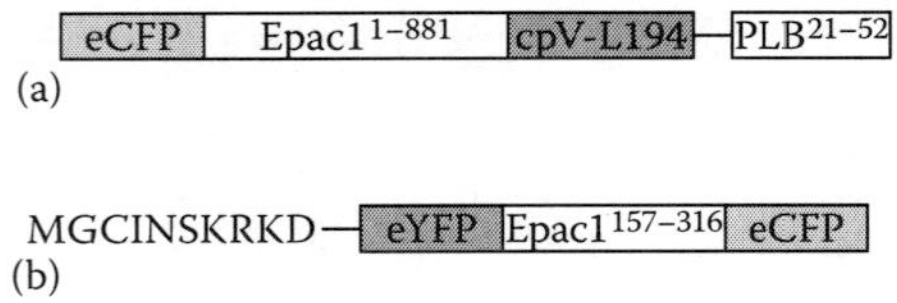

FIGURE 8.1 Design of the targeted FRET cyclic nucleotide biosensors. (a) the first sensor contains a sequence of ICUE3 fused on its C-terminus to the transmembrane domain of phospholamban (PLB) for targeting to the sarcoplasmic reticulum (SR) membrane. ICUE3 comprises the full-length Epacl sequence sandwiched between eCFP (the donor) and cpVenus-L194 (acceptor) fluorophores. (b) pmEpacl-camps contains an N-terminal membrane targeting 10 amino acid motive from the Lyn kinase fused to the cytosolic Epacl-camps sensor to achieve sensor localization to caveolin-rich membrane fractions.

circularly permuted Venus fluorescent protein (DiPilato and Zhang 2009). Fusion of this sensor to the sarcoplasmic reticulum protein phospholamban enables the localization and measurements at the membrane of this organelle (Chakir et al. 2011).

Figure 8.1b shows the design of the membrane-targeted cAMP biosensors pmEpacl-camps. It is derived from the cytosolic biosensor Epacl-camps, which contains a single cyclic nucleotide binding domain from Epacl flanked by eCFP and enhanced yellow fluorescent protein (eYFP) and shows a conformational change leading to a decrease of FRET upon activation. To clone pmEpacl-camps, digest the Epacl-camps plasmid (Nikolaev et al. 2004) with *Hin*dIII and *Eco*RI restriction enzymes and ligate it with the digested PCR fragment, which has a sequence encoding the N-terminal 10 amino acid tag MGCINSKRKD (see Figure 8.2).

Materials needed:

1. Epacl-camps plasmid vector (Nikolaev et al. 2004)
2. Pfu polymerase and buffer (Promega)
3. *Hin*dIII and *Eco*RI restriction enzymes (New England Biolabs)

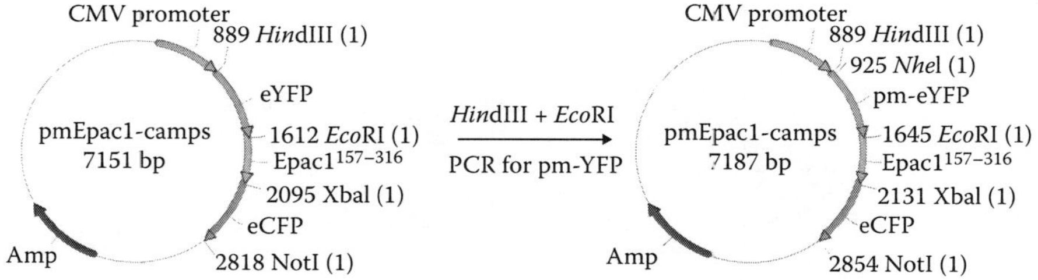

FIGURE 8.2 Cloning of the pmEpacl-camps sensor. The previously described pcDNA3.0-Epacl-camps vector was cut with *Hin*dIII and *Eco*RI, and the pm-eYFP sequence amplified using PCR and digested with the same enzymes was subcloned into the vector. The resulting construct contained a sequence for the above-described N-terminal 10 amino acid Lyn kinase tag MGCINSKRKD, followed by the *Nhe*I restriction site (resulting in additional two amino acids Ala-Ser) and the Epacl-camps sensor sequence.

4. T4 DNA ligase (New England Biolabs)
5. QIAquick Gel Extraction Kit (Qiagen)
6. QIAfilter Plasmid Midi Kit (Qiagen)
7. Primers (ordered from MWG Eurofins or alternative sources, the linker sequence is in italic, the tag sequence is underlined):
   forward: 5′ *AAA AAG CTT* <u>ATG GGA TGT ATC AAT AGC AAG CGC AAA GAT</u> GCT AGC ATG GTG AGC AAG GGC 3′
   reverse: 5′ *AAA GAA TTC* CTT GTA CAG CTC GTC CAT GC 3′

8.4.2 Testing the Proper Sensor Localization

To test the proper subcellular localization of the pmEpac1-camps sensor, it can be cotransfected into 293A cells together with a marker protein construct such as red fluorescent protein–tagged caveolin 3 (cav3-RFP) for subsequent confocal microscopy analysis. Alternatively, membrane dyes such as CellMask™ DeepRed (Molecular Probes) can be used to label sensor-transfected cells immediately before microscopy.

8.4.2.1 Cell Culture and Transfection Materials

1. Mammalian expression plasmid encoding the targeted FRET biosensor (e.g., pmEpac1-camps, see above)
2. 293A cells (Invitrogen)
3. Dulbecco's phosphate buffer saline without Ca^{2+} and Mg^{2+} (PBS, Biochrom)
4. 293A cell growth medium:
   Dulbecco's MEM (DMEM without additives, Biochrom)
   10% (v/v) fetal bovine serum (FBS, Biochrom)
   1% (v/v) of 100 U/mL penicillin and 100 μg/mL streptomycin (P/S, Biochrom)
   1% (v/v) of 200 mM L-glutamine solution (Biochrom)
5. 0.05% Trypsin/EDTA solution (Biochrom)
6. Lipofectamine 2000 transfection reagent (Life Technologies)
7. Tissue culture 75 cm² flasks and six-well plates
8. 25 mm round glass coverslides (Thermo Scientific)

8.4.2.2 Cell Culture and Transfection

8.4.2.2.1 Thawing and Initial Culture Procedure

1. Rapidly thaw the cells at 37°C in a water bath
2. Gently pipette the cells in a 50 mL falcon tube with 29 mL growth medium in it
3. Centrifuge at 800 rpm for 5 min
4. After aspirating the supernatant, resuspend the cells with 12 mL medium and transfer the cell suspension in a 75 cm² flask for proliferation until they are 90% confluent
5. When the cells have reached the appropriate density, aspirate the medium from the flask, wash the cells with 10 mL of PBS and add 2 mL of 37°C

0.05% Trypsin to dissociate them by tilting the flask or briefly incubating at 37°C
6. Check under the microscope for sufficient trypsinization
7. Add 9 mL of DMEM-growth medium and resuspend the cells thoroughly
8. Prepare a six-well plate by placing autoclaved 24 mm cover-slides in each well
9. Split the cell suspension 1:12 and plate them onto the cover-slide
10. Transfect the cells after 24 h when they are 70% to 90% confluent

8.4.2.2.2 Transfection

1. Dilute the DNA construct and the Lipofectamine 2000 reagent each with 150 μL DMEM without additives. Use 2.33 μL of Lipofectamine 2000 per 1.0 μg of DNA. Add 7 μL of Lipofectamine into one tube with 150 μL of DMEM and do the same for 3 μg DNA in another tube with 150 μL of DMEM
2. Add diluted DNA to the diluted Lipofectamine 2000, mix gently and incubate for 20 min at room temperature
3. Pipette 50 μL per well DNA–lipid complex dropwise to the cells
4. Incubate the cells for 24 to 48 h and analyze the expression and FRET responses. Make sure that cells have not reached confluency at this point, as this might influence the activity of several cell surface receptors

8.4.2.3 Confocal Microscopy

First, the FRET biosensor should be tested for its expression and localization in cultured 293A cells. To characterize the cellular localization of the targeted FRET biosensor, 293A cells are cotransfected with the construct and the cav3-RFP membrane marker construct (as described previously).

Confocal images of transfected cells can be taken using any suitable commercially available confocal microscope in YFP (excitation at 488 or 514 nm) and RFP (excitation at 561 nm) emission channels. The images of the separately recorded emission channels should show a high degree of overlay (see Figure 8.3a).

8.4.3 FRET Measurements in Live Cells

8.4.3.1 Materials and Instrumentation

1. Microscope: a Nikon Ti inverted microscope equipped with 60× oil immersion objective and a single-wavelength light-emitting diode (CoolLED, 440 nm) selectively exciting the donor protein eCFP. Images were captured using a CCD camera (e.g., ORCA 03G, Hamamatsu). CV2 DualView beam splitter with emission filter set containing the 505dcxr dichroic mirror plus ET480/30M and ET535/40M emission filters for simultaneous detection of CFP and YFP emissions and ratiometric imaging (Sprenger et al. 2012)
2. Fluorescence-free immersion oil (Zeiss)
3. Attofluor® cell chamber (A-7816, Life Technologies)
4. Imaging software: for example, MicroManager freeware (download from: https://micro-manager.org/wiki/Micro-Manager)

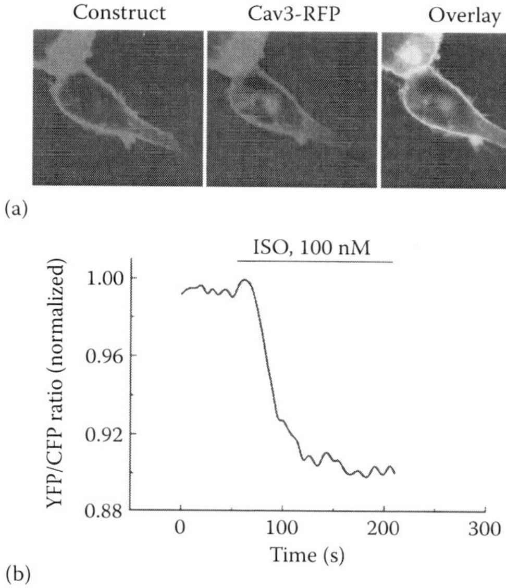

FIGURE 8.3 **(See color insert.)** Testing the targeted FRET construct in 293A cells. (a) pmEpac1-camps was cotransfected together with cav3-RFP plasmid and imaged by confocal microscopy. The biosensor shows good colocalization with cav3-RFP at the plasma membrane, suggesting the successful targeting to the desired compartment. (b) FRET analysis of local cAMP levels measured in transfected cells using the pmEpac1-camps sensor. Cells were stimulated with 100 nM of the β-adrenergic agonist isoproterenol, and the changes of the YFP/CFP (acceptor/donor) ratio were monitored online. A decrease of the ratio indicated an increase in cAMP measured with the sensor at the membrane.

5. FRET buffer: 144 mM NaCl, 5.4 mM KCl, 1 mM $MgCl_2$, 2 mM $CaCl_2$, 10 mM HEPES in deionized water, the pH is adjusted to 7.3 with NaOH. All compounds to be used in the imaging experiments should be diluted with the FRET buffer
6. Isoproterenol (I6504, Sigma)

8.4.3.2 FRET Microscopy

After 24 to 48 h transfection of the 293A cells with the targeted FRET biosensor, they should be tested for the appropriate FRET response.

1. Mount a coverslide with adherent cells in a cell chamber
2. After washing the cells once with FRET buffer, add 400 μL of the buffer into the chamber
3. Place some immersion oil onto the objective and transfer the imaging chamber onto the microscope
4. Mount the chamber on the microscope stage, switch on the fluorescent light, and quickly choose a cell with optimal sensor expression and fluorescence

5. Adjust exposure time (5–10 ms), set an image acquisition time interval (e.g., 5 s) and start the FRET measurement
6. As soon as the FRET ratio has reached a stable baseline, apply isoprenaline at 100 nM final concentration by accurately pipetting its solution into the chamber, and continue monitoring the FRET ratio change

8.4.3.3 Offline Data Analysis

Refer to the previously described protocol (Sprenger et al. 2012), and split the acquired images into donor and acceptor channels to measure fluorescent intensities in multiple regions of interest. These intensities can further be used for the corrected FRET ratio. Binding of cAMP to the biosensor increases the distance between the fluorophores, leading to reduced FRET signal (see Figure 8.3b).

FRET signal is calculated as a corrected acceptor/donor ratio:

$$\text{Ratio} = (\text{YFP} - B \times \text{CFP})/\text{CFP}$$

where B is the correction factor for the bleed-through of the donor fluorescence into the acceptor channel. It can be determined by transfecting cells with eCFP plasmid alone and measuring a fraction of the donor fluorescence in the YFP channel (B = YFP/CFP).

8.5 DISCUSSION AND FUTURE PERSPECTIVES

There are several ways of analyzing local compartmentalized cyclic nucleotide signaling, each having its advantages and disadvantages. Although the classic methods such ELISA and radioimmunoassays have a limited applicability due to low spatial resolution, new biosensors are promising candidates for direct real-time visualization of microdomain-specific cAMP and cGMP dynamics. In particular, differentially targeted FRET reporters are of high interest because, by using versatile targeting motifs, they can be adapted to measuring cyclic nucleotides specifically in a wide variety of subcellular locations. Thereby, it is essential to first ensure an exclusive localization of the probe to the microdomain to provide the specificity of response. Second, it is important to exclude that the presence of the biosensor at this location per se might interfere with the normal physiological function and with the composition of the microdomain. In the future, the spectrum of the available targeted biosensors will definitely be further increased. Apart from using simple tags, which target the reporter to certain organelles or just simply to the cell membrane, further more refined strategies such as fusions of the biosensors to particular individual proteins or their domains (Sin et al. 2011; Di Benedetto et al. 2008) should result in many more targeted sensors. Because a number of such proteins form individual microdomains due to direct or indirect interaction with PKA, PDE, and other molecules, one can expect the development of many new targeted biosensors and imaging tools. These probes can be applied in different cellular systems and under disease conditions to directly understand exact molecular mechanisms and the role of compartmentalized cyclic nucleotide signaling in physiology and pathophysiology.

ACKNOWLEDGMENTS

The work in the authors' laboratory is supported by the Deutsche Forschungsgemeinschaft (grants NI 1301/1-1, SFB 1002 TP A01, IRTG 1816, FOR 2060) and the University of Göttingen Medical Center (*pro futura* grant).

REFERENCES

Baillie, G. S. 2009. Compartmentalized signalling: Spatial regulation of cAMP by the action of compartmentalized phosphodiesterases. *FEBS J* 276 (7):1790–1799.

Beavo, J. A., and L. L. Brunton. 2002. Cyclic nucleotide research—still expanding after half a century. *Nat Rev Mol Cell Biol* 3 (9):710–718.

Brooker, G., J. F. Harper, W. L. Terasaki, and R. D. Moylan. 1979. Radioimmunoassay of cyclic AMP and cyclic GMP. *Adv Cyclic Nucleotide Res* 10:1–33.

Brunton, L. L., J. S. Hayes, and S. E. Mayer. 1981. Functional compartmentation of cyclic AMP and protein kinase in heart. *Adv Cyclic Nucleotide Res* 14:391–397.

Buxton, I. L., and L. L. Brunton. 1983. Compartments of cyclic AMP and protein kinase in mammalian cardiomyocytes. *J Biol Chem* 258 (17):10233–10239.

Castro, L. R., I. Verde, D. M. Cooper, and R. Fischmeister. 2006. Cyclic guanosine monophosphate compartmentation in rat cardiac myocytes. *Circulation* 113 (18):2221–2228.

Castro, L. R., J. Schittl, and R. Fischmeister. 2010. Feedback control through cGMP-dependent protein kinase contributes to differential regulation and compartmentation of cGMP in rat cardiac myocytes. *Circ Res* 107 (10):1232–1240.

Chakir, K., C. Depry, V. L. Dimaano et al. 2011. Galphas-biased beta2-adrenergic receptor signaling from restoring synchronous contraction in the failing heart. *Sci Transl Med* 3 (100):100ra88.

Conti, M., and J. Beavo. 2007. Biochemistry and physiology of cyclic nucleotide phosphodiesterases: Essential components in cyclic nucleotide signaling. *Annu Rev Biochem* 76:481–511.

Di Benedetto, G., A. Zoccarato, V. Lissandron et al. 2008. Protein kinase A type I and type II define distinct intracellular signaling compartments. *Circ Res* 103 (8):836–844.

DiPilato, L. M., and J. Zhang. 2009. The role of membrane microdomains in shaping [small beta]2-adrenergic receptor-mediated cAMP dynamics. *Molecular BioSystems* 5 (8):832–837.

DiPilato, L. M., X. Cheng, and J. Zhang. 2004. Fluorescent indicators of cAMP and Epac activation reveal differential dynamics of cAMP signaling within discrete subcellular compartments. *Proc Natl Acad Sci U S A* 101 (47):16513–16518.

Diviani, D., K. L. Dodge-Kafka, J. Li, and M. S. Kapiloff. 2011. A-kinase anchoring proteins: Scaffolding proteins in the heart. *Am J Physiol Heart Circ Physiol* 301 (5):H1742–H1753.

Dodge-Kafka, K. L., L. Langeberg, and J. D. Scott. 2006. Compartmentation of cyclic nucleotide signaling in the heart: The role of A-kinase anchoring proteins. *Circ Res* 98 (8):993–1001.

Fischmeister, R., L. R. Castro, A. Abi-Gerges et al. 2006. Compartmentation of cyclic nucleotide signaling in the heart: The role of cyclic nucleotide phosphodiesterases. *Circ Res* 99 (8):816–828.

Harper, J. F., and G. Brooker. 1975. Femtomole sensitive radioimmunoassay for cyclic AMP and cyclic GMP after 2′0 acetylation by acetic anhydride in aqueous solution. *J Cyclic Nucleotide Res* 1 (4):207–218.

Hofmann, F., and J. W. Wegener. 2013. cGMP-dependent protein kinases (cGK). *Methods Mol Biol* 1020:17–50.

Lefkimmiatis, K., and M. Zaccolo. 2014. cAMP signaling in subcellular compartments. *Pharmacol Ther* 143 (3):295–304.

Lehnart, S. E., X. H. Wehrens, S. Reiken et al. 2005. Phosphodiesterase 4D deficiency in the ryanodine-receptor complex promotes heart failure and arrhythmias. *Cell* 123 (1):25–35.

Lygren, B., C. R. Carlson, K. Santamaria et al. 2007. AKAP complex regulates Ca2+ re-uptake into heart sarcoplasmic reticulum. *EMBO Rep* 8 (11):1061–1067.

Mika, D., W. Richter, R. E. Westenbroek, W. A. Catterall, and M. Conti. 2014. PDE4B mediates local feedback regulation of beta(1)-adrenergic cAMP signaling in a sarcolemmal compartment of cardiac myocytes. *J Cell Sci* 127 (Pt 5):1033–1042.

Nikolaev, V. O., M. Bunemann, L. Hein, A. Hannawacker, and M. J. Lohse. 2004. Novel single chain cAMP sensors for receptor-induced signal propagation. *J Biol Chem* 279 (36):37215–37218.

Perera, R. K., and V. O. Nikolaev. 2013. Compartmentation of cAMP signalling in cardiomyocytes in health and disease. *Acta Physiol (Oxf)* 207 (4):650–662.

Piggott, L. A., K. A. Hassell, Z. Berkova, A. P. Morris, M. Silberbach, and T. C. Rich. 2006. Natriuretic peptides and nitric oxide stimulate cGMP synthesis in different cellular compartments. *J Gen Physiol* 128 (1):3–14.

Rich, T. C., K. A. Fagan, H. Nakata, J. Schaack, D. M. Cooper, and J. W. Karpen. 2000. Cyclic nucleotide-gated channels colocalize with adenylyl cyclase in regions of restricted cAMP diffusion. *J Gen Physiol* 116 (2):147–161.

Rich, T. C., T. E. Tse, J. G. Rohan, J. Schaack, and J. W. Karpen. 2001. In vivo assessment of local phosphodiesterase activity using tailored cyclic nucleotide-gated channels as cAMP sensors. *J Gen Physiol* 118 (1):63–78.

Richter, W., P. Day, R. Agrawal et al. 2008. Signaling from beta1- and beta2-adrenergic receptors is defined by differential interactions with PDE4. *EMBO J* 27 (2):384–393.

Rochais, F., G. Vandecasteele, F. Lefebvre et al. 2004. Negative feedback exerted by cAMP-dependent protein kinase and cAMP phosphodiesterase on subsarcolemmal cAMP signals in intact cardiac myocytes: An in vivo study using adenovirus-mediated expression of CNG channels. *J Biol Chem* 279 (50):52095–52105.

Rochais, F., A. Abi-Gerges, K. Horner et al. 2006. A specific pattern of phosphodiesterases controls the cAMP signals generated by different Gs-coupled receptors in adult rat ventricular myocytes. *Circ Res* 98 (8):1081–1088.

Sin, Y. Y., H. V. Edwards, X. Li et al. 2011. Disruption of the cyclic AMP phosphodiesterase-4 (PDE4) –HSP20 complex attenuates the beta-agonist induced hypertrophic response in cardiac myocytes. *J Mol Cell Cardiol* 50 (5):872–883.

Sprenger, J. U., and V. O. Nikolaev. 2013. Biophysical techniques for detection of cAMP and cGMP in living cells. *Int J Mol Sci* 14 (4):8025–8046.

Sprenger, J. U., R. K. Perera, K. R. Götz, and V. O. Nikolaev. 2012. FRET microscopy for real-time monitoring of signaling events in live cells using unimolecular biosensors. *J Vis Exp* (66):e4081.

Wachten, S., N. Masada, L. J. Ayling et al. 2010. Distinct pools of cAMP centre on different isoforms of adenylyl cyclase in pituitary-derived GH3B6 cells. *J Cell Sci* 123 (Pt 1):95–106.

Williams, C. 2004. cAMP detection methods in HTS: Selecting the best from the rest. *Nat Rev Drug Discov* 3 (2):125–135.

Zaccolo, M., and M. A. Movsesian. 2007. cAMP and cGMP signaling cross-talk: Role of phosphodiesterases and implications for cardiac pathophysiology. *Circ Res* 100 (11): 1569–1578.

9 Identifying Complexes of Adenylyl Cyclase with A-Kinase Anchoring Proteins

Yong Li and Carmen W. Dessauer

CONTENTS

9.1 INTRODUCTION

9.1.1 Physiological Significance of Adenylyl Cyclase: A-Kinase Anchoring Protein Complex Formation in Heart

Cyclic AMP (cAMP) is a second messenger that regulates many aspects of cardiac physiology including contraction rate and force of contraction, in addition to the pathophysiology of hypertrophy and heart failure. cAMP is synthesized by adenylyl cyclase (AC), a family of nine membrane-bound enzymes that serve as nodes for cross-talk between sympathetic and parasympathetic signaling, in addition to inputs by calcium channels and other G protein–coupled receptors. In heart, cAMP activates cAMP-gated channels (HCN), cAMP-activated Rap exchange proteins (EPAC), phosphodiesterases, and most notably protein kinase A (PKA), which leads to the increased phosphorylation of numerous proteins involved in contraction.[1,2] Although most ACs are readily detected in cardiac fibroblasts or sino-atrial node, in cardiac myocytes, the major isoforms include AC5 and AC6 and, to a lesser extent, AC4 and AC9.[1,3] Deletion of either AC5 or AC6 leads to major alterations in sympathetic and parasympathetic control of ejection fraction, baroreflexes, calcium handling, and stress responses.[4]

It is becoming increasingly clear that localization of AC to specific effector complexes is an important aspect of its function. For example, AC9 is associated with the I_{Ks} channel,[3] whereas AC5/6 is found in complexes containing TRPV1,[5] L-type Ca^{2+} channel,[6,7] and AMPA-type glutamate receptor.[8] These complexes are assembled by A-kinase anchoring proteins (AKAPs), a family of highly divergent scaffolds. Approximately 20 AKAPs are expressed in heart and disruption of PKA binding to AKAPs results in altered cardiac contraction.[2,9] AKAPs are defined by the presence of a unique PKA regulatory subunit docking motif but also contain anchoring sites for a diverse set of regulatory molecules and downstream effectors to promote spatial and temporal regulation. For instance, the AKAP Yotiao anchors both AC9 and I_{Ks} to facilitate PKA phosphorylation of the pore-forming KCNQ1 channel subunit.[3] PKA phosphorylation increases I_{Ks} current and shortening of the action potential duration in cardiac myocytes.[10–14] The anchoring of AC5 to TRPV1 is mediated by interactions with AKAP5 (also known as AKAP79/150). The formation of an AC5-AKAP5-TRPV1-PKA complex increases the sensitivity of TRPV1 to local cAMP production by approximately 100-fold.[5] Similar anchoring of AC to AMPA receptors by AKAP5 is an important aspect of cAMP-PKA regulation of β-adrenergic–stimulated AMPA receptor activity.[8] Finally, the mislocalization of AC5 from yet another AKAP anchor in the heart, mAKAP, leads to the loss of proper cAMP regulation and an increase in cardiac myocyte hypertrophy.[15]

A detailed understanding of signalosomes that are present in native systems is crucial to understanding the physiology and pathophysiology associated with these molecules. Therefore, analysis of the composition of AKAP-mediated complexes, including the identification of the associated AC isoforms and downstream effector targets is a prerequisite for uncovering their physiological roles and potential as therapeutic targets.

9.1.2 Analysis of AC Interaction with Regulators and Scaffolding Proteins

This review will examine methods to determine if AC is present in an AKAP complex or associated with an effector of choice. The major technique used is an IP-AC assay, which allows one to immunoprecipitate (IP) an AKAP or protein of choice and measure the associated AC activity that copurifies with the given complex. A similar method using glutathione-S-transferase (GST) pull-down allows for quick mapping of interaction sites. Although we focus on AKAP-mediated complexes, these techniques will work for any AC-containing macromolecular complex and can be applied to any effector system, assuming quality antibodies against the target of interest are available.

The following will address the advantages as well as pitfalls of various approaches with respects to certain aspects of AC complex analysis. Although classic methods will be introduced including purification of AC by forskolin-agarose, focus will be placed on the limitations of these methods and why an emphasis on measuring AC activity is preferred. For the key techniques, detailed step-by-step working protocols and technical requirements will be provided.

9.2 ISSUES WITH CLASSIC METHODS

9.2.1 Purification on Forskolin-Agarose

The classic approach to purify AC and associated G proteins took advantage of forskolin-sepharose or forskolin-agarose chromatography.[16,17] Forskolin is a natural plant diterpene that directly activates nearly all isoforms of AC.[18] Enrichment of AC on forskolin-sepharose yielded a 2000-fold purification and was ultimately used to sequence the protein and clone the first AC isoform.[19] When a nonhydrolyzable GTP analogue was added to the initial tissue extracts, an activated Gαs copurified with AC.[16] Moreover, larger complexes containing AC, AKAP150, and RII subunits of PKA have been enriched using forskolin-based chromatography.[20] Thus, this technique can potentially be quite useful in analyzing AC-containing complexes.

9.2.1.1 Advantages

The main advantage of this method is its relative selectivity for AC. The protocol starts with the homogenization and preparation of crude membranes from cells or tissue. The membrane proteins are then solubilized with Lubrol PX (solubilization occurs with 4 mg/mL protein and 0.6% Lubrol) and the insoluble material is removed by centrifugation. Lubrol PX is a nonionic detergent that does not greatly inhibit AC enzymatic or forskolin-binding activity. However, its production was discontinued more than 15 years ago. Today, the major components of the original Lubrol, $C_{12}E_9$ or $C_{12}E_{10}$, are used instead. Solubilized extracts are then incubated overnight with forskolin-agarose. The resin is washed with a series of high salt buffers and the AC eluted with 0.2 mM forskolin in buffer containing 0.05% $C_{12}E_9$ or $C_{12}E_{10}$. The beauty of this technique is that active AC (based on forskolin interactions) is isolated. Moreover, nearly all isoforms of AC (except AC9) and their interacting proteins can be analyzed using the same method.

9.2.1.2 Requirements, Limitations, and Problems

The major issue with this method is the cost and the need to synthesize the resin. Unfortunately, forskolin-agarose is no longer commercially available. The protocol for synthesis uses as the starting material 7-deacetyl-7-*O*-hemisuccinyl-forskolin (CAS 83797-56-2; available from Santa Cruz Biotechnology) and is briefly outlined below.[16,17,21] The starting material is activated by treatment of 7-deacetyl-7-*O*-hemisuccinyl-forskolin (10 μmol, 4.7 mg sufficient for 5 mL resin) with *N*-hydroxysuccinimide (15 μmol) and 3.09 mg dicyclohexylcarbodiimide (15 μmol) in 125 μL of anhydrous acetonitrile for 4 h at 22°C. The dicyclohexylurea that is formed is removed by centrifugation and the solution is extracted three times with 125 μL of acetonitrile. The supernatants are added to 6.25 g (wet weight) of activated resin in 9.4 mL dimethylformamide. One can either use aminoethyl agarose (available from ABT in an activated form) or activate Affigel 10 (Bio-Rad), which has a longer spacer and tends to be better for many applications. The latter is activated by reacting Affigel 10 (6.25 mL) with 0.2 mL ethylenediamine for 12 h at 22°C, followed by washing with 50 mL each of 0.1 M HC1 and distilled water. Activated resin is then transferred to anhydrous dimethylformamide and reacted with 2 μmol/mL of activated 7-succinyl-7-deactylforskolin for 15 to 24 h at 22°C. Details surrounding washing conditions and the choice of resins are discussed by Pfeuffer.[17] The cost of synthesis is also a factor when using forskolin-agarose (7-deactyl-7-*O*-hemisuccinyl-forskolin is ~$875 for 5 mg yielding ~5 mL resin; plus the cost of resins, chemicals, and forskolin for elution).

A minor issue surrounding the use of forskolin-agarose to pull down AC-containing complexes is the off-target binding partners of forskolin. Forskolin is known to interact with and inhibit the galactose symport protein GalP from *Escherichia coli*,[22] in addition to a number of mammalian transporters, including the glucose transporter GLUT1 and the P-glycoprotein multidrug transporter (ABCB1).[23,24] Forskolin is also known to modulate voltage-gated K^+ channels.[25] An analogue of forskolin, 1,9-dideoxyforskolin, does not stimulate AC activity but does interact with and modulate these membrane transporters and channels with a similar potency as forskolin, serving as an effective control for cellular assays. However, one might expect that these membrane proteins would copurify with AC on forskolin-affinity columns. An elution with 1,9-dideoxyforskolin to remove contaminants prior to elution of AC with forskolin is one method to increase selectivity. Finally, the inability to differentiate between AC isoforms can be a disadvantage of this technique if one is trying to isolate a complex containing a particular AC isoform.

9.2.2 Coimmunoprecipitation and Western Blot Analysis

Coimmunoprecipitation (Co-IP) of protein complexes has become one of the most common techniques to analyze protein–protein interactions. It has the distinct advantage over forskolin-agarose of being able to identify a given AC isoform and other members of the complex by Western blotting. However, for AC, this technique is largely impractical due to the poor quality of antibodies available for most AC isoforms. Except for antibodies against AC3 and, to a lesser extent, AC5/6 (which recognize AC6 much better than AC5), most AC antibodies yield multiple bands by Western blotting and it is often difficult to identify the correct band corresponding to AC without proper controls. We often include membranes from Sf9 cells expressing a given AC isoform as a marker of the correct size, but comparison to cells or tissue from a knockout of the AC isoform is best. In addition, most AC antibodies can only reasonably detect AC isoforms overexpressed in cell lines and are unable to recognize endogenous AC in membranes or whole cell extracts. Moreover, many of the RIPA-based buffers used in IP experiments are not compatible with keeping AC in an active conformation. Therefore, we prefer to rely on AC activity as the more accurate readout for the presence of AC.

9.3 COIMMUNOPRECIPITATION OF AC ACTIVITY

Coimmunoprecipitation of AC activity (termed IP-AC assay) is well suited for determining if AC is associated with macromolecular complexes. It is similar in design to an IP kinase assay, except for measuring the associated kinase activity that is pulled down in the IP, the associated AC activity is measured instead. This assay has been particularly useful for determining the AC selectivity for interactions with a given AKAP upon overexpression in a cell line. For example, this technique was used to determine the AC isoform selectivity of the AKAP Yotiao (Figure 9.1).[26] HEK293 cells were transfected with various AC isoforms in the presence or absence of V5-tagged Yotiao. Yotiao was then immunoprecipitated from cell lysates followed by measurement of associated AC activity present in the washed beads of the IP. The in vitro AC assay used radiolabeled [^{32}P]ATP as substrate and the AC present in the IP was stimulated by exogenously added forskolin and purified GTPγS-bound Gαs (Figure 9.1a). AC9 is largely insensitive to forskolin, so a higher concentration of Gαs was used for this AC isoform (Figure 9.1b). The cAMP generated was separated by column chromatography and detected by scintillation counting. When compared with cells lacking Yotiao expression, it is easy to determine that Yotiao associates with AC1, AC2, AC3, and AC9, but not AC4, AC5, and AC6 (Figure 9.1a and b). Although not shown here, AC activity in the starting cell lysates can be used to confirm that each AC isoform is expressed over background AC activity in control HEK293 cells.[26] The details of this protocol are found in Section 9.3.3 and modifications for use in tissue or with disrupting peptides are discussed in Sections 9.3.5 and 9.3.6.

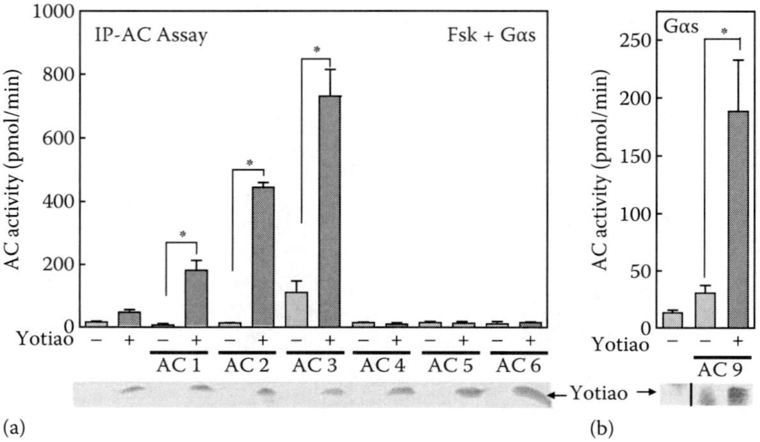

FIGURE 9.1 IP-AC assay used to screen for Yotiao-interacting AC isoforms. (a) Immunoprecipitations of V5-tagged Yotiao from transfected HEK293 cells were assayed for AC activity by stimulation with 50 nM Gαs plus 100 μM forskolin. Yotiao expression was confirmed by Western blot analysis. (b) Cells transfected with AC9 plus or minus Yotiao were immunoprecipitated with anti-Yotiao serum and then stimulated for associated AC activity with 400 nM Gαs. (This figure was originally published in Piggott, L.A., A.L. Bauman, J.D. Scott, and C.W. Dessauer. The A-kinase anchoring protein Yotiao binds and regulates adenylyl cyclase in brain. *Proc Natl Acad Sci U S A* 105: 13835–13840. Copyright 2008 National Academy of Sciences, U.S.A.)

9.3.1 Advantages

The major advantage of an IP-AC assay is its sensitivity. Measurement of AC activity is far more sensitive than Western blotting for any AC isoform. Antibodies against scaffolding proteins (AKAP) or another protein associated with the complex of interest can be used for IP. Besides the limitations of antibody quality, IP is a rather selective method to enrich a complex. For example, antibodies against nesprin have been used to purify mAKAP-containing complexes, whereas antibodies against effector proteins such as AMPA receptor or TRPVI can enrich for AKAP79-associated complexes. Similarly, antibodies against KCNQ1 were used to purify the IKs–Yotiao complex. In addition, co-IP can be performed with sufficiently accurate antibodies within primary cells or tissue, and thus does not require heterologous (over)expression of a protein and avoids artifacts produced by distorted relative concentrations of the interacting proteins (Figure 9.2). Finally, differences in AC regulation allow one to probe different classes of AC isoforms.

9.3.2 Requirements, Limitations, and Problems

As with any IP method, good quality antibodies are the limiting factor. Controls include using IgG or preimmune serum, or if performing experiments in cultured cell lines using epitope-tagged AKAPs, including controls that lack expression of

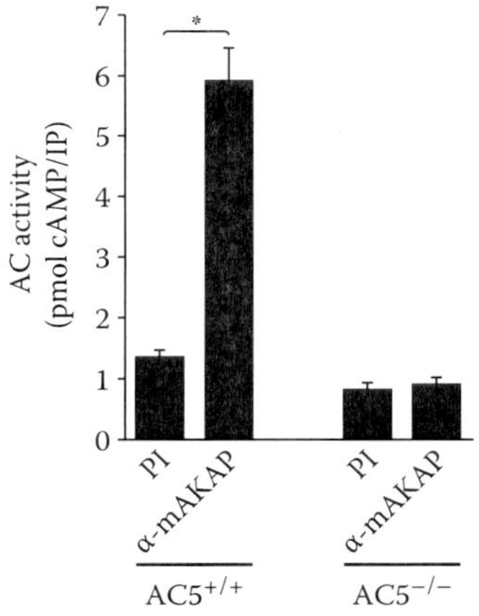

FIGURE 9.2 IP-AC assay using heart extracts to probe mAKAP–AC interactions. Heart-solubilized extracts were prepared from wild-type ($AC5^{+/+}$) or AC5 knockout ($AC5^{-/-}$) mice. AC activity associated with mAKAP or preimmune immunoprecipitates was measured in the presence of 50 nM Gαs + 100 μM forskolin. (This research was originally published in Kapiloff, M.S. et al., *J Biol Chem* 284: 23540–23546, 2009. The American Society for Biochemistry and Molecular Biology.)

the tagged AKAP. Controls must also be able to distinguish nonspecific binding of AKAPs or effector proteins to certain resins. For example, AKAP79 nonspecifically interacts with Dynabeads. Suitable antibodies are commercially available for many AKAPs (AKAP79, BD Transduction Laboratories; AKAP150, Santa Cruz Biotechnology) and effectors (TRPV1, EMD Biosciences), but for others, it was necessary to develop these antibodies (e.g., Yotiao[26] and mAKAP[15,27,28]). It is also important to remember that not all antibodies that work well for Western blotting are appropriate for IPs. For example, the TRPV1 antibody from BD Transduction Labs works great for IP but it does not work well in a Western blot, whereas the antibody for TRPV1 from Alomone Labs works great for Western blotting but not IP.

The other consideration for IP-AC assays is the use of detergents. As discussed above, $C_{12}E_9$ and $C_{12}E_{10}$ are the major detergent components of Lubrol PX and are generally good at maintaining AC activity. Although softer tissues such as brain can be solubilized with $C_{12}E_{10}$ (0.6%), the heart requires the use of Triton X-100 (1%). Although still a nonionic detergent, Triton X-100 is not as mild as $C_{12}E_{10}$ and it is best to perform all subsequent washes with $C_{12}E_{10}$ to remove the Triton X-100 prior to assay of AC activity. Ionic detergents such as cholate should be avoided completely.

9.3.3 Step-by-Step Showcase Protocol

9.3.3.1 Preparation of $C_{12}E_{10}$ Detergent

Incubate the polyoxyethylene 10 lauryl ether ($C_{12}E_{10}$; Sigma-Aldrich) at 42°C overnight to melt it out of the wax state. Fill a 500 mL bottle with 100 mL dH_2O. Tare a balance with the bottle and weigh out 50 g of $C_{12}E_{10}$ into the bottle. The wax will resolidify in the water. Intermittently swirl the bottles until the wax goes into solution. Adjust final volume to 500 mL with dH_2O. Store at 4°C.

9.3.3.2 Preparation of Cell Lysates from Transiently Transfected HEK293 Cells

1. HEK293 cells are plated in 10-cm dishes 1 day before transient transfection to obtain a cell density of 70% to 80% at the time of transfection.
2. The cells are transfected using Lipofectamine 2000 (Invitrogen). Plasmid DNA (10 μg total) is used to transfect cells in 10-cm dishes.
3. Cells at 36 to 40 h posttransfection are washed once with 3 to 5 mL ice-cold PBS, removing PBS by aspiration (aspiration must be done very well to keep volumes down).
4. Lysis buffer is added to the plate (150 μL; 50 mM Hepes, pH 7.5, 1 mM EDTA, 1 mM $MgCl_2$, 150 mM NaCl, 0.5% $C_{12}E_{10}$ plus protease inhibitors). Cells are scraped with a plastic policeman and collected in an Eppendorf tube.
5. Using a 23-gauge needle and 1 mL syringe, homogenize cells by going up and down in a syringe three to five times. Avoid creating bubbles, as AC activity is sensitive to oxidation.
6. Spin down the cell lysate at 13,000 × *g* for 10 min at 4°C to remove cellular debris.
7. Transfer the supernatant (cleared lysate) to a new 1.5 mL Eppendorf tube and determine the protein concentration using a method compatible with detergents. Save 20 μL of supernatant for AC assay. *Note*: It is important to save an aliquot of the starting extract to measure the total AC activity in step 17. This allows one to determine what percentage of total AC activity is associated with the AKAP of interest. In addition, measurement of total AC activity allows one to determine if an AC isoform transfected in a cell line of choice was actually expressed as compared with the activity present in the nontransfected control. Larger volumes can be saved for additional Western blotting of samples if necessary.
8. Lysates thus obtained are used in the following IP-AC assay.

9.3.3.3 Prepare Protein A or G Beads

9. Wash the protein A or protein G sepharose beads (30 μL of 50% slurry per sample) with ice-cold lysis buffer, centrifuge them with a maximum of 500 × *g* for 2 min at room temperature, and repeat this twice. *Note*: The choice of either protein A or protein G beads is based on the affinity of the antibodies for protein A or protein G.

9.3.3.4 Incubate to Form Antibody-Antigen Complexes

10. Add 1 to 2 μg of appropriate antibody to each vial, using 1 to 2 μg of IgG or preimmune serum as a control, and gently rotate the tubes for 1 h at 4°C.
11. Add 30 μL of washed beads and incubate them with rotation for one more hour at 4°C.

9.3.3.5 Purify Protein Complexes

12. After incubation, centrifuge the sample and remove the supernatant. Next, wash the beads three times (300 μL) with wash buffer (lysis buffer with 0.05% $C_{12}E_{10}$). After the third wash, fully remove buffer from each sample with Hamilton gel-loading syringe. *Tip*: Centrifuge gently (500 × *g*, 2 min) each time to prevent any damage to the beads.
13. Resuspend samples in 35 μL of resuspension buffer (50 mM Hepes, pH 7.5, 1 mM EDTA, 1 mM $MgCl_2$, 0.05% $C_{12}E_{10}$, and protease inhibitors). *Note*: Save 5 μL for Western blotting.
14. Make the AC assay mix:

Basic AC Assay Components

Ro 20-1724[a]	100 μM
EDTA (pH 7.0)	0.6 mM
Bovine serum albumin (BSA)	100 μg/mL
HEPES (pH 8.0)	50 mM
Dipotassium phosphoenolpyruvate	3 mM
Pyruvate kinase	10 μg/mL
[α-^{32}P]ATP	1–2 × 10^6 cpm/tube
ATP, unlabeled[b]	50–200 μM
$MgCl_2$	5 mM

[a] Ro 20-1724 is a phosphodiesterase (PDE) inhibitor selective for PDE4. If needed, a more general PDE inhibitor, such as IBMX (1 mM), can be added.

[b] The K_m of most AC isoforms for ATP is approximately 22 μM. We generally prefer to use ATP concentrations that are greater than 10 times the K_m, but this greatly lowers the specific activity of the ATP and the sensitivity of the assay. It is also important not to use more than 10% of the substrate in the assay (at 200 μM ATP, this is rarely a problem).

15. The final volume is 120 μL. *Note*: The reactions also include forskolin (50–100 μM) or purified regulators of AC, such as Gαs or Gαi, depending on which AC isoform will be examined (see Section 9.3.4). These are added to the AC assay mix just before use. The assay is usually started with the addition of 90 μL of AC mix to 30 μL beads and proceeds for 10 min at 30°C. The starting lysate (~20 μL) should be assayed at the same time. The reaction is stopped with 0.8 mL of 0.25% (w/v) sodium dodecyl sulfate (SDS), 5 mM ATP, and 0.175 mM cAMP. Approximately 10,000 cpm of [^{3}H]cAMP (100 μL) is then added to monitor column recovery of sequential chromatography on Dowex and alumina columns (see below). *Tips*: A stock solution containing Ro20-1724, HEPES, EDTA, BSA, and phosphoenolpyruvate can be made ahead of time and aliquoted for storage at −20°C. Make sure to adjust the pH of phosphoenolpyruvate stock to 7.0 with KOH. A 10× stock of rabbit muscle pyruvate kinase (1 mg/mL) and 10× STOP

solution can be aliquoted and stored at –20°C. Diluted 1× STOP solution can be stored at 4°C for direct use in AC assays.

16. The reactions are subjected to centrifugation (500 × *g*, 2 min) and the supernatants are applied to prepared Dowex and alumina columns for separation of cAMP.

9.3.3.6 Dowex and Alumina Chromatography

cAMP produced in the reaction is separated by sequential chromatography on Dowex and alumina columns using Dowex 50 H^+ form (Bio-Rad: AG50-XS, 100–200 mesh; 1 mL packed resin) and neutral alumina (Sigma WN-3 A9003; 0.8 g resin per column). For convenience, columns are arranged in racks that form a 10 × 10 array that can be stacked to allow elution from Dowex to drip directly onto the alumina column. Care must be taken to properly contain the flow-through from the Dowex columns as most of the radioactive ATP substrate is eluted in this fraction. For detailed protocols on the preparation and regeneration of Dowex and alumina columns, see Johnson and Salomon.[29] Briefly, 1 mL reactions were applied to Dowex columns. Once the sample had dripped through, the resin was washed with 2 mL dH_2O. The Dowex columns are then stacked on top of the alumina columns and the cAMP eluted from the Dowex onto the alumina resin with two washes of 2 mL dH_2O. The Dowex columns were then removed and the alumina columns placed over scintillation vials. The cAMP is eluted from the alumina columns with 3 mL of imidazole solution (200 mM NaCl and 20 mM imidazole). Scintillation fluid is added and the radioactivity measured by dual channel counting using ^3H to adjust for individual column recoveries. *Note*: Columns are regenerated after every use and, once prepared, can be reused repeatedly for several years.

One can also measure cAMP production without the use of radioactivity and column chromatography by taking advantage of one of the many enzyme-immunoassay (EIA) kits or homogeneous time-resolved fluorescence (HTRF) that have been commercially developed for cAMP. These kits and the necessary alterations to the AC assay are discussed in Section 9.3.5.

9.3.4 Activation of AC: Isoform-Specific Considerations

In general, if the specific AC isoform bound to the complex of interest is unknown, we probe for AC activity using 50 nM GTPγS-bound Gαs and 100 μM forskolin. The purification and activation of Gαs is discussed below. Most AC isoforms display additive or synergistic activity with Gαs and forskolin and this combination often yields maximal AC activity. There are exceptions of course. AC9 displays little sensitivity to forskolin and maximal activity is generally achieved at a higher Gαs concentration (400 nM). One can use only forskolin alone to stimulate AC activity but the sensitivity of the assay is generally much lower. Note, AKAPs have been shown to regulate a subset of ACs in membranes and cells.[20,26,30–32] However, we find that many inhibitory effects are lost upon detergent solubilization. Inhibition is not due to direct binding, but rather is often due to PKA-dependent phosphorylation of anchored AC isoforms (particularly AC5, AC6, and AC8).[20,30,33] However, in the absence of phosphatase inhibitors, this regulation is not readily observed in detergent-solubilized extracts.

Isoform-selective activation conditions can be very useful to probe for AC isoforms present in a complex of interest. For example, when probing for Yotiao interactions with AC2 in brain, AC activity was stimulated with Gαs plus Gβγ.[26] This helped to distinguish between possible AC1-Yotiao interactions because Gβγ conditionally stimulates AC2 by approximately 7-fold over Gαs alone, whereas AC1 is inhibited by Gβγ. AC1 and AC8 complexes can be examined using Ca^{2+} and calmodulin (Ca^{2+}/CaM). For low signals with AC1 and AC8, stimulation with Ca^{2+}/CaM plus forskolin can greatly increase the activity due to the high synergy between these two activators. Although the hallmark of AC5 and AC6 regulation is inhibition by Gαi,[4] the conditions for observing inhibition can be tricky in the detergent-based assay necessary for the IP. In this case, stimulation with Gαs alone must be compared with Gαs plus GTPγS-bound myristoylated Gαi; however, these are submaximal activities and can be difficult to observe.

9.3.4.1 Purification and Activation of Gαs

Hexahistidine tagged Gαs can be purified from *E. coli*. The H_6-tag is inserted at exon 3 before the αA helix in the helical domain.[34] Methods for expression in BL21(DE3) cells cotransformed with pREP4 and purification by metal affinity chromatography have been described previously.[35] Briefly, cells are grown in T7 medium at 30°C and induced with 30 μM isopropyl-1-thio-β-D-galactopyranoside at an A_{600} of 0.4. Cells are harvested after 12 to 16 h. After incubation with lysozyme and then DNase to facilitate cell lysis, cellular debris is removed by centrifugation and the supernatant applied to a metal-affinity column (Ni-NTA). The column is washed with buffer containing 400 mM NaCl and 10 mM imidazole. The column is then briefly washed in buffer lacking NaCl and the protein eluted with 100 mM imidazole. To stabilize Gαs, 10 μM GDP is present in all buffers except for the elution buffer. The protein is then immediately applied to an ion exchange MonoQ fast protein liquid chromatography column and eluted with a linear gradient of NaCl (0–400 mM). Peak fractions are pooled and concentrated and stored in aliquots at −80°C. To activate the G protein, purified Gαs is incubated with 50 mM Hepes (pH 8.0), 5 mM $MgSO_4$, 1 mM EDTA, 2 mM dithiothreitol, and 400 mM [^{35}S]GTPγS at 30°C for 30 min. Free GTPγS is removed by rapid gel filtration through a 1.8 mL desalting column using Sephadex G-25 or G-50 resin. An aliquot of the starting reaction mixture and the radioactive protein obtained after gel filtration is measured by scintillation counting to determine the concentration of GTPγS-bound Gαs (concentration of Gαs (μM) = cpm per microliter of sample × 400 μM GTPγS/cpm per microliter of mix). This is compared with the total protein concentration; typically, 70% to 80% of the protein is present in the active state.

9.3.5 Additional Considerations for Application in Tissues

To process tissues for IP-AC assays, a homogenization step is required prior to the addition of detergents. Conditions vary depending on the tissue but generally disruption with a Polytron is sufficient for most samples. In addition, many tissues can be solubilized using $C_{12}E_{10}$ by Dounce homogenization of the disrupted tissue. However, this is not true for fibrous tissues such as heart, which requires harsher disruption methods and Triton X-100 detergent. Moreover, depending on the amount of available starting tissue, higher sensitivity methods using EIA kits may be necessary for the detection of cAMP production. Outlined below is a general protocol for IP-AC assays from heart.

9.3.5.1 Preparation of Lysates from Mouse Hearts

1. Rapidly thaw frozen mouse heart in water bath at 30°C. Incubate the heart in a water bath for only a few seconds. *Note*: Use of fresh tissue is recommended when possible.
2. Rinse the heart in ice-cold PBS buffer with protease inhibitors.
3. Quickly quarter the heart and transfer into 3 mL ice-cold homogenization buffer (50 mM HEPES pH 7.4, 5 mM EDTA, 100 mM NaCl, DTT, PT, Leu/Lima, 50 mM NaF, 150 μM orthovanadate, 10% glycerol).
4. Homogenize the heart on ice with Polytron for 4 to 5 s.
5. Add Triton X-100 to 1% and homogenize the heart using a Dounce homogenizer.
6. Centrifuge homogenates for 15 min with 13,000 × *g* at 4°C.
7. Save 100 to 200 μL of supernatant for AC assay and Western blotting and divide remaining supernatant into four samples of approximately 600 μL each.
8. Incubate each sample with 1 to 2 μg antibody or preimmune serum for 1 h at 4°C.
9. Prewash protein A beads by adding 300 to 500 μL of homogenization buffer to 30 μL of beads, centrifuge them at a maximum of 500 × *g* for 2 min at room temperature and repeat this twice.
10. Add 30 μL of prewashed protein A beads to each sample and incubate for 2 h at 4°C.
11. Wash beads three times with 300 μL wash buffer (50 mM HEPES, pH 7.4, 1 mM EDTA, 1 mM $MgCl_2$, 150 mM NaCl, 0.05% $C_{12}E_{10}$, protease inhibitors) being careful not to aspirate the resin.
12. Remove all supernatants with a Hamilton syringe.
13. Resuspend pellet in 30 μL resuspension buffer (50 mM HEPES, 1 mM EDTA, 1 mM $MgCl_2$, 0.05% $C_{12}E_{10}$, protease inhibitors).

9.3.5.2 AC Assay

The basic AC assay mix is the same as that shown in Section 9.3.3. However, if using EIA or HTRF to detect cAMP production, the [α-^{32}P]ATP is eliminated and the final ATP concentration is 200 μM. The more sensitive but expensive EIA detection method is recommended if small amounts of starting material (tissue or primary cells) are used for the IP-AC assay. The major difference in the AC assay protocol is the lack of radioactivity and the use of a different stop solution.

14. Add 90 μL of AC assay mix to 30 μL of IP reaction or starting extracts. After 10 min, the reaction is stopped by adding 120 μL of 0.2 N HCl.
15. Reactions are subjected to centrifugation (500 × *g*, 2 min) and the supernatant (120 μL) is removed from the protein A beads.

16. The cAMP generated is measured by direct cAMP EIA kit (ADI-900-066, Enzo Life Sciences) according to the manufacturer's directions. *Note*: We generally use the acetylation step to improve the sensitivity of the EIA kit, particularly when using small amounts of starting material. Many options are commercially available for cAMP detection by EIA or by HTRF, which combines FRET with time-resolved measurement.[36] However, all of these are very expensive to run, costing approximately $3 to $4 per well. It is important to use the appropriate amount of the sample to stay within the standard curve of the assay. This will vary greatly depending on the starting material and assay conditions and will have to be determined empirically.

9.3.6 Use of Disrupting Peptides to Further Define Complexes

The use of disrupting peptides to further define the components of macromolecular complexes has proved quite useful in IP-AC assays. Once an interacting domain has been mapped between two members of a complex, this region can be used to specifically disrupt that interaction, leaving other members of the complex intact. This has been particularly powerful in the case of PKA–AKAP interactions. Peptides based on the amphipathic helix that binds the RII or RI regulatory domain of PKA are often used to displace PKA from an AKAP complex.[37–40] Similar peptides have been used for effector–AKAP interactions such as TRPV1–AKAP79.[41,42] Slightly larger disrupting agents have also been developed for particular AC–AKAP interactions, including AC2–Yotiao, AC5–mAKAP, and AC5/6–AKAP79.[15,26,31] Therefore, one can often confirm that a unique AC–AKAP interaction is associated with an effector or other component of the complex used for immunoprecipitation.

9.3.6.1 IP-AC Assay with Competition Proteins (Using Transfected HEK293 Cells)

The protocol is the same as described in Section 9.3.3 except that disrupting peptides/proteins are added during step 4. It is important to add the peptides prior to the homogenization of the sample. We generally use approximately five times the IC_{50} for the disrupting peptide/protein. After incubating with disrupting protein for 5 to 10 min, the cells are lysed using a 23-gauge needle and 1 mL syringe. The remaining steps (#6–18) are the same as described in Section 9.3.3.

9.3.6.2 IP-AC Assay from Tissue with Competition Proteins (Mouse Heart)

Once again, disrupting peptides/proteins are added prior to the solubilization of the sample. In this case, the sample is first homogenized using a Polytron (see step 4 of Section 9.3.5). If needed, the sample is divided into equal tubes. The disrupting peptides/proteins are subsequently added to the appropriate tubes and incubated on ice for 5 to 10 min prior to addition of Triton X-100 (1%). The tissue samples are then subjected to Dounce homogenization. The remaining steps (#6–16) are the same as described in Section 9.3.5.

9.4 PULL-DOWN ASSAYS USING GST-TAGGED PROTEINS

GST fusion proteins are a common tool for researchers due to the ease of synthesis and purification of the recombinant proteins in bacteria. GST pull-down experiments can be used to identify interactions between a probe protein and unknown targets. The probe protein is generally fused to the C-terminus of GST; this fusion protein is expressed in bacteria and purified by affinity chromatography on glutathione-agarose column. Target proteins are usually cell lysates, tissue extracts, or purified proteins. For example, we have previously used GST-based pull-down assays to map binding sites for AC on mAKAP (Figure 9.3). Mouse heart extracts were used as the source of the target AC. Purified GST-mAKAP fusion proteins were incubated with heart extracts prior to pull-down (PD) with glutathione resin and measurement of Gαs/Fsk-stimulated AC assay. Although, at the time, it was not known which AC isoform was contributing to mAKAP complexes, we were able to map the cardiac AC interacting domain of mAKAP to the first 245 amino acids using this technique (Figure 9.3). To demonstrate direct protein–protein interactions between AC and mAKAP, this method was used with GST-tagged fusions of mAKAP and purified hexahistadine-tagged AC domains.[15] In this case, the interaction was detected by Western blotting for the tagged AC domains.

9.4.1 Advantages, Requirements, and Limitations

GST pull-down assays are excellent for mapping sites of interactions, even for very large scaffolds such as mAKAP or Yotiao. If AC activity is being measured as a readout

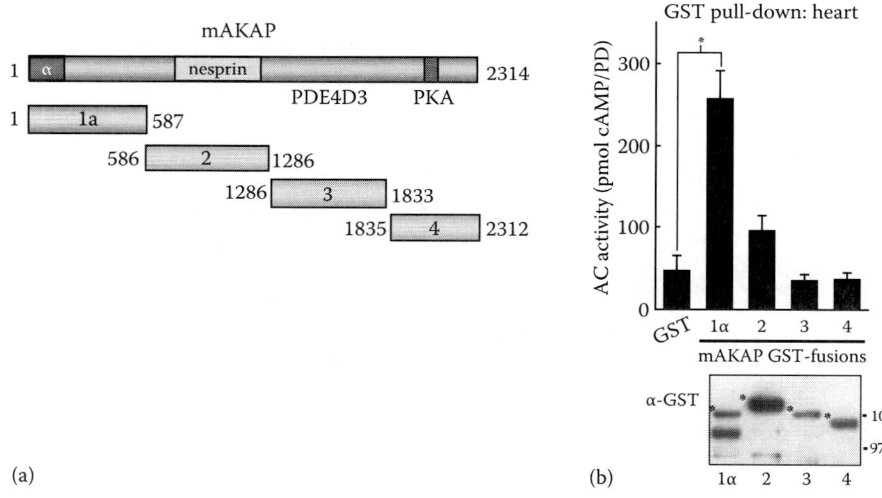

FIGURE 9.3 GST pull-down assays to map AKAP–AC interaction sites. (a) schematic of mAKAP domains used in this study. (b) GST-mAKAP fusion proteins were incubated with heart extracts prior to pull-down with glutathione resin and measurement of Gαs/forskolin-stimulated AC assay. Western blots of GST proteins (starred bands) are shown below. (This research was originally published in Kapiloff, M.S. et al., *J Biol Chem* 284: 23540–23546, 2009. The American Society for Biochemistry and Molecular Biology.)

for AC–AKAP interactions, it is not necessary to know a priori which AC isoform is associated with the AKAP of interest in your particular tissue type. Generally, this method works well for high-affinity interactions (<1 μM kD). Assuming a cloned gene is available, standard molecular biology methods can be employed to subclone the entire coding region or fragments of the scaffolding protein in frame with GST using one of the many available pGEX vectors. In general, the size of the fusion should be limited to approximately less than 100 kDa to ensure good expression of the fusion protein. Recombinant clones can be overexpressed and easily purified, resulting in an abundance of GST fusion protein for use in pull-down assays. For difficult proteins, expression can be increased by reducing the temperature to 30°C to help in folding of the protein or limiting the time of induction to 3 h to reduce degradation problems.

Difficulties mapping interaction sites can be incurred if multiple ACs can associate with the AKAP of choice using unique binding sites. Alternatively, some ACs have multiple sites of interaction with a given AKAP (Yotiao and AC9, for example).[3] It is also important to remember that the association of AC activity from a cellular lysate with a GST fusion protein does not necessarily indicate a direct interaction. To determine if direct protein–protein interactions occur, pull-downs must be performed with purified proteins.

9.4.2 Step-by-Step Showcase Protocol

The protocol for the preparation of lysates from cultured cells or tissue (mouse heart) is the same as described in steps 1 through 7 from Sections 9.3.3 and 9.3.5.

8. Add to the cell or tissue extract 50 μg of GST (as a negative control) or GST fusion protein and incubate for 30 min at 4°C. *Note*: For mouse heart, we use approximately 1/3 mouse heart per pull-down. *Note*: For cell culture, determine the protein concentration by Bradford protein assay and use 100 μg of cell lysate per pull-down. The final volume of each sample should be brought to 100 to 150 μL with lysis buffer (see Section 9.3.3, step 4).
9. To prepare glutathione-agarose (G4510-10 ML, Sigma; 30 μL of slurry per sample) wash the resin three times with 300 μL lysis or homogenization buffer (Section 9.3.5, step 3).
10. Add glutathione-agarose to each sample (30 μL of 50% slurry per sample) and rotate for an additional 2 h at 4°C. The remaining steps are the same as described in Section 9.3.3 for use in cultured cells (#12–16) or Section 9.3.5 for use with tissue (#11–16). *Note*: Generally, the signal is sufficient that it is not necessary to detect AC assay by EIA.

9.5 PERSPECTIVES

We have outlined techniques to identify AC–AKAP complexes from cultured cells or native tissues. Complementary strategies can be used to further characterize these complexes, including imaging techniques such as fluorescence resonance energy transfer (FRET) microscopy. One striking advantage of FRET is the possibility for real-time observation of protein–protein interaction dynamics in an intact cellular system. Once appropriate fluorescent fusion proteins are generated (either with CFP-YFP or GFP-RFP pairs), FRET results can be obtained relatively quickly using standard cell culture expression. However, one limitation of FRET approaches is that false-negative results may be produced simply by the distance constraints of the complex. This is particularly true for AC–AKAP complexes. FRET requires that the donor and acceptor proteins be within 100 Å. However, the distance across AC cytoplasmic domains alone can be as much as 65 Å. Combined with size of the fluorescent proteins and scaffolds, the measurement of FRET signals can be problematic for complexes scaffolded by the larger AKAPs. Thus, only a positive FRET signal represents potentially conclusive information. Moreover, FRET requires overexpression of the molecules in question, which may perturb the very complex that one is trying to measure. Thus, such FRET-based imaging within cells is a complementary method. Measurement of complex formation by IP-AC assay allows for direct measurement of cAMP signaling from a variety of systems, including endogenous complexes.

ACKNOWLEDGMENTS

The authors thank Cameron Brand and Dr. Alexis Bavencoffe for helpful comments and discussions. This work was supported by the National Institutes of Health (GM060419) and the United States–Israel Binational Science Foundation.

REFERENCES

1. Efendiev, R., and C.W. Dessauer. 2011. A kinase-anchoring proteins and adenylyl cyclase in cardiovascular physiology and pathology. *Journal of Cardiovascular Pharmacology* 58: 339–344.
2. Scott, J.D., C.W. Dessauer, and K. Tasken. 2013. Creating order from chaos: Cellular regulation by kinase anchoring. *Annual Review of Pharmacology and Toxicology* 53: 187–210.
3. Li, Y., L. Chen, R.S. Kass, and C.W. Dessauer. 2012. The A-kinase anchoring protein Yotiao facilitates complex formation between type 9 adenylyl cyclase and the IKs potassium channel in heart. *Journal of Biological Chemistry* 287: 29815–29824.
4. Sadana, R., and C.W. Dessauer. 2009. Physiological roles for G protein-regulated adenylyl cyclase isoforms: Insights from knockout and overexpression studies. *NeuroSignals* 17: 5–22.
5. Efendiev, R., A. Bavencoffe, H. Hu, M.X. Zhu, and C.W. Dessauer. 2013. Scaffolding by A-kinase anchoring protein enhances functional coupling between adenylyl cyclase and TRPV1 channel. *Journal of Biological Chemistry* 288: 3929–3937.
6. Davare, M.A., V. Avdonin, D.D. Hall, E.M. Peden, A. Burette et al. 2001. A beta2 adrenergic receptor signaling complex assembled with the Ca^{2+} channel Cav1.2. *Science* 293: 98–101.
7. Nichols, C.B., C.F. Rossow, M.F. Navedo, R.E. Westenbroek, W.A. Catterall et al. 2010. Sympathetic stimulation of adult cardiomyocytes requires association of AKAP5 with a subpopulation of L-type calcium channels. *Circulation Research* 107: 747–756.
8. Zhang, M., T. Patriarchi, I.S. Stein, H. Qian, L. Matt et al. 2013. Adenylyl cyclase anchoring by a kinase anchor protein AKAP5 (AKAP79/150) is important for postsynaptic beta-adrenergic signaling. *Journal of Biological Chemistry* 288: 17918–17931.

9. McConnell, B.K., Z. Popovic, N. Mal, K. Lee, J. Bautista et al. 2009. Disruption of protein kinase A interaction with A-kinase-anchoring proteins in the heart in vivo: Effects on cardiac contractility, protein kinase A phosphorylation, and troponin I proteolysis. *Journal of Biological Chemistry* 284: 1583–1592.
10. Kass, R.S., and S.E. Wiegers. 1982. The ionic basis of concentration-related effects of noradrenaline on the action potential of calf cardiac purkinje fibres. *The Journal of Physiology* 322: 541–558.
11. Walsh, K.B., and R.S. Kass. 1988. Regulation of a heart potassium channel by protein kinase A and C. *Science* 242: 67–69.
12. Terrenoire, C., C.E. Clancy, J.W. Cormier, K.J. Sampson, and R.S. Kass. 2005. Autonomic control of cardiac action potentials—Role of potassium channel kinetics in response to sympathetic stimulation. *Circulation Research* 96: E25–E34.
13. Sampson, K.J., and R.S. Kass. 2010. Molecular mechanisms of adrenergic stimulation in the heart. *Heart Rhythm* 7: 1151–1153.
14. Chen, L., and R.S. Kass. 2011. A-kinase anchoring protein 9 and IKs channel regulation. *Journal of Cardiovascular Pharmacology* 58: 459–464.
15. Kapiloff, M.S., L.A. Piggott, R. Sadana, J. Li, L.A. Heredia et al. 2009. An adenylyl cyclase-mAKAPbeta signaling complex regulates cAMP levels in cardiac myocytes. *Journal of Biological Chemistry* 284: 23540–23546.
16. Pfeuffer, E., R.M. Dreher, H. Metzger, and T. Pfeuffer. 1985. Catalytic unit of adenylate cyclase: Purification and identification by affinity crosslinking. *Proceedings of the National Academy of Sciences of the United States of America* 82: 3086.
17. Pfeuffer, T. 1991. Synthesis of forskolin-agarose affinity matrices. *Methods in Enzymology* 195: 44–51.
18. Seamon, K.B., W. Padgett, and J.W. Daly. 1981. Forskolin: Unique diterpene activator of adenylate cyclase in membranes and in intact cells. *Proceedings of the National Academy of Sciences of the United States of America* 78: 3363.
19. Krupinski, J., F. Coussen, H.A. Bakalyar, W.J. Tang, P.G. Feinstein et al. 1989. Adenylyl cyclase amino acid sequence: Possible channel- or transporter-like structure. *Science* 244: 1558–1564.
20. Bauman, A.L., J. Soughayer, B.T. Nguyen, D. Willoughby, G.K. Carnegie et al. 2006. Dynamic regulation of cAMP synthesis through anchored PKA-adenylyl cyclase V/VI complexes. *Molecular Cell* 23: 925–931.
21. Pfeuffer, E., S. Mollner, and T. Pfeuffer. 1985. Adenylate cyclase from bovine brain cortex: Purification and characterization of the catalytic unit. *EMBO Journal* 4: 3675.
22. Spooner, P.J., N.G. Rutherford, A. Watts, and P.J. Henderson. 1994. NMR observation of substrate in the binding site of an active sugar-H^+ symport protein in native membranes. *Proceedings of the National Academy of Sciences of the United States of America* 91: 3877–3881.
23. Morris, D.I., J.D. Robbins, A.E. Ruoho, E.M. Sutkowski, and K.B. Seamon. 1991. Forskolin photoaffinity labels with specificity for adenylyl cyclase and the glucose transporter. *Journal of Biological Chemistry* 266: 13377–13384.
24. Morris, D.I., L.A. Speicher, A.E. Ruoho, K.D. Tew, and K.B. Seamon. 1991. Interaction of forskolin with the P-glycoprotein multidrug transporter. *Biochemistry* 30: 8371.
25. Hoshi, T., S.S. Garber, and R.W. Aldrich. 1988. Effect of forskolin on voltage-gated K^+ channels is independent of adenylate cyclase activation. *Science* 240: 1652–1655.
26. Piggott, L.A., A.L. Bauman, J.D. Scott, and C.W. Dessauer. 2008. The A-kinase anchoring protein Yotiao binds and regulates adenylyl cyclase in brain. *Proceedings of the National Academy of Sciences of the United States of America* 105: 13835–13840.
27. Kapiloff, M.S., R.V. Schillace, A.M. Westphal, and J.D. Scott. 1999. mAKAP: An A-kinase anchoring protein targeted to the nuclear membrane of differentiated myocytes. *Journal of Cell Science* 112: 2725–2736.

28. Pare, G.C., A.L. Bauman, M. McHenry, J.J. Michel, K.L. Dodge-Kafka, and M.S. Kapiloff. 2005. The mAKAP complex participates in the induction of cardiac myocyte hypertrophy by adrenergic receptor signaling. *Journal of Cell Science* 118: 5637–5646.
29. Johnson, R.A., and Y. Salomon. 1991. Assay of adenylyl cyclase catalytic activity. *Methods in Enzymology* 195:3–21.
30. Dessauer, C.W. 2009. Adenylyl cyclase—A-kinase anchoring protein complexes: The next dimension in cAMP signaling. *Molecular Pharmacology* 76: 935–941.
31. Efendiev, R., B.K. Samelson, B.T. Nguyen, P.V. Phatarpekar, F. Baameur et al. 2010. AKAP79 interacts with multiple adenylyl cyclase (AC) isoforms and scaffolds AC5 and -6 to alpha-amino-3-hydroxyl-5-methyl-4-isoxazole-propionate (AMPA) receptors. *Journal of Biological Chemistry* 285: 14450–14458.
32. Willoughby, D., N. Masada, S. Wachten, M. Pagano, M.L. Halls et al. 2010. AKAP79/150 interacts with AC8 and regulates Ca2+-dependent cAMP synthesis in pancreatic and neuronal systems. *Journal of Biological Chemistry* 285: 20328–20342.
33. Willoughby, D., M.L. Halls, K.L. Everett, A. Ciruela, P. Skroblin et al. 2012. A key phosphorylation site in AC8 mediates regulation of Ca(2+)-dependent cAMP dynamics by an AC8-AKAP79-PKA signalling complex. *Journal of Cell Science* 125: 5850–5859.
34. Kleuss, C., and A.G. Gilman. 1997. Gsa contains an unidentified covalent modification that increases its affinity for adenylyl cyclase. *Proceedings of the National Academy of Sciences of the United States of America* 94: 6116.
35. Lee, E., M.E. Linder, and A.G. Gilman. 1994. Expression of G protein a subunits in *Escherichia coli*. *Methods in Enzymology* 237: 146–164.
36. Conley, J.M., T.F. Brust, R. Xu, K.D. Burris, and V.J. Watts. 2014. Drug-induced sensitization of adenylyl cyclase: Assay streamlining and miniaturization for small molecule and siRNA screening applications. *Journal of Visualized Experiments* 83: e51218.
37. Carr, D.W., R.E. Stofko-Hahn, I.D. Fraser, S.M. Bishop, T.S. Acott et al. 1991. Interaction of the regulatory subunit (RII) of cAMP-dependent protein kinase with RII-anchoring proteins occurs through an amphipathic helix binding motif. *Journal of Biological Chemistry* 266: 14188–14192.
38. Herberg, F.W., A. Maleszka, T. Eide, L. Vossebein, and K. Tasken. 2000. Analysis of A-kinase anchoring protein (AKAP) interaction with protein kinase A (PKA) regulatory subunits: PKA isoform specificity in AKAP binding. *Journal of Molecular Biology* 298: 329–339.
39. Carlson, C.R., B. Lygren, T. Berge, N. Hoshi, W. Wong et al. 2006. Delineation of type I protein kinase A-selective signaling events using an RI anchoring disruptor. *Journal of Biological Chemistry* 281: 21535–21545.
40. Alto, N.M., S.H. Soderling, N. Hoshi, L.K. Langeberg, R. Fayos et al. 2003. Bioinformatic design of A-kinase anchoring protein-in silico: A potent and selective peptide antagonist of type II protein kinase A anchoring. *Proceedings of the National Academy of Sciences of the United States of America* 100: 4445–4450.
41. Btesh, J., M.J. Fischer, K. Stott, and P.A. McNaughton. 2013. Mapping the binding site of TRPV1 on AKAP79: Implications for inflammatory hyperalgesia. *Journal of Neuroscience* 33: 9184–9193.
42. Fischer, M.J., J. Btesh, and P.A. McNaughton. 2013. Disrupting sensitization of transient receptor potential vanilloid subtype 1 inhibits inflammatory hyperalgesia. *Journal of Neuroscience* 33: 7407–7414.

10 Assessing Cyclic Nucleotide Binding Domain Allostery and Dynamics by NMR Spectroscopy

Bryan VanSchouwen, Madoka Akimoto, Stephen Boulton, Kody Moleschi, Rajanish Giri, and Giuseppe Melacini

CONTENTS

10.1 INTRODUCTION

10.1.1 What Is Dynamics in the Context of Cyclic Nucleotide Binding Domains?

Cyclic nucleotide binding domains (CBDs) regulate proteins that are involved in a wide range of cellular functions, including protein kinase A (PKA), protein kinase G (PKG), the exchange protein directly activated by cAMP (EPAC), and the hyper-polarization-activated cyclic nucleotide–modulated (HCN) and cyclic nucleotide–gated (CNG) ion channels. Although it is known that cyclic nucleotide binding to CBDs leads to perturbations that promote the activation of key functions within the respective host proteins, the role played by dynamics in the cyclic nucleotide–dependent activation is currently not fully understood. Therefore, an in-depth analysis of dynamics is essential to achieve a more complete understanding of CBD function and to fully exploit the potential of CBDs as therapeutic targets.

Dynamics in CBDs occur at multiple levels encompassing several timescales and length scales. First of all, nuclear magnetic resonance (NMR) studies have suggested that apo-CBDs exists in an autoinhibitory equilibrium between inactive (i.e., inhibition-competent) and active (i.e., inhibition-incompetent) conformational states (Figure 10.1). The inactive versus active introversion is fast in the chemical shift NMR timescale and therefore the observed chemical shifts are an average of the isolated states. The active and inactive states differ not only in structure but also in their profiles of structural fluctuations affecting both the peptide backbone and the side chains (internal dynamics). Such fluctuations can occur on short (ps–ns) or longer (ms–μs) timescales, and in some cases, they can result in a transient local unfolding with the exposure of protein functional groups to the aqueous solvent. The active versus inactive state differences both at the level of structure and of internal dynamics are critical to explain how cyclic nucleotides or mutations (or both) regulate protein activity.

Dynamics are paramount not only for the CBD but also within the cyclic nucleotide ligand itself, which exists in equilibrium between *syn* and *anti* conformational states (Figure 10.1). In the *syn* conformation, the cyclic nucleotide base points toward the ribose moiety, whereas in the *anti* conformation, it is oriented in the opposite direction. It has been found that CBDs typically bind cyclic nucleotides in one of these two conformations, and the *syn*/*anti* conformational tendencies of cyclic nucleotides are thus expected to play a key role in CBD activation. Overall, the allosteric control of eukaryotic CBDs investigated thus far seem to conform to a general model of *reciprocal conformational selection*, whereby it is not only the cyclic nucleotide that selects for a specific state presampled by the *apo* CBD, but it is also the CBD that selectively binds one of the conformations adopted by the *apo* ligand (Figure 10.1).[1–5]

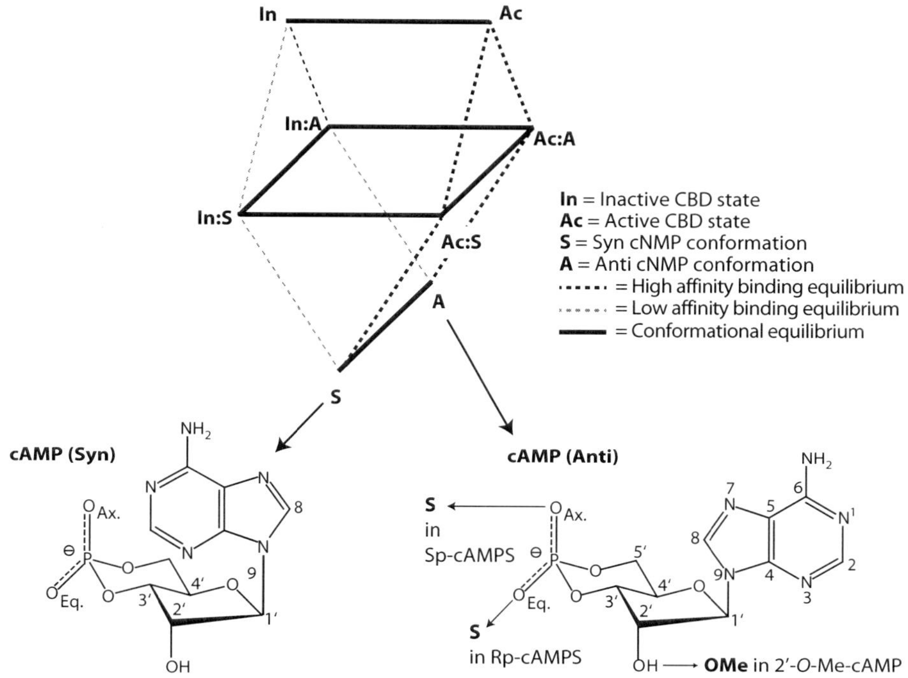

FIGURE 10.1 Eight-state thermodynamic cycle for the allosteric control of CBDs through reciprocal protein-ligand conformational selection. Symbols are defined in the figure legend, and arrows in the equilibria are omitted for simplification purposes. Boldface fonts denote states with higher populations. The apo CBD samples both inactive (autoinhibitory) and active states, which differ in structure or dynamics (or both). In the absence of ligands, the inactive state is typically, although not always (Figure 10.2), more populated than the active state (top solid line). However, in the holo CBD, the reverse applies (square with solid lines), due to the active versus inactive selectivity of the cyclic nucleotide. Similarly, unbound cAMP samples primarily the *anti* conformation (bottom solid line), whereas it transitions to the *syn* conformation upon binding to the CBDs of PKA or EPAC.[6] The covalent structure of *anti* cAMP is used to illustrate the library of cAMP analogues originally used as a perturbation set for CHESCA.[7]

10.1.2 Why Are CBD Dynamics Functionally Significant?

Dynamics play a key role in protein allostery and this is particularly evident in the context of CBDs, as proven by several recent applications of NMR to CBDs. An example of the relevance of dynamics in allostery is provided by PKA, in which a flexible linker region N-terminal to the regulatory subunit CBD-A domain was found to serve as a key allosteric element controlling PKA activity, despite the fact that this linker is only partially structured.[8] Specifically, the linker forms weak, conformation-selective interactions with the active conformation of CBD-A, thus tuning the CBD-A inactive/active conformational equilibrium of the *apo* regulatory subunit to a near-degenerate state that facilitates an optimal response of PKA to cAMP

(Figure 10.2).[8] Furthermore, the linker/catalytic subunit interactions formed in the inactive state occur at the expense of disrupting the active conformation–selective linker/CBD-A interactions and vice versa (Figure 10.2), resulting in a structural frustration within the linker that facilitates the disruption of the large regulatory subunit/catalytic subunit interface in response to the low molecular weight (MW) ligand cAMP.[8]

Dynamics are also a key entropic determinant of the free energy changes that control allosteric processes (Figure 10.1). This general concept has been elegantly illustrated by the group of Kalodimos et al. through a series of pioneering publications on the catabolite activator protein (CAP), which is a dimer including two CBDs. In CAP, dynamics control not only the free energy of allosteric coupling between the two cAMP-binding sites[9] but also the free energy of DNA binding by CAP. In this respect, a striking example illustrating the relevance of dynamics in allostery is provided by a recently studied point mutant of CAP, where cAMP binding to the mutant promotes its binding to DNA—forming a complex with a structure very similar to that of the wild-type protein—despite an apparent failure to induce the active, DNA-binding–competent conformation in the DNA-binding domain (DBD) of the mutant.[10] NMR analyses revealed that despite the apparent lack of a full shift to the active state, cAMP binding to the mutant CAP resulted in a dynamic exchange between an inactive DBD conformation and a weakly populated active DBD conformation.[10] The latter binds DNA with high affinity due to a DNA-induced enhancement of dynamics, which promotes an entropic stabilization of the DNA-bound form—that is, whereas binding of wild-type CAP to DNA is driven by enthalpy, binding of the mutant to DNA is driven by entropy resulting from an increase in structural fluctuations of the mutant upon DNA binding.[10] Because the DNA binding affinity of the active DBD conformation is several orders

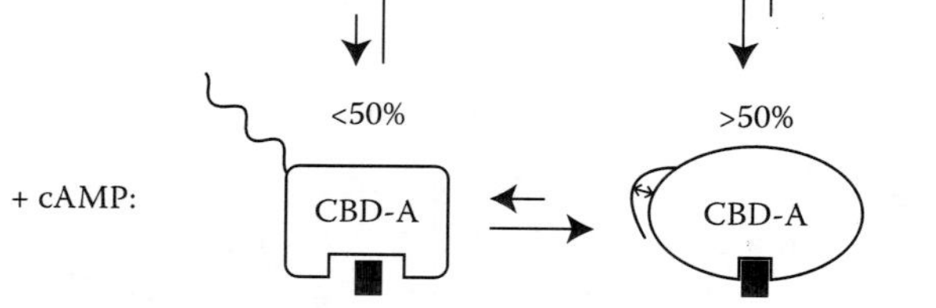

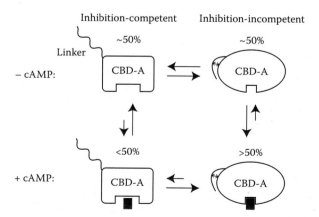

FIGURE 10.2 Simplified four-state thermodynamic cycle for the allosteric control of the N-terminal cyclic nucleotide binding domain of PKA (CBD-A). Single-headed arrows denote conformational or binding equilibria, whereas the double-headed arrow denotes state-selective linker/CBD interactions. The inhibition-competent state binds the catalytic subunit of PKA (C) with higher affinity than the inhibition-incompetent state. In the absence of cAMP, the two states exhibit nearly degenerate free energies.

of magnitude higher than that of the inactive DBD conformation, it was postulated that DNA would selectively bind to the weakly populated active DBD conformation, thus shifting the conformational equilibrium from the inactive to the active conformation.[10] Conversely, suppression of transient conformational states is useful for eliciting allosteric inhibition.[11] Dynamically driven allosteric phenomena, although first experimentally proven for the prokaryotic CAP,[9–12] are also common in eukaryotic CBDs. For example, cAMP binding to the CBD of EPAC was found to result in an increase in the internal dynamics of the domain's N-terminal α-helical bundle (NTHB), with only a minimal change in the secondary or tertiary structure of the NTHB.[13] Such dynamic enhancement was proposed to entropically promote the dissociation of a known autoinhibitory interface formed between the NTHB and the catalytic domain of EPAC, thereby promoting EPAC activation.[13]

Besides its involvement in protein allostery, dynamics have also been implicated as a key factor in cyclic nucleotide binding and selectivity of CBDs. For example, the higher cAMP binding affinity of PKA, compared with other eukaryotic CBDs such as those of EPAC and HCN, was recently attributed to the relative positions of the inactive/active conformational equilibria of the apo state CBDs.[14] Specifically, whereas the apo state CBDs of EPAC and HCN have been proposed to exist predominantly in their inactive conformations, the apo state CBD-A of PKA exists in an equilibrium containing a greater population of the active conformation.[8,14] Because the active CBD conformation is expected to have a higher binding affinity than the inactive conformation, the greater apo state population of active CBD was postulated to result in the greater observed binding affinity of the PKA CBD-A, relative to EPAC and HCN.[14] Indeed, C-terminal deletion mutations of EPAC that were found to increase its apo-active population also significantly increased the observed binding affinity of the EPAC CBD for cAMP,[15] providing further evidence of the relevance of CBD conformational equilibria for binding. However, it is notable that in the prokaryotic CBD of CAP, mutation-induced changes in the CBD inhibitory equilibrium do not always correlate with the observed changes in DNA affinity.[12] This apparent paradox was solved by considering that mutations that modulate the inhibitory equilibrium are also often effective modulators of the entropy of DNA binding, even in the absence of apparent structural changes.[12]

The role of dynamics in cyclic nucleotide selectivity is illustrated by the effects of cAMP and cGMP on the EPAC CBD.[6] Unlike cAMP, cGMP is an EPAC antagonist. This cAMP versus cGMP functional difference with respect to EPAC activation is explained by the observation that whereas cAMP binds to EPAC in a *syn* conformation, cGMP binds in an *anti* conformation. The alternative binding mode of cGMP leads to a decrease in the internal dynamics of the CBD NTHB, compared with cAMP-bound EPAC.[6] The reduced NTHB dynamics were proposed to hinder the entropically induced NTHB/catalytic domain dissociation normally promoted by cAMP, thus highlighting a selection against EPAC activation by cGMP at the level of CBD dynamics.[6] Last but not least, it should be noted that in addition to improving our understanding of CBD function, analysis of CBD dynamics is also useful as an ancillary tool to other experimental techniques. For instance, analysis of flexible N- and C-termini by NMR can be used to accurately identify the domain boundaries of a minimal CBD construct to be used in further structural characterizations.

The examples outlined above prove the pivotal role of dynamics in cyclic nucleotide–dependent modulation of function and highlight the need to map the dynamic profiles of CBDs across different timescales and length scales. NMR spectroscopy is ideally suited to serve this purpose and the following sections summarize the methods we have previously used to prepare, validate, and investigate NMR samples of eukaryotic CBD constructs. Although experimental details can be found in the original publication references, here we will try to highlight critical aspects of NMR-related protocols. Knowledge of the critical strength and weakness of the NMR-based approaches will allow the reliable application and adaptation of previously published protocols to new systems.

10.2 NMR SAMPLE PREPARATION AND VALIDATION

10.2.1 Selection of Construct Boundaries

Cyclic nucleotide–dependent proteins are often multidomain systems, and meaningful NMR analyses of cyclic nucleotide–dependent dynamics requires the use of a family of protein constructs, which includes both single and multiple-domain fragments of the wild-type system under investigation. Longer constructs, to the ideal limit of the full-length integral wild-type protein, are obviously preferable in terms of biological relevance, but often pose technical challenges in terms of resolution and sensitivity of solution NMR spectra. The large size of the intact host proteins slows the tumbling in solution, accelerating transverse relaxation and, consequently, reducing the efficiency of coherence transfer in multidimensional NMR experiments and increasing line broadening. As a result, both sensitivity and resolution are typically reduced as the MW increases. This is a major drawback because high MW proteins often exhibit limited solubility and extensive peak overlap due to the large number of signals, which can obscure features of interest in the NMR spectra. Although recent advances in isotopic labeling schemes and pulse sequence design[16–20] elegantly circumvent the technical challenges traditionally associated with the NMR investigation of large MW proteins, it is often advantageous to complement the NMR analyses of multidomain constructs with studies of shorter fragments, including one-domain segments.

Lower MW constructs not only provide enhanced sensitivity and resolution in NMR spectra but also often simplify sample preparation through the deletion of flexible elements that promote self-association, and potentially aggregation, or of elements that normally reside in nonsolution environments, such as the transmembrane regions of HCN and CNG ion channels.[21–23] Low MW constructs also serve as a necessary stepping-stone in the studying of the intact, full-length system. Therefore, NMR analyses of CBDs have often focused on fragments of the respective host multidomain proteins, usually containing the CBD(s) of interest,[8,13,24–26] and occasionally, additional structural elements N- and C-terminal to the CBD(s).[8,26] Considering that protein NMR investigations often rely on a *divide-and-conquer* experimental strategy involving one- or multidomain fragments, a well-designed protein construct with carefully selected domain boundaries and containing the region(s) of interest is a necessity for successful NMR experiments. These points are well illustrated by the regulatory subunit (R) of PKA, which is one of the eukaryotic CBDs first studied by NMR.[25]

The shortest PKA R construct examined so far, referred to as *PKA RIα (119–244)*, consists of residues 119 to 244 of the CBD-A domain of the PKA RIα regulatory subunit, and binds cAMP with nanomolar affinity.[6,25,27] PKA RIα (119–244) is therefore a minimal binding unit for cAMP, which has proven useful for studying conformational and dynamic changes upon binding of cAMP and various analogues. However, this PKA construct lacks the catalytic subunit-binding inhibitory region as well as the adjacent linker, and therefore lacks several allosteric features present in the PKA regulatory subunit. To address this problem, a PKA RIα (91–244) construct containing the inhibitory and linker regions was created, and was verified as the minimal regulatory construct that can bind to both the catalytic subunit and cAMP. As such, it is an excellent model for the biological interactions involving PKA. However, the inhibitory region and the adjacent linker cause the PKA RIα (91–244) construct to undergo self-association at high concentrations.[8] Therefore, a shorter PKA RIα (96–244) construct was used to obtain concentrated protein samples for use in low-sensitivity NMR experiments, such as triple resonance or {^{1}H-^{15}N} nuclear overhauser effect (HN NOE) NMR experiments. Although the PKA RIα (96–244) construct does not bind the catalytic subunit with high affinity, it has a similar HSQC pattern compared to PKA RIα (91–244). Furthermore, to study inter-CBD interactions, a PKA construct was used which contains both the CBD-A and CBD-B domains of PKA RIα, referred to as *RIα (119–379)*. Indeed, PKA RIα (119–379) has proven to be an excellent construct for understanding CBD domain cross-talk, and also for studying the role of the CBD-A lid, which is located in the N-terminal segment of the CBD-B domain. These examples illustrate the importance of tailoring the NMR construct design to the specific questions to be addressed. Once a construct is designed, the next step focuses on the preparation of amounts sufficient for NMR analyses.

10.2.2 Optimization of Protein Yields

High yields of sample preparations are desirable because they allow multiple protein solution aliquots for NMR analysis to be obtained from a single purified batch of protein, thus minimizing the expenditure of unnecessary time and costs on repetitive protein preparations, and permitting execution of multiple NMR experiments under consistent and comparable solution conditions. These considerations are paramount especially when expensive isotopically labeled precursors are required. In this respect, it is critical to improve the protein yield by first optimizing the expression conditions used for sample preparation. This can be achieved by such means as codon optimization of the protein-encoding DNA sequence, incorporation of a cleavable solubility tag in the encoded protein construct, and ensuring that a suitable plasmid vector and *Escherichia coli* strain have been used for expressing the protein. For example, expression of the PKA RIα (91–244) construct initially had a low protein yield (~0.5 mg/L of M9 medium), but by using a new DNA clone whose sequence was codon-optimized by replacing rare codons and optimizing GC content to stabilize the mRNA, the protein yield was increased 20-fold compared with the initial protein preparation.[8] Furthermore, to enhance protein solubility during expression, the PKA constructs were expressed with an N-terminal small ubiquitin-related modifier (SUMO) tag.[8] Similarly, the EPAC1$_h$ (149–318) CBD construct was expressed with an N-terminal GST tag.[13] Additional means of optimizing the protein expression conditions include adjustments of the cell culture concentration at which protein expression is induced, the concentration of inducing reagent added to the cell culture to induce protein expression (IPTG), and the incubation temperature and time for which protein expression is allowed to proceed after induction.

10.2.3 Preparation of NMR Samples

The ideal NMR sample has long-lasting stability and high solubility, and for some NMR experiments, such sample qualities are a necessity. For example, acquisition of ^{15}N T_1, T_2 and {^{1}H-^{15}N}-NOE relaxation data with adequate signal-to-noise ratios often requires approximately 1 week. Reduction of acquisition time for these experiments by concentrating the samples is often not an option because even a minimal degree of self-association can lead to abnormal increases in the R_2 relaxation rates. Therefore, due to the long acquisition times for these experiments, the sample must remain stable under the NMR experimental conditions for an extended time. Furthermore, a high sample concentration, typically 0.4 to 0.5 mM, is advisable for some of the most informative triple resonance experiments due to their low sensitivity, and so the samples used for these experiments must have high solubility under the NMR experimental conditions to remain stable for the duration of the experiments.

The buffer conditions used for NMR samples are often dictated by the need to optimize probe performance. For instance, because the high ionic mobility of the buffer reduces the signal-to-noise ratio drastically in cryogenic NMR probes required for dilute samples, low-conductivity buffers such as MES or MOPS are preferred over higher-conductivity buffers such as sodium phosphate buffer. However, the latter offers other advantages, including a lack of nonexchangeable protons that might interfere with signals of interest. If phosphate buffer is needed, low concentrations (~20 mM) are advisable, and the samples should be checked for possible pH drifts. In addition, salt concentrations have to be sufficiently low to maintain NMR pulse performance and sensitivity, but high enough to enhance protein solubility. A sodium chloride concentration in the 50 to 150 mM range typically meets both requirements in CBD NMR samples. Other common additives for NMR samples include DTT, EDTA and NaN_3. DTT is used to disrupt disulfide bridges, if needed, and to prevent methionine oxidation. Chelating reagents such as EDTA are added to bind metal ions and inactivate residual metalloproteases, as well as to prevent undesired relaxation enhancements caused by potential paramagnetic metal ions. NaN_3 is used to inhibit microbial growth in the NMR sample. For example, 50 mM MES buffer with pH 6.5, 100 mM NaCl, 2 mM EGTA, 2 mM EDTA, 5 mM DTT, and 0.02% NaN_3 was used for triple-resonance experiments performed on PKA samples.[28]

10.2.4 Removal of cAMP and Monitoring of cAMP Binding

The cAMP-free (*apo*) form of CBDs is a key component of comparative NMR analysis aimed at understanding the molecular basis of cAMP-dependent binding and allostery. The apo CBD often closely represents the autoinhibited state of the CBD (Figure 10.1) and it provides a reference spectrum to evaluate the perturbations caused by cAMP. However, the preparation of apo CBDs poses experimental challenges when the K_d is less than a few micromolar because cAMP is endogenous to *E. coli*, and CBDs that are overexpressed in *E. coli* can easily bind with cAMP. This is a potential complication because the residual cAMP-bound CBD will shift the apo/bound state equilibrium enough to alter the NMR spectral analysis results. It is therefore critical to establish, before further NMR analyses, whether cAMP-bound CBD is present in the samples purified from *E. coli*.

One means of detecting the presence of residual cAMP-bound CBD is by acquiring {^{1}H,^{15}N}-HSQC spectra. For example, in the case of EPAC, where the cAMP binding affinity is in the micromolar range,[24] the equilibrium between apo and cAMP-bound states exhibits a mixture of fast exchange and slow exchange dynamics, depending on the residues and the magnitude of the respective cAMP-dependent chemical shift change. Here, we can use the slow-exchanging residues to detect the presence of residual cAMP-bound CBD in the apo sample by checking for the presence of a second set of {^{1}H,^{15}N}-HSQC cross-peaks that would be generated for these residues by the residual cAMP-bound CBD. Any endogenous cAMP-bound EPAC can then be eliminated by removing the cAMP through dialysis, which is effective due to the micromolar binding affinity of EPAC for cAMP. For the PKA R-subunit, however, the CBD constructs examined by NMR typically exhibit high binding affinities for cAMP (K_d << μM), and as a result, the purified PKA construct is always a mixture of cAMP-bound and apo states even after extensive dialysis or size exclusion purification. Therefore, to remove residual cAMP from the PKA R-subunit, it is necessary to unfold the constructs using 8 M of urea and remove the residual cAMP through size exclusion column (SEC) or PD10 desalting column.[8,25] The unfolded PKA is then refolded by extensive dialysis, and any aggregated protein is eliminated by further SEC purification. Alternatively, cAMP-free PKA can be prepared by expressing the protein in an *E. coli* strain that lacks endogenous cAMP, such as the TP2000 strain devoid of adenylate cyclase activity described in Chapter 11.[29] To confirm that the purified PKA is indeed in its apo state, {^{1}H,^{15}N}-HSQC analysis is then performed to check the slow-exchanging residues for the absence of {^{1}H,^{15}N}-HSQC cross-peaks arising from the cAMP-bound state. Alternatively, residual cAMP-bound CBD can be detected using 260/280 nm absorbance ratio measurements.[30]

10.3 VALIDATION OF CBD CONSTRUCTS SELECTED FOR NMR ANALYSES

Although low MW protein segments benefit from increased NMR sensitivity and resolution relative to higher MW constructs, and hence have the potential to provide access to a wealth of dynamic and structural information on CBDs, it is critical to validate the functional relevance of short protein fragments early on in the NMR investigation to verify whether the selected NMR constructs will provide a suitable representation of the functional properties normally exhibited in the context of the respective intact proteins. Specifically, the validation process is typically aimed at ensuring that N- and C-terminal truncations do not significantly perturb binding and allosteric properties relative to the intact protein.

10.3.1 Validation of Binding Affinities Measured by NMR

One means of validating a selected NMR construct is by measuring its binding affinity for a particular ligand(s), and comparing the measured affinity with that determined for the intact protein. In the case of CBDs, this has been accomplished by NMR either by saturation transfer difference (STD) measurements, if the ligand binding exhibits micromolar affinity, or by differential unfolding free energy ($\Delta\Delta G_{unfolding}$) measurements, if the ligand binding exhibits nanomolar affinity. For example, in an NMR study of the EPAC CBD performed by Mazhab-Jafari et al.,[24] an $EPAC1_h$ fragment composed of the N-terminal helical bundle (NTHB), β-subdomain and hinge helix of the $EPAC1_h$ CBD (i.e., $EPAC1_h$ residues 149–305) was used to study the effect of cAMP binding on the CBD. Measurement of the dissociation constant by STD resulted in a binding isotherm exhibiting a clear dose–response pattern and a well-defined plateau, indicating that the EPAC construct bound cAMP specifically and with an affinity in the micromolar range, comparable to that reported previously for full-length EPAC1. This observation provided evidence that despite the truncation, the selected EPAC1 construct preserved key cAMP-responsive features present in the full-length protein.[24] This example also illustrates the potential of STD NMR experiments for micromolar K_d measurements using the intact cNMP without the need of ad-hoc added tags, which may perturb the binding affinities. However, accurate K_d determinations require that the protein concentration into which a ligand is titrated be in the same order of magnitude as the K_d value. The requirement of protein concentrations that are approximately in the micromolar range often imposes limits on the sensitivity of the STD NMR spectra, which are however effectively circumvented through the use of cryogenic probes with enhanced sensitivity.

Another example of binding affinity–based construct validation is illustrated in the study of Das et al.,[6] in which the effects of cAMP and cGMP binding on the PKA CBD-A domain were examined using a PKA RIα fragment composed of the isolated CBD-A domain (i.e., PKA RIα residues 119–244). The difference between the free energies of unfolding for the cAMP-bound and cGMP-bound states of the construct (i.e., $\Delta\Delta G_{unfolding}$) was computed based on the average H/D exchange protection factors (PFs) for the buried inner β-strand residues of the β-subdomain, whose H/D exchange results primarily from transient global unfolding of the domain, as explained in Section 10.4.2.2.[6] This value of $\Delta\Delta G_{unfolding}$ was then compared with a value computed based on previously determined dissociation constants for cAMP and cGMP binding to the full-length regulatory subunit, and the two values were found to agree within error with one another ($\Delta\Delta G_{unfolding} = 1.65 \pm 0.70$ kcal/mol computed from the PFs versus $\Delta\Delta G_{unfolding} = 1.5$ kcal/mol computed from the dissociation constants), providing evidence that despite the truncation, the selected PKA RIα construct preserved key ligand-responsive features present in the full-length protein.[6] This example also shows how differential global unfolding free energies provide a direct assessment of relative K_d values for different ligand-bound states of a given protein construct, without the need for isolating the apo form, which may be poorly soluble and unstable, as in the case of the regulatory subunit of PKA. However, it should be noted that the measurement of global unfolding free energies from H/D exchange protein factors assumes that an EX2 exchange mechanism applies and that accurate PFs are measured for the buried inner strands of the CBD β-subdomain (e.g., strands β3, 4, 7, and 8). As these PFs are typically high and correspond to slow decays after exposure to ^2H_2O, it is important to avoid excessively high free-ligand concentrations, which may otherwise result in NMR intensities with negligible decays within the time frame of the NMR sample life.

10.3.2 Validation of Allosteric Responses by the NMR Chemical Shift Projection Analysis

Another critical phase of the construct validation process pertains to verifying that the extent of activation measured by NMR for a given protein fragment in response to selected perturbations correlates with functional changes observed for the full-length system in response to the same perturbations. Perturbations typically arise from either mutations or covalent modifications of the allosteric effector ligand. The relative extent of activation of the perturbed CBD constructs is effectively quantified using a recently developed method called *NMR chemical shift projection analysis*, or CHESPA.[31] The CHESPA approach probes the position of the inactive-versus-active equilibrium of a perturbed CBD relative to the apo and holo (i.e., cAMP-bound) wild-type forms of the same CBD construct, which approximate the inactive and active states, respectively (Figure 10.1). For this purpose, chemical shifts are used to compute through simple vectorial algebra a fractional activation (X) on a per residue basis for a perturbed CBD. An X value close to 0 (1) indicates that the position of the autoinhibitory equilibrium of the perturbed CBD is similar to that of the wild-type apo (holo) CBD. An X value ≤ 0 points to inactivation, whereas $X \geq 1$ points to activation or superactivation, with X values in between 0 and 1 corresponding to partial activation.

The usefulness of the CHESPA approach in the validation of a CBD construct was illustrated in a recent NMR study of the EPAC CBD performed by Selvaratnam et al.,[31] in which an $EPAC1_h$ CBD construct (i.e., $EPAC1_h$ residues 149–318) was examined using CHESPA. The CHESPA was used to compare apo wild-type, cAMP-bound wild-type, and apo L273W mutant forms of the selected $EPAC1_h$ segment.[31] Notably, whereas cAMP binding promoted a shift of the construct toward its active state, the L273W mutant promoted a shift further toward the inactive state, in agreement with previous bioassay results for full-length EPAC, demonstrating that the L273W mutant promotes inactivation of full-length EPAC.[31] Furthermore, analysis of the cAMP-bound L273W mutant indicated that the apo state inactivation by the L273W mutant was preserved upon cAMP binding, in agreement with the previously observed inability of cAMP to activate the L273W mutant of full-length EPAC.[31]

Another example of construct validation through the comparison of NMR-based assessments of autoinhibitory equilibria and functional assays on intact systems is illustrated by the study of Akimoto et al.,[8] in which a PKA RIα fragment composed of the CBD-A domain and an N-terminal linker (i.e., PKA RIα residues 91–244) was examined by chemical shift covariance analyses. Specifically, a cluster of residues was identified in the RIα (91–244) fragment for which the relative {^{1}H,^{15}N} chemical shifts across five states correlated with previously measured activation profiles of the same states obtained for the full-length regulatory subunit. These states include the kinase-inhibiting apo form and four bound forms, in which the ligands were the allosteric effector (i.e., cAMP), an inverse agonist (i.e., R_p-cAMPS), a partial agonist (i.e., 2′-*O*-Me-cAMP) and a full agonist (i.e., S_p-cAMPS). The correlation between the positions of the inactive-versus-active equilibrium as probed by NMR chemical shifts and the relative kinase activation potencies measured through bioassays on the integral PKA system provided evidence that despite the truncation, the selected PKA RIα (91–244) construct preserves key cAMP-responsive features present in the full-length protein.[8]

10.3.3 Validation of Structural Integrity

The validation process can be further extended to comparative structural assessments, if the structure of the full-length protein is known. For instance, a simple structural validation is obtained by verifying that the secondary structure of a protein fragment is similar to that observed in the integral system. Overall secondary structure content is effectively measured in solution by circular dichroism (CD), and a residue-resolution map of secondary structure elements is reliably obtained through the analysis of secondary chemical shifts.[32,33] Once an NMR-amenable construct is validated at the level of binding, allosteric responses and, when possible, structural integrity, further NMR studies on the same construct are typically directed at in-depth explorations of dynamics, interactions, and their mutual coupling—that is, allostery.

10.4 METHODS FOR ASSESSING CBD DYNAMICS AND ALLOSTERY BY NMR

10.4.1 NMR Relaxation Experiments

Direct assessment at atomic resolution of ps–ns and ms–μs structural fluctuations within proteins is possible through the use of classical NMR relaxation techniques.[34,35] In particular, T_1 and T_2 ^{15}N relaxation and {^{1}H-^{15}N}-NOE experiments are frequently used to probe dynamics in the peptide backbone on a per residue basis and in N-H containing side chains. The ^{15}N T_1, T_2 and {^{1}H-^{15}N}-NOE relaxation data is typically analyzed in terms of either the model-free formalism or reduced spectral density maps to provide information on both ps–ns and ms–μs timescales.[34,35] The slower (ms–μs) motions are best analyzed through measurements at multiple static fields and through nuclear magnetic relaxation dispersion (NMRD) specifically designed to examine ms–μs dynamics arising from the exchange between ground and excited conformations of proteins.[36,37] The ^{15}N relaxation and dispersion pulse sequence and acquisition protocols have already been extensively reviewed elsewhere,[34–40] and additional NMR relaxation experiments are discussed in Section 10.4.5. Hence, here, we will focus primarily on how to use these experiments to extract functionally relevant dynamic profiles of CBDs.

To establish correlations between CBD dynamics and CBD function, it is critical to comparatively analyze the dynamic profiles of a given CBD in several forms that span different degrees of functional activation. This is why NMR relaxation experiments are best used in conjunction with sample manipulations designed to trap the protein in a particular functional state. For example, in a recent study of the CBD of EPAC, it was possible to separate cAMP-dependent variations in dynamics due to cAMP binding versus allostery by performing a comparative analysis not only of the apo and cAMP-bound CBDs, which approximate the wild-type inactive and active states, respectively, but also of an antagonist-bound CBD. In this respect, a useful antagonist of cAMP in relation to PKA and EPAC activation is the phosphorothioate cAMP analogue R_p-cAMPS. The antagonism of R_p-cAMPS arises from its inactive-versus-active selectivity pattern, which is different from cAMP. Whereas cAMP binds more tightly to the active rather than the inactive state (Figure 10.1), R_p-cAMPS elicits higher affinity for the inactive state of the PKA and EPAC CBDs due to its bulky exocyclic equatorial sulfur atom (Figure 10.1).[13] Therefore, the R_p-cAMPS antagonist is a useful tool to trap the inactive-bound state of the CBD and dissect contributions to dynamics arising from binding versus purely allosteric effects, which would have otherwise been convoluted together had the comparative dynamic analyses been limited exclusively to the pairwise *apo* versus cAMP-bound forms of the CBD. These comparative analyses used to correlate dynamics to function are easily extended to several NMR methods, which have proven useful for studying CBD structural fluctuations and allostery. Such methods are not limited to NMR relaxation experiments, and include also hydrogen exchange and chemical shift analyses, which are discussed in the following sections.

10.4.2 Hydrogen Exchange NMR Experiments: H/D and H/H

The dynamic structural fluctuations that define the native ensemble accessible to a protein often result in transient structural unfolding, which permits hydrogen exchange between the aqueous solvent and the backbone amides of the affected residues. The higher the population of the state in which a given residue is transiently solvent exposed, the more rapidly hydrogen exchange at that residue will occur. Therefore, measurement of backbone hydrogen exchange rates provides an additional means of assessing conformational dynamics within the protein. Hydrogen exchange is probed with excellent sensitivity through mass spectrometry, as shown by several elegant applications to the PKA and EPAC systems.[41,42] However, when hydrogen exchange is monitored by mass spectrometry, it is not always possible to go beyond peptide resolution. In these cases, useful complementary techniques to ensure that hydrogen exchange is probed at residue resolution are H/D NMR and H/H NMR. In H/D NMR, hydrogen exchange rates are measured through real-time monitoring of $\{^1H,^{15}N\}$-HSQC cross-peak intensity decays resulting from deuterium incorporation into the backbone amides for a sample dissolved in ^2H_2O.[24] However, H/D exchange rates cannot be quantified for residues whose backbone amides undergo full deuterium incorporation within the dead time of the H/D experiment (usually ~20 min). To further characterize these fast-exchanging residues, H/H NMR pulse sequences are used to identify residues that undergo hydrogen exchange on the 10^0–10^2 ms timescale,[24] with the remaining residues classified as exchanging on a seconds-to-minutes timescale.

10.4.2.1 Transient Local Unfolding Probed by Hydrogen Exchange NMR

Residues that are predicted to be highly solvent-exposed, for example, based on high solvent-accessible surface areas calculated from the protein structure, generally tend to exhibit rapid hydrogen exchange, whereas residues predicted to be buried within the protein based on low solvent-accessible surface areas, tend to exhibit slower hydrogen exchange. Indeed, upon examination of the apo CBDs of EPAC and PKA, residues of the solvent-exposed N-terminal α-helical elements exhibited rapid exchange, as evidenced by fast H/D exchange, and the observation of H/H NMR signals, whereas residues of the inner strands of the CBD β-barrels exhibited slower exchange, as evidenced by slower H/D exchange, and a lack of H/H NMR signals.[24,25] However, notable exceptions to this general trend have been observed for the CBDs of EPAC and PKA, and such exceptions have highlighted critical allosteric features of these proteins.[24,25] For example, despite the prediction of low solvent exposure for several residues in the phosphate binding cassette (PBC), hinge helix, and β2–β3 loop elements of the apo state EPAC CBD, the apo state exchange rates were mostly fast for these residues, suggesting that the PBC, hinge helix, and β2–β3 loop elements are subject to significant local structural fluctuations in the apo state.[24] Meanwhile, upon cAMP binding, a dramatic decrease in exchange rate was observed for several residues in these elements, highlighting a quenching of local structural fluctuations within these regions.[24] However, these residues still exhibited more rapid exchange than residues in the inner β-barrel strands, and this was attributed to the existence of residual structural fluctuations in the cAMP-bound state.[24]

Additional changes in local structural fluctuations observed in the PKA and EPAC CBDs by H/D and H/H NMR proved to be significant also at the level of autoinhibitory interfaces formed by these CBDs. For example, cAMP removal from the PKA CBD resulted in faster hydrogen exchange for several residues throughout the CBD α-helical subdomain (α-subdomain), suggesting an increased solvent exposure within the affected regions of the α-subdomain.[25] Because the α-subdomain regions in question were known to form key components of the autoinhibitory interface with the PKA C-subunit in the inactive state, the increased solvent exposure in these regions upon cAMP removal was postulated to render these regions more available for establishment of interactions with the C-subunit, with cAMP-dependent CBD perturbations hindering the formation of the C-subunit interface through increased sequestration of key interaction sites in the CBD.[25]

In the case of EPAC, cGMP binding to the EPAC CBD resulted in slower hydrogen exchange for several residues of the CBD NTHB compared with cAMP binding, which was attributed to reduced local structural fluctuations in the cGMP-bound state that were also noted from NMR relaxation experiments.[6] Because the NTHB was known to form key autoinhibitory interactions with the catalytic domain, and because an enhancement of NTHB dynamics upon cAMP binding had been previously postulated to promote NTHB/catalytic domain dissociation,[13] the reduced NTHB dynamics in the cGMP-bound state were proposed to hinder entropically induced NTHB/catalytic domain dissociation, thus contributing to the selectivity of the EPAC CBD for cAMP versus cGMP at the level of activation.[6]

10.4.2.2 Transient Global Unfolding Probed by Hydrogen Exchange NMR

Hydrogen exchange NMR experiments are not only sensitive to local unfolding, but also to transient global unfolding events. For instance, residues in the inner β-barrel strands of both EPAC and PKA exhibited slower hydrogen exchange upon cAMP binding, despite the very low solvent exposure and slow hydrogen exchange already exhibited by these residues in the apo state.[6,24,25] This observation was attributed to the occurrence of transient global unfolding in the apo CBD, which was quenched upon cAMP binding, suggesting an overall stabilization of the CBD structure upon cAMP binding.[6,24,25] The identification of global unfolding H/D exchange pathways for the inner strands of the β-barrel of PKA was also independently confirmed by the agreement between the ΔG of unfolding estimated from the corresponding PFs and that measured through urea unfolding monitored by intrinsic fluorescence.[25]

The ability to measure free energies of global unfolding ($\Delta G_{unfolding}$) by hydrogen exchange is useful in several applications involving ligand binding, allostery, and interdomain communication. As discussed in Section 10.3.1, ligand-dependent changes in $\Delta G_{unfolding}$ provide an assessment of both absolute and relative ligand affinities (K_d values). Furthermore, mutation-induced changes in $\Delta G_{unfolding}$ are useful to probe the free energy landscape of allostery through the application of mutant cycles designed to identify and quantify long-range allosteric couplings.[28] Another application of $\Delta G_{unfolding}$ measurements from H/D exchange pertains to interdomain interactions. For instance, a PKA construct composed of both tandem CBDs of PKA RIα (referred to here as *PKA RIα (119–379)*) was examined using a combination of hydrogen exchange NMR and $\{^1H, ^{15}N\}$ compounded chemical shifts to gain insight into the mechanism of allosteric cross-talk between the two adjacent CBDs.[43] The wild-type PKA RIα (119–379) construct and two mutants, R209K and R333K, with reduced cAMP binding affinity in the CBD-A or CBD-B, respectively, were examined in both the presence and absence of excess cAMP. Comparative analysis of the wild-type and mutant constructs revealed previously unknown details of both the intradomain and interdomain allosteric networks, including notable differences between the two domains despite their sequence and structure homology.[43]

In particular, H/D exchange analysis revealed that CBD-A exerts multiple levels of control on CBD-B, as suggested by losses of H/D exchange protection in both CBDs when cAMP binding to the CBD-A is reduced through the R209K mutation.[43] Specifically, a loss of H/D exchange protection for highly buried residues in both CBDs suggested that CBD-B is coupled to CBD-A at the level of global unfolding. In addition, a loss of H/D exchange protection for more exposed residues in both CBDs suggested that CBD-B is also coupled to CBD-A at the level of local unfolding, whereby local unfolding of the CBD-A phosphate-binding cassette (PBC) leads to increased local unfolding not only for the CBD-A αB and αC helices, but also for the N3A motif and αB and αC helices of CBD-B. The observed coupling of CBD-B to CBD-A was rationalized by the existence of a *tandem transdomain lid* whereby the CBD-B N3A motif, which is structurally and functionally part of the CBD-B, also forms the cAMP-capping lid region for CBD-A and is covalently linked to the αB-αC region of CBD-A, thereby establishing a pathway for allosteric signal transmission between the α-helical subdomains of the two CBDs.[43]

Interestingly, whereas the H/D exchange analysis suggested that global unfolding originating in CBD-A, as triggered by the R209K mutation, promotes global unfolding in CBD-B, global unfolding originating in CBD-B, as triggered by the R333K mutation, did not seem to promote an appreciable global unfolding in CBD-A.[43] Meanwhile, local unfolding at the CBD-B PBC, as prompted by the R333K mutation, seemed to promote *local* unfolding of the CBD-A αB and αC helices as well as the N3A motif and αB and αC helices of CBD-B, in accordance with the *tandem transdomain lid* model, although the CBD-A local unfolding seemed to occur to a lesser degree than the CBD-B local unfolding that was observed for the R209K mutant.[43] The apparent asymmetry in interdomain communication at the level of global unfolding was concluded to have evolved to ensure robustness in both PKA activation and deactivation, whereby the stability of CBD-A irrespective of CBD-B ensures that PKA activation is only minimally susceptible to dynamic release of apo state CBD-B from the catalytic subunit, whereas the instability of CBD-B as a result of CBD-A global unfolding ensures a prompt deactivation of PKA with the release of cAMP from both CBDs before new cAMP molecules need to bind to transduce new incoming signals.[43] These conclusions were reached without the need to isolate the pure *apo* form of PKA RIα (119–379), which would have required a challenging unfolding and refolding of the two-domain construct. The preparation of *apo* RIα (119–379) was circumvented through the use of a technique called *equilibrium perturbation NMR*, which expands the scope of application for both relaxation and hydrogen exchange NMR measurements, as discussed in the next section.

10.4.3 Equilibrium Perturbation NMR

Although understanding the structural basis of allostery often relies on the comparative analysis of the *apo* and ligand-bound forms of a protein, a direct comparison of the *apo* and ligand-bound states is often experimentally difficult due to the potential instability of one of these states in solution.[44] For example, the CBD-A domain of PKA is poorly soluble in its *apo* form, thus hindering examination by NMR. To circumvent this problem, an experimental strategy based on equilibrium perturbation NMR was implemented.[44] In this approach, the direct comparison of apo and ligand-bound states is replaced by a comparison between a solution of the stable form, and another solution in which the apo/bound state equilibrium is partially shifted toward the unstable state, whereby a reduced effective concentration of the unstable state and continuous ligand binding/dissociation serve to stabilize the equilibrium-shifted solution.[44] The two solutions are then analyzed by techniques such as NMRD or hydrogen exchange NMR, taking advantage of the ability of such methods to probe minimally populated states within dynamic conformational equilibria.[44]

To study the CBD-A domain of PKA, the equilibrium perturbation approach was applied to a pair of solutions of the PKA RIα (119–244) construct: one in which the ligand-bound state was stabilized by the addition of a 10-fold excess of cAMP, and one in which a minor population of the apo state was created by removing excess unbound cAMP through dialysis under nondenaturing conditions.[44] Using this approach, allosteric network elements that had been previously postulated based on mutagenesis and sequence analysis studies were successfully identified, including residues of the phosphate-binding cassette (PBC), β2–β3 region, and αB and αC helices.[44] More importantly, even though no detectable {^{1}H,^{15}N} chemical shift differences were observed between the two solutions, the minor difference in the relative apo/ligand-bound state populations generated marked changes in the $\Delta\Delta R_2^{eff}$ parameters and hydrogen exchange rates measured by NMRD and hydrogen exchange NMR, respectively, proving that it is possible to probe cAMP-associated perturbations even without a directly observable fully apo state of the protein.[44] Furthermore, the equilibrium perturbation approach was found to increase the number of fast-exchanging amide protons detectable by H/D exchange that would otherwise have fully exchanged within the dead time of the H/D exchange analysis of apo PKA RIα (119–244), thus highlighting a significant expansion of the scope of applicability of hydrogen exchange experiments through the use of equilibrium perturbation.[44]

10.4.4 Chemical Shift Covariance Analysis

The fast-exchanging inactive-versus-active conformational equilibria exhibited by CBDs involves long-range allosteric perturbations that are propagated by changes in structure or dynamics (or both), and whereas the end points of these allosteric perturbations can be effectively characterized by comparative analyses of the structural and dynamic profiles of the apo and cyclic nucleotide–bound CBD forms, it is often difficult to define the networks of residues involved in allosteric signal propagation using such methods.[7] This is especially true if the allosteric signal propagation involves very subtle changes in residues of the allosteric network, as such changes, although potentially of functional relevance, may fall below the resolution of the aforementioned comparative analysis methods.[7]

To identify otherwise elusive allosteric networks of residues, a method based on NMR chemical shift covariance analysis (CHESCA) was recently developed and applied to the CBDs of EPAC and PKA.[7,8] In this method, {^{1}H,^{15}N} chemical shifts are measured for five or more states of the protein, which are selected to represent the protein at varying degrees of activation—that is, at different positions along its intrinsic inactive-versus-active equilibrium.[7] For examination of CBDs, such states may include residue mutations or ligand-bound states with chemical modifications at the cyclic nucleotide ligand. Ideally, the perturbations within each CHESCA library are chosen to be confined within a limited spatial region, such that the overall extent of chemical shift variations due to nearest-neighbor effects, as opposed to inactive-versus-active equilibrium shifts, is minimized.[7] In addition, careful referencing of chemical shifts is vital to accurate chemical shift measurements, especially in cases of small chemical shift variations, and is achieved by using an internal referencing compound in the samples (e.g., ^{15}N-labeled acetylglycine), coupled with analysis of the {^{1}H,^{15}N}-HSQC cross-peaks through Gaussian line-fitting.[7] The measured chemical shifts are then probed for correlated variations among the examined perturbation states. Groups of residues that exhibit closely correlated chemical shift variations among the examined states (referred to as residue *clusters*) are identified as undergoing a concerted perturbation in response to the selected set of mutations or ligands. The residues within a given cluster are thus identified as belonging to a common allosteric network.[7] The CHESCA clusters are identified through a combination of correlation matrix and principal component statistical analyses of the *residue* × *perturbation* chemical shift matrix.[7]

The CHESCA method was first implemented to identify the network of residues involved in cAMP-associated allosteric and dynamically driven signal propagation within the CBD of EPAC.[7] Previously, it had been determined that cAMP binding to the EPAC CBD results in a significant structural shift of the cAMP phosphate binding cassette (PBC) and the adjacent C-terminal helix, referred to as the *hinge helix*, as well as an enhancement of dynamics at NTHB residues that form key autoinhibitory interactions with the EPAC catalytic region. However, the means by which the allosteric signal propagates from the PBC and hinge helix to the NTHB residues exhibiting dynamic enhancement had remained elusive, as little structural change had been previously noted in regions of the CBD between these sites. Through CHESCA analysis based on the apo state and four ligand-bound states of the CBD (Figure 10.1), an allosteric network of residues was identified that links the PBC and hinge helix to the NTHB residues exhibiting dynamic enhancement, providing a pathway for allosteric signal propagation between these sites.[7] Specifically, the pathway originates at the PBC, where cAMP binds, and extends through the hinge helix and NTHB C-terminal helix, ultimately reaching the NTHB N-terminal helices, where the residues exhibiting dynamic enhancement are located.[7] This pathway was independently validated by several known mutants of EPAC, for which the k_{max} for guanine exchange had been assessed in full-length EPAC.[7]

Although the first CHESCA performed on EPAC used a series of ligand-bound states as a perturbation library for the statistical analysis,[7] a more recent implementation of CHESCA took advantage of a series of C-terminal deletion mutants of EPAC.[15] Specifically, a series of progressive truncations within the C-terminal portion of the hinge helix were designed to simulate hinge helix C-terminus unwinding, as observed upon cAMP binding and activation. The wild-type and mutant constructs were then analyzed in their apo states to permit examination of the allosteric network controlled by the hinge helix C-terminus in the apo-inactive versus apo-active equilibrium.[15] The results indicated that hinge helix C-terminal truncation leads to a greater stabilization of the active CBD conformation even in the absence of cAMP, and that the perturbation propagates to all known allosteric sites within the CBD, including sites identified in the previous CHESCA analysis of EPAC. Therefore, it was concluded that the hinge helix C-terminus plays a key role in EPAC autoinhibition, and is tightly coupled to the other allosteric elements of the CBD even in the absence of cAMP.[15]

Besides being implemented as a stand-alone method, CHESCA can also be effectively used in conjunction with other experimental data to generate or test specific hypotheses of interest. For example, the CHESCA analysis of a PKA fragment containing the CBD-A and N-terminal linker revealed that the network of residues involved in allosteric signal propagation within the PKA CBD-A also includes residues of the linker, suggesting that the linker, although quite flexible, forms a key component of the CBD-A allosteric network.[8] Based on these results, it was hypothesized that the N-terminal linker selectively interacts with the active rather than the inactive state of the adjacent CBD. Although consistent with the CHESCA analysis, this hypothesis seemed counterintuitive and nonobvious based on the inspection of the x-ray structures of the regulatory subunit of PKA bound to the catalytic subunit (inactive state) or to cAMP (active state). In the former structure, the linker is visible and in the vicinity of the CBD, whereas in the latter structure, the linker is only partially visible due to its dynamics, and involved in crystal contacts. Hence, the hypothesis of active state-selective linker/CBD interactions needed to be tested by additional independent data in solution. For this purpose, paramagnetic relaxation enhancement (PRE) experiments were designed whereby a spin-label was introduced in the linker. The location of the spin-label within the linker was critically guided by the CHESCA results, which identified a site that was not part of the CHESCA allosteric network, but at the same time was in the vicinity of the linker residues involved in CBD interactions. This CHESCA-identified site allowed us to noninvasively and reliably probe the linker/CBD interactions through PREs, which indicated that the linker significantly interacts with the PKA CBD-A only when the latter is in its active conformation, as predicted based on the CHESCA. Furthermore, mutations were designed based on CHESCA and confirmed the existence of active state–selective interactions of the linker with the CBD-A.[8] Together, these results suggested that the flexible linker is not simply a passive covalent thread, but rather is an active allosteric element involved in the conversion of the PKA R-subunit between inhibition-competent and incompetent states.[8]

10.4.5 Probing Dynamics beyond the Single Isolated CBD Using Methyl-TROSY and Hybrid NMR-MD Approaches

Although NMR has yielded valuable information about CBD allostery in the context of isolated CBDs or CBD-containing protein fragments, as illustrated in Sections 10.4.1 through 10.4.4, such an analysis is only a starting point toward gaining a full understanding of CBD protein function. In particular, analyzing CBDs in the context of the full-length protein as well as capturing cAMP-dependent variations in structure and dynamics in other domains of CBD-containing proteins are critical aspects of understanding CBD-associated protein function.[41,45] However, as discussed in Section 10.2.1, examination of full-length CBD-containing proteins by NMR has often proven experimentally challenging. In addition, whereas the endpoint states of CBD protein allostery (i.e., the apo/inactive conformation and ligand-bound/active conformation states; Figure 10.1) are readily examined by NMR, it is generally difficult to probe the transient intermediate states that are critical to allostery, such as the ligand-bound/inactive conformation and apo/active conformation states, which serve as intermediates for the classic induced-fit and conformational-selection allosteric pathways, respectively (Figure 10.1).[45]

One means of examining larger proteins by NMR is to substitute protein backbone–based NMR methods (i.e., $\{^{1}H,^{15}N\}$-HSQC, etc.) with methyl-TROSY-based NMR experiments. In this approach, the protein to be examined is selectively $\{^{1}H,^{13}C\}$-labeled at the side-chain methyl groups of selected amino acids, but fully deuterated elsewhere in the protein, and the labeled methyl groups are probed through $\{^1H,^{13}C\}$-HMQC correlation spectra.[46] Methyl groups are chosen because they yield intense signals due to their three degenerate ^1H spins as well as their internal dynamics, and are typically well dispersed in $\{^1H,^{13}C\}$-correlation spectra.[46] Furthermore, methyl-based multiple quantum coherences benefit from desirably slow transverse relaxation, thanks to the destructive interference between relaxation mechanisms of the ^1H and ^{13}C spins (*methyl TROSY*), as well as to the scarcity of ^1H spins achieved by deuteration. The combination of methyl TROSY and deuteration dramatically reduces line broadening and enhances both the resolution and signal-to-noise ratio of the methyl-based NMR spectra.[46] In addition, by probing only the selected side-chain methyl groups, methyl-TROSY–based methods alleviate problems due to peak overlap that often plague protein backbone–based analyses of large proteins. The usefulness of methyl NMR in the investigation of CBDs is well illustrated by the CAP studies of the Kalodimos group[9–12] mentioned in Section 10.1.

In some cases, it may be desirable to complement methyl NMR studies with examinations of backbone dynamics, which also probe residues devoid of methyls, as well as with the examination of transient intermediate states that remain elusive to experimental investigations. In this respect, it is useful to consider an alternative approach to examine dynamics within full-length CBD-containing proteins, which is based on a hybrid methodology combining NMR and molecular dynamics (MD) simulations.[45] In this hybrid NMR-MD approach, a comparative assessment of results obtained from the MD simulations and from NMR is first performed to determine whether key CBD dynamic features observed from experiments can also be captured by the MD simulations. Once the MD simulations have been validated, the MD-based analysis is then extended to the examination of dynamics in the full-length protein and in the transient intermediate states.

The application of the hybrid NMR-MD approach to CBDs is illustrated by the recent residue-resolution examination of dynamics in a fully functional EPAC2 construct containing the essential regulatory CBD domain (i.e., CBD-B) and the full-length catalytic region of EPAC2—i.e., EPAC2 (280–990).[45] To gain insights into the variations of dynamics in full-length EPAC along the thermodynamic cycle of cAMP-dependent activation (Figure 10.1), MD simulations were performed starting from all four states of the EPAC2 (280–990) construct (apo/inactive, apo/active, holo/inactive, and holo/active; Figure 10.1).[45] The MD simulations were first validated by comparing CBD backbone order parameters for the apo/inactive conformation and cAMP-bound/active conformation states, as obtained from both NMR and MD simulations, which indicated that key experimentally observed trends in CBD dynamics were successfully captured by the MD simulations.[45] More significantly, the comparative simulation analyses not only confirmed the experimentally observed trends in CBD dynamics, but they also revealed unanticipated dynamic attributes of EPAC, thereby rationalizing previously unexplained aspects of EPAC activation and autoinhibition.[45] Specifically, the simulations showed that cAMP binding causes an extensive perturbation of dynamics in the distal catalytic region, possibly assisting recognition of the Rap1b substrate. In addition, analysis of the intermediate states pointed to a possible hybrid mechanism of EPAC allostery incorporating elements of both the induced fit and conformational selection models, resulting in a low-free-energy pathway by which EPAC is effectively and reversibly activated through an entropy compensation strategy. Finally, simulations revealed that the autoinhibitory interactions of EPAC were more dynamic than had previously been anticipated, leading to a revised model of autoinhibition in which dynamics fine-tune the stability of the autoinhibitory interactions, which become optimally sensitized to cAMP while at the same time ensuring that they are stable enough to avoid constitutive EPAC activation in the absence of cAMP.[45]

Besides probing dynamics beyond the experimentally observable CBD construct or states, the hybrid NMR-MD approach can also be used for exploring dynamic attributes of the CBD itself that may not be as readily detected by NMR alone. For example, in a more recent study of the $EPAC1_h$ CBD,[31] a combination of MD and NMR comparative mutational analyses was used to dissect the determinants for the EPAC apo-inactive versus apo-active autoinhibitory equilibrium. It was found by NMR that mutating a highly conserved glycine residue (i.e., G238) in the CBD β2–3 loop to alanine induced a partial shift of the apo CBD toward the active conformation, while producing a relatively negligible effect on the cAMP-bound CBD. It was concluded that G238 contributed to the autoinhibition of $EPAC1_h$ CBD, exerting an inhibitory control on all key allosteric sites of the CBD, but it remained unclear from NMR alone how G238 exerted such an effect.[31] Therefore, dynamics within the β2–3 loop and adjacent phosphate binding cassette (PBC) were further probed by comparative analyses of MD simulations performed on the wild-type apo/inactive conformation, the G238A-mutant apo/inactive conformation and the wild-type cAMP-bound/active conformation states of the CBD.

The MD simulations were first validated by computing residue-specific secondary structure probabilities for the simulated states and comparing these results with trends observed in NMR experiments.[31] Further analysis of the simulations revealed that a CH-π interaction between the Cα of G238 and the guanidinium moiety of R279 in the PBC is tightened by both the G238A mutation and cAMP binding, as reflected by a similar shortening of G238 Cα/R279 guanidinium interaction distances in both the G238A-mutant apo/inactive conformation and wild-type cAMP-bound/active conformation states relative to the wild-type apo/inactive conformation state. Furthermore, Procrustean rotation analyses of the MD simulations indicated that the dynamics of the PBC and β2–3 loop relative to one another were quenched by both the G238A mutation and cAMP binding, and NMR-derived backbone order parameters indicated a quenching of dynamics within the PBC and the β2–3 loop. Together, these results suggested that G238 controls EPAC autoinhibition through its CH-π interaction with R279, which in turn modulates the dynamics of the PBC and β2–3 loop, leading to a loss of conformational entropy upon inactive-to-active conformational transition within the apo form.[31] Based on this idea, it was further concluded that the G238A mutation stabilizes the active conformation of the apo state by prequenching dynamics within the PBC/β2–3 loop region, thereby reducing the conformational entropy loss that occurs upon inactive-to-active transition, and explaining the partial shift of the apo CBD toward the active conformation that was observed from NMR.[31] Overall, this example illustrates the potential of combined comparative NMR/MD mutational analyses for dissecting the determinants of autoinhibitory equilibria.

10.4.6 Assessment of Cyclic Nucleotide Conformational Propensities

Besides examining dynamics within the CBD itself, it is also important to consider dynamics within the cyclic nucleotide ligand, which exists in equilibrium between *syn* and *anti* conformational states (Figure 10.1). Indeed, given the selectivity of CBDs for one of these two conformations, the *syn/anti* conformational tendencies of cyclic nucleotides are expected to play a potentially functionally significant role in CBD activation. Cyclic nucleotide *syn/anti* conformational tendencies are typically probed using two-dimensional off-resonance ROESY experiments for the unbound cyclic nucleotide, or two-dimensional transfer-NOESY experiments for the CBD-bound cyclic nucleotide.[6] If the presence of the protein causes background signals that obscure the cyclic nucleotide resonances, transfer-NOESY can be combined with isotope filtering methods, which dramatically reduce contributions from isotopically labeled proteins. Once the cyclic nucleotide signals are clearly detectable, the resulting two-dimensional spectra are analyzed to check for the presence of characteristic ^{1}H-^{1}H NOE cross-peaks that arise due to the proximity of specific pairs of cyclic nucleotide hydrogen atoms.[6] For example, the H8 and H1′ cyclic nucleotide hydrogen atoms are significantly closer together in the *syn* conformation than in the *anti* conformation, whereas the H8 and H3′ hydrogen atoms are further apart in the *syn* conformation than in the *anti* conformation (Figure 10.1).[6] Therefore, the *syn* conformation would give rise to a stronger H8/H1′ NOE cross-peak and a weaker H8/H3′ NOE cross-peak than the *anti* conformation, thus providing an NOE signature that can be used to identify the predominant conformation of a cyclic nucleotide.[6] A challenging aspect of these applications is the overlap between the cAMP H8 and H2 ^{1}H NMR lines. A simple, but effective way to circumvent this problem is to preincubate the cAMP in $^{2}H_2O$, which promotes the exchange of H8 into D8. The deuteration-dependent change in the intensity of the H8 line is then exploited to separate NOE/ROE contributions from H8 versus H2. Using these approaches, it was determined that the EPAC CBD binds cAMP in a *syn* conformation, but binds cGMP in an *anti* conformation, thus revealing a previously unknown feature of EPAC selectivity for cAMP versus cGMP.[6]

10.5 CONCLUDING REMARKS AND FUTURE PERSPECTIVES

The analysis of CBDs through a combination of multiple NMR-based methods has significantly advanced our understanding of the role that dynamics play in allostery and in turn in the function of cyclic nucleotide–dependent proteins. These studies have revealed that although eukaryotic CBDs share a structurally conserved fold, they exhibit a wide range of dynamic profiles. Such differences provide a starting point for the development of ligands that are selective for specific eukaryotic CBDs, as needed to fully exploit the therapeutic potential of CBD targeting. Hence, we envision that the methods outlined here will be extended to map how drug-leads selected through screening perturb the multiple states of the allosteric cycle for the cyclic nucleotide–dependent activation of cyclic nucleotide–dependent proteins (Figure 10.1).

ACKNOWLEDGMENT

This work was funded by grant MOP-68897 from CIHR to G.M.

REFERENCES

1. Gunasekaran, K., Ma, B., and Nussinov, R. 2004. Is allostery an intrinsic property of all dynamic proteins? *Proteins: Structure, Function, and Bioinformatics* 57: 433–443.
2. Nussinov, R., and Tsai, C.J. 2013. Allostery in disease and in drug discovery. *Cell* 153: 293–305.
3. Nussinov, R., Tsai, C.J., and Ma, B. 2013. The underappreciated role of allostery in the cellular network. *Annual Review of Biophysics* 42: 169–189.
4. Ma, B., Shatsky, M., Wolfson, H.J., and Nussinov, R. 2002. Multiple diverse ligands binding at a single protein site: A matter of pre-existing populations. *Protein Science* 11: 184–197.
5. Andrusier, N., Mashiach, E., Nussinov, R., and Wolfson, H.J. 2008. Principles of flexible protein–protein docking. *Proteins: Structure, Function, and Bioinformatics* 73: 271–289.
6. Das, R., Chowdhury, S., Mazhab-Jafari, M.T., SilDas, S., Selvaratnam, R., and Melacini, G. 2009. Dynamically driven ligand selectivity in cyclic nucleotide binding domains. *Journal of Biological Chemistry* 284: 23682–23696.
7. Selvaratnam, R., Chowdhury, S., VanSchouwen, B., and Melacini, G. 2011. Mapping allostery through the covariance analysis of NMR chemical shifts. *Proceedings of the National Academy of Sciences of the United States of America* 108: 6133–6138.
8. Akimoto, M., Selvaratnam, R., McNicholl, E.T., Verma, G., Taylor, S.S., and Melacini, G. 2013. Signaling through dynamic linkers as revealed by PKA. *Proceedings of the National Academy of Sciences of the United States of America* 110: 14231–14236.
9. Popovych, N., Sun, S., Ebright, R.H., and Kalodimos, C.G. 2006. Dynamically driven protein allostery. *Nature Structural & Molecular Biology* 13: 831–838.
10. Tzeng, S.R., and Kalodimos, C.G. 2009. Dynamic activation of an allosteric regulatory protein. *Nature* 462: 368–372.
11. Tzeng, S.R., and Kalodimos, C.G. 2013. Allosteric inhibition through suppression of transient conformational states. *Nature Chemical Biology* 9: 462–465.
12. Tzeng, S.R., and Kalodimos, C.G. 2012. Protein activity regulation by conformational entropy. *Nature* 488: 236–240.
13. Das, R., Mazhab-Jafari, M.T., Chowdhury, S., Das, S.S., Selvaratnam, R., and Melacini, G. 2008. Entropy-driven cAMP-dependent allosteric control of inhibitory interactions in exchange proteins directly activated by cAMP. *Journal of Biological Chemistry* 283: 19691–19703.
14. Boulton, S., Akimoto, M., VanSchouwen, B., Moleschi, K., Selvaratnam, R., and Melacini, G. 2014. Tapping the translation potential of cAMP signalling: Molecular basis for selectivity in cAMP agonism and antagonism as revealed by NMR. *Biochemical Society Transactions* 42: 302–307.
15. Selvaratnam, R., Mazhab-Jafari, M.T., Das, R., and Melacini, G. 2012. The auto-inhibitory role of the EPAC hinge helix as mapped by NMR. *PLoS One* 7: e48707.
16. Goto, N.K., and Kay, L.E. 2000. New developments in isotope labeling strategies for protein solution NMR spectroscopy. *Current Opinion in Structural Biology* 10: 585–592.
17. Tugarinov, V., Hwang, P.M., and Kay, L.E. 2004. Nuclear magnetic resonance spectroscopy of high-molecular-weight proteins. *Annual Review of Biochemistry* 73: 107–146.

18. Marion, D., Driscoll, P.C., Kay, L.E., Wingfield, P.T., Bax, A., Gronenborn, A.M., and Clore, G.M. 1989. Overcoming the overlap problem in the assignment of proton NMR spectra of larger proteins by use of three-dimensional heteronuclear proton-nitrogen-15 Hartmann-Hahn-multiple quantum coherence and nuclear Overhauser-multiple quantum coherence spectroscopy: Application to interleukin 1.beta. *Biochemistry* 28: 6150–6156.
19. Ikura, M., Kay, L.E., and Bax, A. 1990. A novel approach for sequential assignment of 1H, 13C, and 15N spectra of larger proteins: Heteronuclear triple-resonance three-dimensional NMR spectroscopy. Application to calmodulin. *Biochemistry* 29: 4659–4667.
20. Gardner, K.H., and Kay, L.E. 1998. The use of 2H, 13C, 15N multidimensional NMR to study the structure and dynamics of proteins. *Annual Review of Biophysics and Biomolecular Structure* 27: 357–406.
21. Craven, K.B., and Zagotta, W.N. 2006. CNG and HCN channels: Two peas, one pod. *Annual Review of Physiology* 68: 375–401.
22. Zagotta, W.N., Olivier, N.B., Black, K.D., Young, E.C., Olson, R., and Gouaux, E. 2003. Structural basis for modulation and agonist specificity of HCN pacemaker channels. *Nature* 425: 200–205.
23. Biel, M., Wahl-Schott, C., Michalakis, S., and Zong, X. 2009. Hyperpolarization-activated cation channels: From genes to function. *Physiological Reviews* 89: 847–885.
24. Mazhab-Jafari, M.T., Das, R., Fotheringham, S.A., SilDas, S., Chowdhury, S., and Melacini, G. 2007. Understanding cAMP-dependent allostery by NMR spectroscopy: Comparative analysis of the EPAC1 cAMP-binding domain in its apo and cAMP-bound states. *Journal of the American Chemical Society* 129: 14482–14492.
25. Das, R., Esposito, V., Abu-Abed, M., Anand, G.S., Taylor, S.S., and Melacini, G. 2007. cAMP activation of PKA defines an ancient signaling mechanism. *Proceedings of the National Academy of Sciences of the United States of America* 104: 93–98.
26. Huang, G., Kim, J.J., Reger, A.S., Lorenz, R., Moon, E.W., Zhao, C., Casteel, D.E., Bertinetti, D., VanSchouwen, B., Selvaratnam, R., Pflugrath, J.W., Sankaran, B., Melacini, G., Herberg, F.W., and Kim, C. 2014. Structural basis for cyclic-nucleotide selectivity and cGMP-selective activation of PKG I. *Structure* 22: 116–124.
27. Das, R., and Melacini, G. 2007. A model for agonism and antagonism in an ancient and ubiquitous cAMP binding domain. *Journal of Biological Chemistry* 282: 581–593.
28. Boyer, J.A., Clay, C.J., Luce, K.S., Edgell, M.H., and Lee, A.L. 2010. Detection of native-state non-additivity in double mutant cycles via hydrogen exchange. *Journal of the American Chemical Society* 132: 8010–8019.
29. Kim, J.J., Huang, G.Y., Rieger, R., Koller, A., Chow, D.C., and Kim, C. 2014. A protocol for expression and purification of cyclic-nucleotide free protein in *Escherichia coli*. In *Cyclic Nucleotide Signaling*, ed. Cheng, X. CRC Press, Boca Raton, FL.
30. Schünke, S., Stoldt, M., Lecher, J., Kaupp, U.B., and Willbold, D. 2011. Structural insights into conformational changes of a cyclic nucleotide-binding domain in solution from *Mesorhizobium loti* K1 channel. *Proceedings of the National Academy of Sciences of the United States of America* 108: 6121–6126.
31. Selvaratnam, R., VanSchouwen, B., Fogolari, F., Mazhab-Jafari, M.T., Das, R., and Melacini, G. 2012. The projection analysis of NMR chemical shifts reveals extended EPAC autoinhibition determinants. *Biophysical Journal* 102: 630–639.
32. Gillard, R.D. 1963. Circular dichroism: A review. *Analyst* 88: 825–828.
33. Eghbalnia, H.R., Wang, L., Bahrami, A., Assadi, A., and Markley, J.L. 2005. Protein energetic conformational analysis from NMR chemical shifts (PECAN) and its use in determining secondary structural elements. *Journal of Biomolecular NMR* 32: 71–81.
34. Palmer, A.G. 2004. NMR characterization of the dynamics of biomacromolecules. *Chemical Reviews* 104: 3623–3640.

35. Farrow, N.A., Muhandiram, R., Singer, A.U., Pascal, S.M., Kay, C.M., Gish, G., Shoelson, S.E., Pawson, T., Forman-Kay, J.D., and Kay, L.E. 1994. Backbone dynamics of a free and a phosphopeptide-complexed Src Homology 2 domain studied by ^{15}N NMR relaxation. *Biochemistry* 33: 5984–6003.
36. Kleckner, I.R., and Foster, M.P. 2010. An introduction to NMR-based approaches for measuring protein dynamics. *Biochimica et Biophysica Acta* 1814: 942–968.
37. Korzhnev, D.M., and Kay, L.E. 2008. Probing invisible, low-populated states of protein molecules by relaxation dispersion NMR spectroscopy: An application to protein folding. *Accounts of Chemical Research* 41: 442–451.
38. Kay, L.E., Nicholson, L.A., Delaglio, F., Bax, A., and Torchia, D.A. 1992. Pulse sequences for removal of the effects of cross correlation between dipolar and chemical-shift anisotropy relaxation mechanisms on the measurement of heteronuclear T_1 and T_2 values in proteins. *Journal of Magnetic Resonance* 97: 359–375.
39. Kay, L.E. 1998. Protein dynamics from NMR. *Biochemistry and Cell Biology* 76: 145–152.
40. Hansen, D.F., Vallurupalli, P., and Kay, L.E. 2008. An improved 15N relaxation dispersion experiment for the measurement of millisecond time-scale dynamics in proteins. *Journal of Physical Chemistry B* 112: 5898–5904.
41. Li, S., Tsalkova, T., White, M.A., Mei, F.C., Liu, T., Wang, D., Woods, V.L., and Cheng, X. 2011. Mechanism of intracellular cAMP sensor Epac2 activation: cAMP-induced conformational changes identified by amide hydrogen/deuterium exchange mass spectrometry (DXMS). *Journal of Biological Chemistry* 286: 17889–17897.
42. Hamuro, Y., Anand, G.S., Kim, J.S., Juliano, C., Stranz, D.D., Taylor, S.S., and Woods, V.L. 2004. Mapping intersubunit interactions of the regulatory subunit (RIα) in the type I holoenzyme of protein kinase A by amide hydrogen/deuterium exchange mass spectrometry (DXMS). *Journal of Molecular Biology* 340: 1185–1196.
43. McNicholl, E.T., Das, R., SilDas, S., Taylor, S.S., and Melacini, G. 2010. Communication between tandem cAMP binding domains in the regulatory subunit of protein kinase A-Iα as revealed by domain-silencing mutations. *Journal of Biological Chemistry* 285: 15523–15537.
44. Das, R., Abu-Abed, M., and Melacini, G. 2006. Mapping allostery through equilibrium perturbation NMR spectroscopy. *Journal of the American Chemical Society* 128: 8406–8407.
45. VanSchouwen, B., Selvaratnam, R., Fogolari, F., and Melacini, G. 2011. Role of dynamics in the auto-inhibition and activation of the exchange protein directly activated by cyclic AMP (EPAC). *Journal of Biological Chemistry* 286: 42655–42669.
46. Ollerenshaw, J.E., Tugarinov, V., and Kay, L.E. 2003. Methyl TROSY: Explanation and experimental verification. *Magnetic Resonance in Chemistry* 41: 843–852.

11 A Protocol for Expression and Purification of Cyclic Nucleotide–Free Protein in *Escherichia coli*

Jeong Joo Kim, Gilbert Y. Huang,* Robert Rieger, Antonius Koller, Dar-Chone Chow, and Choel Kim*

CONTENTS

11.1 INTRODUCTION

11.1.1 Role of Cyclic Nucleotide Binding Domains

The cyclic nucleotides cyclic adenosine monophosphate (cAMP) and cyclic guanosine monophosphate (cGMP) are ubiquitous second messengers that relay signals in almost all forms of life.[1] Binding of cyclic nucleotides to their receptors results in subsequent modulation of the receptor's activity, which propagates the signal to downstream effectors. Although the rich structural diversity occurring among these receptors includes drastically different folds such as G-Protein-Couped-Receptors (GPCRs) and cGMP-specific phosphodiesterases, adenylyl cyclases and FhlA (GAF) domains,[2,3] the canonical cyclic

* These authors contributed equally.

nucleotide–binding (CNB) domain is the most common and well characterized.[4] CNB domains are ancient domains occurring in diverse multidomain proteins such as kinases,[5] transcription factors,[6–8] ion channels,[9] nucleotidyl cyclases,[10] guanine exchange factors,[11] and acetyltransferases,[12] often functioning as allosteric modulators of enzyme activity.[13] These domains serve as conformational switches that structurally rearrange upon binding of the cyclic nucleotides. In mammalian physiology, cyclic nucleotides are critical second messengers that relay signals through these enzymes, regulating diverse physiological processes such as vision, learning and memory, vasodilation, and metabolism.[1] Consistent with their importance in physiology, mutations in proteins that relay cyclic nucleotide signaling have been linked to a diverse cohort of diseases including achromatopsia,[14,15] epilepsy,[16] Carney complex,[17,18] fibrolamellar hepatocellular carcinoma,[19] thoracic aortic aneurysms, and acute aortic dissections.[20] The universal importance of cyclic nucleotide signaling in the normal function of mammalian cells has stimulated decades of research into the CNB domain-containing proteins that relay these signals and firmly established some of these proteins as drug targets in the treatment of disease.[21,22]

11.1.2 Historical Challenges of Studying PKA and PKG

The most well characterized mammalian receptors of cyclic nucleotides are the homologous cAMP-dependent (PKA) and cGMP-dependent protein kinases (PKG). PKA and PKG were initially purified from rabbit muscle and lobster tail muscle, respectively.[23,24] Later, PKA was shown to consist of an R subunit and a C subunit, which tightly associate in the absence of cAMP, and dissociate upon cAMP binding to the R subunit, allowing the C subunit to phosphorylate downstream substrates.[5] PKG similarly contains a regulatory (R) domain and a catalytic (C) domain that changes conformation in response to cGMP, although these are fused into a single polypeptide chain.

Early efforts to purify these proteins proved challenging and required unconventional methodologies—for example, because of tight association between the R and C subunits in PKA, an early method for purifying isolated R subunits required affinity chromatography using cAMP-cross-linked resin, followed by elution of the protein using free cAMP.[25] This early protocol later became widely used in the community. Unfortunately, once exposed to cAMP, R subunit required denaturation and refolding using 8 M of urea to remove cAMP.[26] Historically, these observations from PKA were some of the first to hint at the difficulty of purifying cyclic nucleotide–free CNB domains—indeed, later studies showed that cAMP could still associate with the full length and various fragments of the R subunit even in the molar concentrations of urea.[27,28] Although understanding the mechanistic details of cyclic nucleotide signaling requires purified protein in both nucleotide-free and nucleotide-bound states, this was challenging because the high affinity of some CNB domains for cyclic nucleotides often meant the sample would copurify with cyclic nucleotides.[29,30]

11.1.3 Past Structural Studies of CNB Domains: Challenges and Limitations

Many groups have pursued structural studies of CNB domains to understand how they allosterically regulate activity, resulting in many ligand-bound structures.[31–39]

In contrast, relatively few ligand-free structures are available due to the problem of cyclic nucleotide copurification, and many of these are stabilized in their inactive state by other domains.[31,40,41] Some groups have acknowledged the problems caused by cyclic nucleotide copurification and sidestepped them by generating mutants that do not bind cyclic nucleotides, modeling unbound structures after the structures of these mutants.[42,43] This is an imperfect solution because it assumes that the mutant constructs mimic wild type in every feature except the ability to bind cyclic nucleotides, which is a difficult assumption to validate.

Another alternative approach was taken in a recent study on the RIα subunit of PKA, which attempted to address this problem by modifying the previously described affinity purification technique so that the RIα bound to the cAMP affinity column was eluted with cGMP rather than cAMP.[44] Hypothetically, because RIα has a lower affinity for cGMP than cAMP, elution with cGMP would allow it to be dialyzed out of the sample before crystallization. However, this was only partly true because the sample ultimately crystallized with a cGMP-occupied A site and an unoccupied B site, suggesting that the A site of RIα still retained enough affinity for cGMP for it to persist over the course of purification. The partial success of this approach reflects the challenges that may occur as a result of using cAMP affinity resin during the purification.

In a more extreme case, our group recently characterized a CNB domain in PKG that had high affinity for both cGMP and cAMP.[30] The affinity for cAMP was so high that the CNB domain crystallized with no added ligand, showing bacterial cAMP occupying the cGMP pocket. These observations not only highlight the difficulty of using cyclic nucleotide affinity resins but also raise the possibility that exposing CNB domains to any cyclic nucleotide at all can result in an occupied domain that persists over the course of purification. Remarkably, the affinity chromatography protocol utilizing cAMP affinity resin was also adapted for use in the PKG system for purification of recombinant enzyme, suggesting a need to reexamine previous studies describing CNB or cyclic nucleotide–dependent activation.[45,46] Taken together, these observations draw attention to the unmet need for sound protocols that can yield purified protein samples that are also free of cyclic nucleotides.

In this chapter, we describe a simple protocol for expressing and purifying protein that is free of cyclic nucleotides. This protocol can be adapted for any protein that could potentially bind cyclic nucleotides produced in an expression host and provides a simple alternative to the current techniques that are associated with low yield and cyclic nucleotide contamination.

11.2 EXPRESSION AND PURIFICATION OF CYCLIC NUCLEOTIDE–FREE SAMPLE

11.2.1 Finding a Cyclic Nucleotide–Free Expression Host

To address the issue of cyclic nucleotide contamination, we searched the literature for possible cell lines that might serve as expression hosts capable of producing cyclic nucleotide–free samples. Fortuitously, we were able to acquire TP2000 cells (a gift from Dr. Carmen Dessauer, UTHSC).[47] The TP2000 strain is an *Escherichia coli* K12 strain lacking adenylate cyclase activity that was initially generated to map the adenylate cyclase locus in the K12 genome and later to study cAMP-dependent gene regulation in *E. coli*, resulting in a number of genetic studies characterizing the function of the cAMP pathway components in this strain.[48–52] Therefore, we generated TP2000 competent cells and tested if TP2000 could express cyclic nucleotide–free CNB domains.

11.2.2 Comparative Expression and Purification of PKG Iβ 92-363 from BL21(DE3) and TP2000

To test the ability of TP2000 to generate cyclic nucleotide–free protein, we cloned PKG Iβ 92-363 into a pQTEV[53] vector and transformed it into both TP2000 and BL21(DE3), a commonly used *E. coli* expression host. PKG Iβ 92-363 is a deletion mutant of PKG Iβ that contains both of PKG I's CNB domains. In both cell hosts, cells were grown to an OD_{600} of 0.8 at 20°C and induced with 0.5 mM IPTG. Cells were then pelleted by centrifugation, resuspended in lysis buffer (10 mM phosphate buffer, pH 7.0; 150 mM NaCl, 1 mM β-mercaptoethanol), and lysed with a cell disrupter. Lysate was cleared by ultracentrifugation and run through a Bio-Rad Ni-NTA cartridge before elution with 250 mM imidazole. Samples were then dialyzed against low-salt buffer (25 mM phosphate buffer, pH 7.0; 100 mM NaCl, 1 mM β-mercaptoethanol) to remove imidazole and loaded onto a Mono Q column (GE Healthcare). All samples were eluted over a gradient with increasing salt concentration.

To test the efficiency of TP2000 in producing cAMP-free samples against traditional methods of cAMP removal, one of the samples expressed in BL21(DE3) was denatured and refolded by the addition of solid guanidium hydrochloride to 6 M and incubated overnight in buffer containing 6 M guanidium hydrochloride. This was followed by serial dialysis against buffer containing 3 M guanidium hydrochloride and no guanidium hydrochloride. Following complete removal of guanidium hydrochloride, the sample was loaded onto a Mono Q column as previously described.

Samples from TP2000 and BL21(DE3) reproducibly produced very different elution profiles during the ion exchange step (Figure 11.1). Specifically, PKG Iβ 92-363 from TP2000 elutes as a single peak, whereas nontreated and refolded PKG Iβ 92-363 from BL21(DE3) each elute as several peaks that have significantly higher absorption at 260 nm. We speculated that the peaks showing high 260 nm absorption might represent differentially cAMP-occupied species. To test this, we collected samples from each peak and analyzed them using mass spectrometry against the TP2000 sample.

11.2.3 Mass Spectrometry Analysis of Cyclic Nucleotide Contamination

Protein samples from each of these peaks were treated with acetonitrile to extract bound cAMP and supernatants were analyzed for the presence of cAMP using liquid chromatography/mass spectrometry (LC/MS). The samples were loaded onto a C18 column and eluted into the mass spectrometer with an acetonitrile gradient. The mass spectrometer (a Thermo TSQ Access triple quadrupole mass spectrometer) was operated in the positive ion mode and set to a transition of m/z 330 to m/z 136 for cAMP analysis. A calibration curve was generated for cAMP from 1 to 50 ng/mL and produced an R^2 of 0.9937. As shown in Figures 11.1 and 11.2, we found that

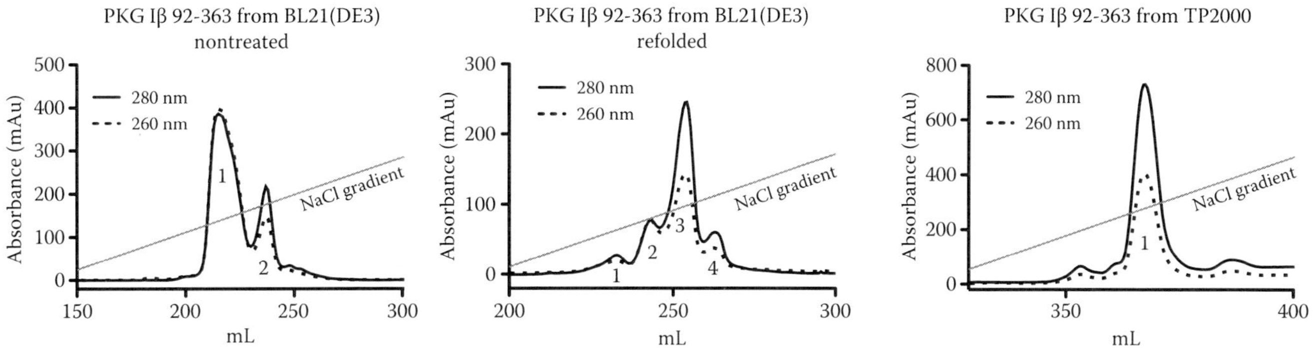

Cell lines	Samples	Peak no.	cAMP concentration (%) vs. protein concentration
BL21(DE3)	Nontreated	1	133.17
		2	66.61
	Refolded	1	148.36
		2	112.94
		3	11.08
		4	8.80
TP2000 Δ*cya*	Nontreated	1	0 .00

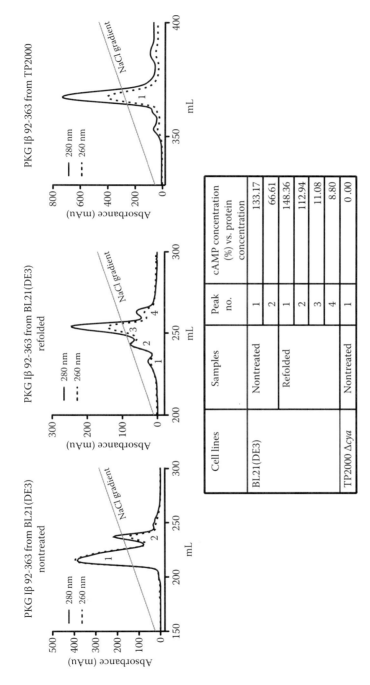

FIGURE 11.1 Ion exchange (Mono Q) profiles of human PKG Iβ 92-363 samples and their cAMP contamination ratios (%). Absorption at 260 nm was used as a qualitative indicator of cAMP contamination during purification, and cAMP concentrations were later measured by mass spectrometry. Samples expressed in BL21(DE3) showed a different elution profile on a Mono Q anion exchange resin, whereas samples expressed in TP2000 produced one major peak that had no detectable cAMP.

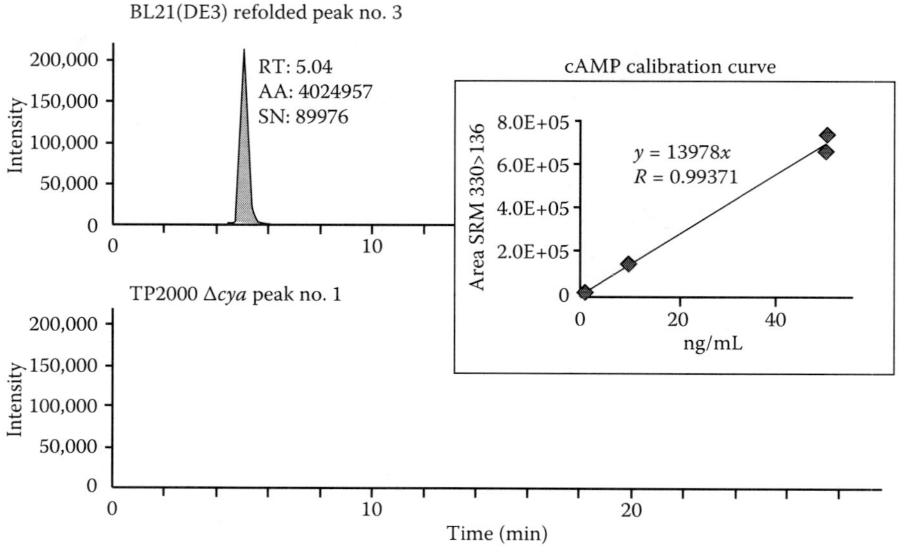

FIGURE 11.2 Representative cAMP elution traces for two samples, BL21(DE3) refolded peak no. 3 (top) and TP2000 Δ*cya* peak no. 1 (bottom) indicating the amount of cAMP in each sample. Both elution traces were calibrated to a top intensity of 200,000. For BL21(DE3) refolded peak no. 3, 50 μL of 36.1 μM protein and for TP2000 Δ*cya* peak no. 1, 50 μL of 35 μM protein was used to measure cAMP concentration. The inset shows duplicates of cAMP standards analyzed with the same method and plotted to show the linearity of cAMP response.

whereas each peak from both native and refolded samples in BL21(DE3) contained variable amounts of cAMP, the major peak from TP2000 contained no detectable cAMP. Thus, we conclude that expression and purification in TP2000 yields a cAMP-free sample.

11.2.4 Isothermal Titration Calorimetry Analysis of PKG Iβ 92-363

We previously attempted to measure binding affinities for PKG Iβ 92-363 purified from BL21(DE3) using isothermal titration calorimetry (ITC) but could not obtain reproducible measurements (data not shown). Our purification and mass spectrometry results suggested that this might be caused by cAMP contamination from the BL21(DE3) expression host. To test if expression/purification from TP2000 could address this issue, we obtained ITC titrations for samples purified from the TP2000 cell. We found that our titrations produced curves that fit a two-site model (Figure 11.3). Consistent with our previous measurements for the isolated domains, the first site binds with high affinity to both cGMP and cAMP, whereas the second site has high affinity only for cGMP compared with cAMP, showing 200-fold selectivity for cGMP.[30,35] Taken together, these ITC measurements validate that our method for expression and purification in TP2000 is a simple protocol for producing a pure,

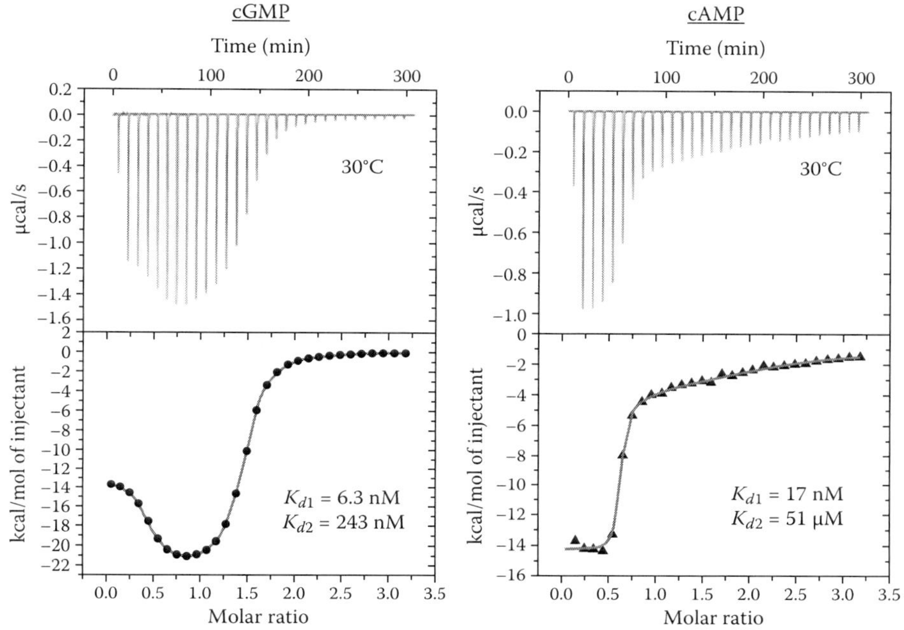

FIGURE 11.3 Isothermal titration calorimetry titration of human PKG Iβ 92-363 samples purified from TP2000. Measurements were performed using a VP-ITC calorimeter (MicroCal LLC) at 30°C. The protein was placed in the sample cell at a concentration of 15 µM in 10 mM Tris (pH 8.0) and 150 mM NaCl. Cyclic nucleotides were dissolved in the same buffer at a concentration of 250 µM and injected 5 µL at a time. Data was processed using Origin.

well-folded sample that retains properties in agreement with the literature, but is free of cyclic nucleotide.

11.3 CONCLUSIONS AND FUTURE DIRECTIONS

To conclude, these data demonstrate that expression in TP2000 is a simple, cost-effective protocol for producing cyclic nucleotide–free samples, and that this protocol is generally useful for any recombinant construct that may bind with high affinity to endogenous cyclic nucleotides. However, limitations remain that require discussion. Many commonly used bacterial expression vectors rely on a T7 promoter and were initially optimized for use in the BL21(DE3) expression strain, which has T7 polymerase incorporated into its genome.[54,55] Importantly, T7 polymerase expression is required for the use of these vectors, so they are not currently compatible with the TP2000 cell line.[56] In contrast, in our protocol described here, the expression vector pQTEV contains a T5-lac promoter that can be recognized by *E. coli* polymerase upon IPTG induction.[53,57] In the future, we anticipate that this protocol can be adapted for new cell lines that are compatible with a broader range of expression vectors, although this requires either generating derivatives of BL21(DE3) that have adenylate cyclase disabled or generating derivatives of TP2000 that have T7 polymerase incorporated into their genomes.

ACKNOWLEDGMENTS

We thank Dr. Carmen Dessauer (UTHSC, Houston, Texas) for the generous gift of the TP2000 cells. This work is supported by NIH grants R01GM090161 and R21HL111953 to Choel Kim. Gilbert Huang is supported by a training grant from the Houston Area Molecular Biophysics Program (NIH T32GM008280, PI Theodore Wensel).

REFERENCES

1. Beavo, J. A., and Brunton, L. L. Cyclic nucleotide research—Still expanding after half a century. *Nat Rev Mol Cell Biol* 3 (9), 710–718, 2002.
2. Prabhu, Y., and Eichinger, L. The *Dictyostelium* repertoire of seven transmembrane domain receptors. *Eur J Cell Biol* 85 (9–10), 937–946, 2006.
3. Heikaus, C. C., Pandit, J., and Klevit, R. E. Cyclic nucleotide binding GAF domains from phosphodiesterases: Structural and mechanistic insights. *Structure* 17 (12), 1551–1557, 2009.
4. Rehmann, H., Wittinghofer, A., and Bos, J. L. Capturing cyclic nucleotides in action: Snapshots from crystallographic studies. *Nat Rev Mol Cell Biol* 8 (1), 63–73, 2007.
5. Pearce, L. R., Komander, D., and Alessi, D. R. The nuts and bolts of AGC protein kinases. *Nat Rev Mol Cell Biol* 11 (1), 9–22, 2010.
6. Cordes, T. J., Worzalla, G. A., Ginster, A. M., and Forest, K. T. Crystal structure of the Pseudomonas aeruginosa virulence factor regulator. *J Bacteriol* 193 (16), 4069–4074, 2011.
7. McKay, D. B., and Steitz, T. A. Structure of catabolite gene activator protein at 2.9 A resolution suggests binding to left-handed B-DNA. *Nature* 290 (5809), 744–749, 1981.
8. Won, H. S., Lee, Y. S., Lee, S. H., and Lee, B. J. Structural overview on the allosteric activation of cyclic AMP receptor protein. *Biochim Biophys Acta* 1794 (9), 1299–1308, 2009.
9. Craven, K. B., and Zagotta, W. N. CNG and HCN channels: Two peas, one pod. *Annu Rev Physiol* 68, 375–401, 2006.
10. An, S. Q., Chin, K. H., Febrer, M., McCarthy, Y., Yang, J. G., Liu, C. L., Swarbreck, D., Rogers, J., Maxwell Dow, J., Chou, S. H., and Ryan, R. P. A cyclic GMP-dependent signalling pathway regulates bacterial phytopathogenesis. *EMBO J* 32 (18), 2430–2438, 2013.
11. Bos, J. L. Epac proteins: Multi-purpose cAMP targets. *Trends Biochem Sci* 31 (12), 680–686, 2006.
12. Lee, H. J., Lang, P. T., Fortune, S. M., Sassetti, C. M., and Alber, T. Cyclic AMP regulation of protein lysine acetylation in *Mycobacterium tuberculosis*. *Nat Struct Mol Biol* 19 (8), 811–818, 2012.
13. Kannan, N., Wu, J., Anand, G. S., Yooseph, S., Neuwald, A. F., Venter, J. C., and Taylor, S. S. Evolution of allostery in the cyclic nucleotide binding module. *Genome Biol* 8 (12), R264, 2007.
14. Wissinger, B., Gamer, D., Jagle, H., Giorda, R., Marx, T., Mayer, S., Tippmann, S., Broghammer, M., Jurklies, B., Rosenberg, T., Jacobson, S. G., Sener, E. C., Tatlipinar, S., Hoyng, C. B., Castellan, C., Bitoun, P., Andreasson, S., Rudolph, G., Kellner, U., Lorenz, B., Wolff, G., Verellen-Dumoulin, C., Schwartz, M., Cremers, F. P., Apfelstedt-Sylla, E., Zrenner, E., Salati, R., Sharpe, L. T., and Kohl, S. CNGA3 mutations in hereditary cone photoreceptor disorders. *Am J Hum Genet* 69 (4), 722–737, 2001.

15. Kohl, S., Marx, T., Giddings, I., Jagle, H., Jacobson, S. G., Apfelstedt-Sylla, E., Zrenner, E., Sharpe, L. T., and Wissinger, B. Total colourblindness is caused by mutations in the gene encoding the alpha-subunit of the cone photoreceptor cGMP-gated cation channel. *Nat Genet* 19 (3), 257–259, 1998.
16. Nava, C., Dalle, C., Rastetter, A., Striano, P., de Kovel, C. G., Nabbout, R., Cances, C., Ville, D., Brilstra, E. H., Gobbi, G., Raffo, E., Bouteiller, D., Marie, Y., Trouillard, O., Robbiano, A., Keren, B., Agher, D., Roze, E., Lesage, S., Nicolas, A., Brice, A., Baulac, M., Vogt, C., El Hajj, N., Schneider, E., Suls, A., Weckhuysen, S., Gormley, P., Lehesjoki, A. E., De Jonghe, P., Helbig, I., Baulac, S., Zara, F., Koeleman, B. P., Euro, E. R. E. S. C., Haaf, T., Leguern, E., and Depienne, C. De novo mutations in HCN1 cause early infantile epileptic encephalopathy. *Nat Genet* 46 (6), 640–645, 2014.
17. Casey, M., Vaughan, C. J., He, J., Hatcher, C. J., Winter, J. M., Weremowicz, S., Montgomery, K., Kucherlapati, R., Morton, C. C., and Basson, C. T. Mutations in the protein kinase A R1alpha regulatory subunit cause familial cardiac myxomas and Carney complex. *J Clin Invest* 106 (5), R31–R38, 2000.
18. Kirschner, L. S., Carney, J. A., Pack, S. D., Taymans, S. E., Giatzakis, C., Cho, Y. S., Cho-Chung, Y. S., and Stratakis, C. A. Mutations of the gene encoding the protein kinase A type I-alpha regulatory subunit in patients with the Carney complex. *Nat Genet* 26 (1), 89–92, 2000.
19. Honeyman, J. N., Simon, E. P., Robine, N., Chiaroni-Clarke, R., Darcy, D. G., Lim, I. I., Gleason, C. E., Murphy, J. M., Rosenberg, B. R., Teegan, L., Takacs, C. N., Botero, S., Belote, R., Germer, S., Emde, A. K., Vacic, V., Bhanot, U., LaQuaglia, M. P., and Simon, S. M. Detection of a recurrent DNAJB1-PRKACA chimeric transcript in fibrolamellar hepatocellular carcinoma. *Science* 343 (6174), 1010–1014, 2014.
20. Guo, D. C., Regalado, E., Casteel, D. E., Santos-Cortez, R. L., Gong, L., Kim, J. J., Dyack, S., Horne, S. G., Chang, G., Jondeau, G., Boileau, C., Coselli, J. S., Li, Z., Leal, S. M., Shendure, J., Rieder, M. J., Bamshad, M. J., Nickerson, D. A., GenTAC Registry Consortium; National Heart, Lung, and Blood Institute Grand Opportunity Exome Sequencing Project, Kim, C., and Milewicz, D. M. Recurrent gain-of-function mutation in PRKG1 causes thoracic aortic aneurysms and acute aortic dissections. *Am J Hum Genet* 93 (2), 398–404, 2013.
21. Schlossmann, J., and Hofmann, F. cGMP-dependent protein kinases in drug discovery. *Drug Discov Today* 10 (9), 627–634, 2005.
22. Postea, O., and Biel, M. Exploring HCN channels as novel drug targets. *Nat Rev Drug Discov* 10 (12), 903–914, 2011.
23. Walsh, D. A., Perkins, J. P., and Krebs, E. G. An adenosine 3′,5′-monophosphate-dependant protein kinase from rabbit skeletal muscle. *J Biol Chem* 243 (13), 3763–3765, 1968.
24. Kuo, J. F., and Greengard, P. Cyclic nucleotide-dependent protein kinases. VI. Isolation and partial purification of a protein kinase activated by guanosine 3′,5′-monophosphate. *J Biol Chem* 245 (10), 2493–2498, 1970.
25. Dills, W. L., Jr., Beavo, J. A., Bechtel, P. J., and Krebs, E. G. Purification of rabbit skeletal muscle protein kinase regulatory subunit using cyclic adenosine-3′:5′-monophosphate affinity chromatography. *Biochem Biophys Res Commun* 62 (1), 70–77, 1975.
26. Builder, S. E., Beavo, J. A., and Krebs, E. G. The mechanism of activation of bovine skeletal muscle protein kinase by adenosine 3′:5′-monophosphate. *J Biol Chem* 255 (8), 3514–3519, 1980.
27. Akimoto, M., Selvaratnam, R., McNicholl, E. T., Verma, G., Taylor, S. S., and Melacini, G. Signaling through dynamic linkers as revealed by PKA. *Proc Natl Acad Sci U S A* 110 (35), 14231–14236, 2013.
28. Leon, D. A., Dostmann, W. R., and Taylor, S. S. Unfolding of the regulatory subunit of cAMP-dependent protein kinase I. *Biochemistry* 30 (12), 3035–3040, 1991.

29. Osborne, B. W., Wu, J., McFarland, C. J., Nickl, C. K., Sankaran, B., Casteel, D. E., Woods, V. L., Jr., Kornev, A. P., Taylor, S. S., and Dostmann, W. R. Crystal structure of cGMP-dependent protein kinase reveals novel site of interchain communication. *Structure* 19 (9), 1317–1327, 2011.
30. Kim, J. J., Casteel, D. E., Huang, G., Kwon, T. H., Ren, R. K., Zwart, P., Headd, J. J., Brown, N. G., Chow, D. C., Palzkill, T., and Kim, C. Co-crystal structures of PKG Ibeta (92–227) with cGMP and cAMP reveal the molecular details of cyclic-nucleotide binding. *PLoS One* 6 (4), e18413, 2011.
31. Kim, C., Xuong, N. H., and Taylor, S. S. Crystal structure of a complex between the catalytic and regulatory (RIalpha) subunits of PKA. *Science* 307 (5710), 690–696, 2005.
32. Su, Y., Dostmann, W. R., Herberg, F. W., Durick, K., Xuong, N. H., Ten Eyck, L., Taylor, S. S., and Varughese, K. I. Regulatory subunit of protein kinase A: Structure of deletion mutant with cAMP binding domains. *Science* 269 (5225), 807–813, 1995.
33. Zagotta, W. N., Olivier, N. B., Black, K. D., Young, E. C., Olson, R., and Gouaux, E. Structural basis for modulation and agonist specificity of HCN pacemaker channels. *Nature* 425 (6954), 200–205, 2003.
34. Weber, I. T., and Steitz, T. A. Structure of a complex of catabolite gene activator protein and cyclic AMP refined at 2.5 A resolution. *J Mol Biol* 198 (2), 311–326, 1987.
35. Huang, G. Y., Kim, J. J., Reger, A. S., Lorenz, R., Moon, E. W., Zhao, C., Casteel, D. E., Bertinetti, D., Vanschouwen, B., Selvaratnam, R., Pflugrath, J. W., Sankaran, B., Melacini, G., Herberg, F. W., and Kim, C. Structural basis for cyclic-nucleotide selectivity and cGMP-selective activation of PKG I. *Structure* 22 (1), 116–124, 2014.
36. Flynn, G. E., Black, K. D., Islas, L. D., Sankaran, B., and Zagotta, W. N. Structure and rearrangements in the carboxy-terminal region of SpIH channels. *Structure* 15 (6), 671–682, 2007.
37. Diller, T. C., Madhusudan, Xuong, N. H., and Taylor, S. S. Molecular basis for regulatory subunit diversity in cAMP-dependent protein kinase: Crystal structure of the type II beta regulatory subunit. *Structure* 9 (1), 73–82, 2001.
38. Rehmann, H., Arias-Palomo, E., Hadders, M. A., Schwede, F., Llorca, O., and Bos, J. L. Structure of Epac2 in complex with a cyclic AMP analogue and RAP1B. *Nature* 455 (7209), 124–127, 2008.
39. Rehmann, H., Prakash, B., Wolf, E., Rueppel, A., de Rooij, J., Bos, J. L., and Wittinghofer, A. Structure and regulation of the cAMP-binding domains of Epac2. *Nat Struct Biol* 10 (1), 26–32, 2003.
40. Rehmann, H., Das, J., Knipscheer, P., Wittinghofer, A., and Bos, J. L. Structure of the cyclic-AMP-responsive exchange factor Epac2 in its auto-inhibited state. *Nature* 439 (7076), 625–628, 2006.
41. Zhang, P., Smith-Nguyen, E. V., Keshwani, M. M., Deal, M. S., Kornev, A. P., and Taylor, S. S. Structure and allostery of the PKA RIIbeta tetrameric holoenzyme. *Science* 335 (6069), 712–716, 2012.
42. Kim, C., Cheng, C. Y., Saldanha, S. A., and Taylor, S. S. PKA-I holoenzyme structure reveals a mechanism for cAMP-dependent activation. *Cell* 130 (6), 1032–1043, 2007.
43. Clayton, G. M., Silverman, W. R., Heginbotham, L., and Morais-Cabral, J. H. Structural basis of ligand activation in a cyclic nucleotide regulated potassium channel. *Cell* 119 (5), 615–627, 2004.
44. Wu, J., Brown, S., Xuong, N. H., and Taylor, S. S. RIalpha subunit of PKA: A cAMP-free structure reveals a hydrophobic capping mechanism for docking cAMP into site B. *Structure* 12 (6), 1057–1065, 2004.
45. Francis, S. H., Wolfe, L., and Corbin, J. D. Purification of type I alpha and type I beta isozymes and proteolyzed type I beta monomeric enzyme of cGMP-dependent protein kinase from bovine aorta. *Methods Enzymol* 200, 332–341, 1991.

46. Richie-Jannetta, R., Busch, J. L., Higgins, K. A., Corbin, J. D., and Francis, S. H. Isolated regulatory domains of cGMP-dependent protein kinase Ialpha and Ibeta retain dimerization and native cGMP-binding properties and undergo isoform-specific conformational changes. *J Biol Chem* 281 (11), 6977–6984, 2006.
47. Roy, A., and Danchin, A. Restriction map of the cya region of the *Escherichia coli* K12 chromosome. *Biochimie* 63 (8–9), 719–722, 1981.
48. Guidi-Rontani, C., Danchin, A., and Ullmann, A. Isolation and characterization of an *Escherichia coli* mutant affected in the regulation of adenylate cyclase. *J Bacteriol* 148 (3), 753–761, 1981.
49. Roy, A., and Danchin, A. The cya locus of *Escherichia coli* K12: Organization and gene products. *Mol Gen Genet* 188 (3), 465–471, 1982.
50. Danchin, A., Guiso, N., Roy, A., and Ullmann, A. Identification of the *Escherichia coli* cya gene product as authentic adenylate cyclase. *J Mol Biol* 175 (3), 403–408, 1984.
51. De Reuse, H., and Danchin, A. The ptsH, ptsI, and crr genes of the *Escherichia coli* phosphoenolpyruvate-dependent phosphotransferase system: A complex operon with several modes of transcription. *J Bacteriol* 170 (9), 3827–3837, 1988.
52. Glaser, P., Roy, A., and Danchin, A. Molecular characterization of two cya mutations, cya-854 and cyaR1. *J Bacteriol* 171 (9), 5176–5178, 1989.
53. Bussow, K., Scheich, C., Sievert, V., Harttig, U., Schultz, J., Simon, B., Bork, P., Lehrach, H., and Heinemann, U. Structural genomics of human proteins—Target selection and generation of a public catalogue of expression clones. *Microb Cell Fact* 4, 21, 2005.
54. Studier, F. W., and Moffatt, B. A. Use of bacteriophage T7 RNA polymerase to direct selective high-level expression of cloned genes. *J Mol Biol* 189 (1), 113–130, 1986.
55. Rosenberg, A. H., Lade, B. N., Chui, D. S., Lin, S. W., Dunn, J. J., and Studier, F. W. Vectors for selective expression of cloned DNAs by T7 RNA polymerase. *Gene* 56 (1), 125–135, 1987.
56. Studier, F. W., Rosenberg, A. H., Dunn, J. J., and Dubendorff, J. W. Use of T7 RNA polymerase to direct expression of cloned genes. *Methods Enzymol* 185, 60–89, 1990.
57. Gentz, R., and Bujard, H. Promoters recognized by *Escherichia coli* RNA polymerase selected by function: Highly efficient promoters from bacteriophage T5. *J Bacteriol* 164 (1), 70–77, 1985.

12 Cyclic Nucleotide Analogues as Chemical Tools for Interaction Analysis

Robin Lorenz, Claudia Hahnefeld, Stefan Möller, Daniela Bertinetti, and Friedrich W. Herberg

CONTENTS

12.1 INTRODUCTION

12.1.1 Cyclic Nucleotide Binding Domains

Cyclic nucleotides play a central role as second messengers in signal transduction in all kingdoms from bacteria to human.[1,2] Proteins involved in cyclic nucleotide signaling need to be able to sense the level of cyclic nucleotides. Therefore, these proteins have specialized domains to bind cyclic nucleotides such as 3′-5′-cyclic adenosine monophosphate (cAMP) or 3′-5′-cyclic guanosine monophosphate (cGMP) and transduce given signals into physiological responses. In prokaryotes, the catabolite activator protein (CAP) acts as a cAMP-regulated master transcriptional regulator of a plethora of genes, including the lactose (lac) operon.[3] The major mammalian effector proteins for cyclic nucleotide signaling encompass the cAMP-dependent protein kinase (PKA),[4,5] the cGMP-dependent protein kinase (PKG),[6,7] the exchange protein directly activated by cAMP (EPAC),[8,9] hyperpolarization-activated cyclic nucleotide–gated cation channels (HCN),[10,11] and cyclic nucleotide–gated ion channels (CNG)[12] (for review, see Berman et al.[2] and Kannan et al.[13]). All these effectors share a highly conserved domain, the so-called cyclic nucleotide binding (CNB) domain.[2,13] The CNB domain consists of 120 to 150 amino acids that form an eight-stranded beta barrel flanked by a variable number of alpha helices at either terminus. The most conserved feature of the CNB domain is the phosphate binding cassette (PBC), which is localized between strand $\beta6$ and $\beta7$. Here, several amino acids contact the ribose and cyclic phosphate moieties of the bound cyclic nucleotide and thus lead to protection from solvent and from phosphodiesterases (PDEs).[14] The CNB domain undergoes major conformational changes upon binding of the cyclic nucleotide and shows two distinct structures for its inactive apo state and active bound state. Generally, the ribose phosphate is nested within the PBC, which tightens whereas the dynamic αC-helix shields the purine base moiety through hydrophobic capping interactions. These underlying conformational rearrangements alter protein–protein interactions and thus generate a biological response.[2]

12.1.2 Synthetic Nucleotide Analogues as Chemical Tools

An important step in cyclic nucleotide research was the synthesis and development of cyclic nucleotide analogues (CNAs).[15] The design of radiolabeled cyclic nucleotides allowed the measurement of cellular cyclic nucleotide levels and gave rise to the first *in vitro* binding assays.[16–18]

Furthermore, advanced CNA synthesis coupled with chemical proteomics has led not only to the identification of new cyclic nucleotide binding proteins but also to a better understanding of cyclic nucleotide signaling interactomes.[19–21]

The development of nonhydrolyzable CNAs further extends the application in *in-cell* assays where degradation by PDEs is problematic for monitoring biological processes in response to fast fluctuating cyclic nucleotide levels.[15] Another landmark was the synthesis of both agonists (activating) and antagonists (inhibiting) of cyclic nucleotide–dependent protein kinases (PKA and PKG) leading to a better understanding of the physiological roles of the respective effector proteins.[22–24] Membrane-permeable CNAs now allow the study of cyclic nucleotide signaling *in vivo*.[25] With some effectors being important drug targets, such analogues have the potential to be applied in a pharmacological context.[15]

Biomolecular interaction analysis (BIA) encompasses a number of biophysical techniques allowing the quantitative analysis of the interaction of biomolecules, for example, protein–protein, protein–ligand, or protein–DNA interactions.[26,27] A number of CNAs are suitable as tools for BIA, where they can be used for specific immobilization (capturing) to a biosensor surface or be modified to carry a fluorophor for *in vitro* or *in-cell* applications, respectively.

In this chapter, we demonstrate how CNAs can be employed in interaction analysis with state-of-the-art techniques. Methods described in this chapter will cover specific strategies for interaction analyses of CNAs with respective binding partners, including the regulatory subunit (R) of PKA and HCN channels.

12.2 BIOMOLECULAR INTERACTION ANALYSIS METHODOLOGY

In this section, we will introduce three established techniques for interaction analysis using CNAs. For each method, examples for assay development and data interpretation are provided.

12.2.1 Surface Plasmon Resonance

Biosensors based on the physical phenomenon of surface plasmon resonance (SPR) have the advantage of real-time monitoring of complex formations. For this technology, one of the interaction partners, the *ligand*, has to be immobilized to a sensor surface. Binding of *analyte* leads to a change in the refractive index close to the sensor surface. These optical changes are proportional to an alteration in the mass concentration, translated to changes in the SPR signal. In SPR, light reflected at an interface between two media with different refractive indices will resonate at a specific angle resulting in a reduction in intensity of the reflected light. This angle is very sensitive to refractive index changes in the medium opposite to the interface of the incident light. In general, the change in refractive index is practically the same for all proteins and peptides and allows the determination of changes in mass concentration with high accuracy. As a rough estimation, a change in the resonance signal of 1000 RU (resonance or response units) corresponds to a change in surface concentration on a sensor chip (Sensor Chip CM5, Biacore™; GE Healthcare) of approximately 1 ng protein/mm^2.[28] For further details regarding the detection principle of optical biosensors, see Hahnefeld et al.[29]

12.2.1.1 Direct Binding Interaction with Immobilized Cyclic Nucleotide Analogue Surfaces

Because SPR instruments are in fact mass detectors, the smaller interaction partner is usually immobilized and the larger partner is detected to achieve a higher signal-to-noise ratio. In the following paragraphs, the coupling of a CNA and the assay development for measurement of a direct binding interaction with such analogue surfaces is described. In this example, the interaction between the analogue 8-(6-aminohexyl) amino-cAMP (8-AHA-cAMP) and the regulatory subunit Iα of PKA was monitored. 8-AHA-cAMP was covalently immobilized through its primary amine function of the 8-AHA-linker to a carboxymethylated sensor chip using standard N-hydroxysuccinimide (NHS)/1-ethyl-3-(3-dimethylaminopropyl)carbodiimide (EDC) chemistry.[30]

12.2.1.1.1 Cyclic Nucleotide Analogue Coupling

8-(6-Aminohexylamino)adenosine-3′,5′-cyclic monophosphate (8-AHA-cAMP)[31] was dissolved in 100 mM HEPES-KOH (pH 8) by gently heating the solution (max. 70°C) and filtered. The concentration was verified through absorption (extinction coefficient $\varepsilon = 17{,}000$ L cm^{-1} mol^{-1} at $\lambda_{max} = 273$ nm) using a Lambda Bio UV/VIS spectrometer (PerkinElmer®). In general, each surface was activated, coupled, and deactivated individually with a flow rate of 5 μL/min at 20°C using a Biacore 2000 or 3000 instrument (Biacore, GE Healthcare) as follows:

1. *Surface activation*: CM5 sensor chip (research grade) surfaces were activated for 10 min with NHS/EDC according to the manufacturer's instructions (amine coupling kit, Biacore, GE Healthcare; see Hahnefeld et al.[29])
2. *Analogue coupling*: 8-AHA-cAMP was injected for 7 min (running buffer: 100 mM HEPES-KOH; pH 8) using various concentrations on surfaces 2 to 4 (surface 2: low density, 0.035 mM; surface 3: medium density, 0.35 mM; surface 4: high density, 3.5 mM)
3. *Surface deactivation*: Surfaces were deactivated with 1 M ethanolamine-HCl (pH 8.5) for 10 min

Surface 1 was used as a reference and was therefore activated and subsequently deactivated without any nucleotide coupled (steps 1 and 3 only).

12.2.1.1.2 Purification of PKA R

The following recombinant mutant PKA R proteins were overexpressed in *E. coli* BL21 (DE3) cells:

- *R dimer*: full-length human RIα with four CNB domains (aa 1–379)[32]
- *R monomer*: bovine RIα Δ1–91 with two CNB domains (aa 92–379)[33]
- *R A-domain*: bovine RIα Δ1–92 260 stop with only CNB domain A (aa 93–259; see Figure 12.1)[34–36]

R dimer and R monomer were purified through an anion exchange chromatography step [diethylaminoethyl (DEAE) cellulose] using a salt gradient for elution. R A-domain was extracted from the pellet fraction after lysis with 8 M urea and subsequently dialyzed. Additionally, purified R dimer and R monomer were treated with urea to remove cAMP (apo cAMP binding proteins can also be prepared using bacterial strain lacking cAMP, as detailed in Chapter 11). The inhibitory activity of the proteins was verified with a phosphotransferase assay using the substrate peptide kemptide (LRRASLG) according to Cook et al.[37] Gel filtration was used to check for purity and to detect aggregated protein.[38]

12.2.1.1.3 Direct Binding Interaction with Analogue Surfaces

All interaction analyses were performed at 20°C in 20 mM MOPS (pH 7.0), 150 mM KCl or NaCl plus 0.005% (v/v) surfactant P20 using a Biacore 2000 or 3000 instrument (Biacore, GE Healthcare) with a flow rate of 30 μL/min. Measurements with cAMP in the dissociation phase were performed with the *Coinject* command. To

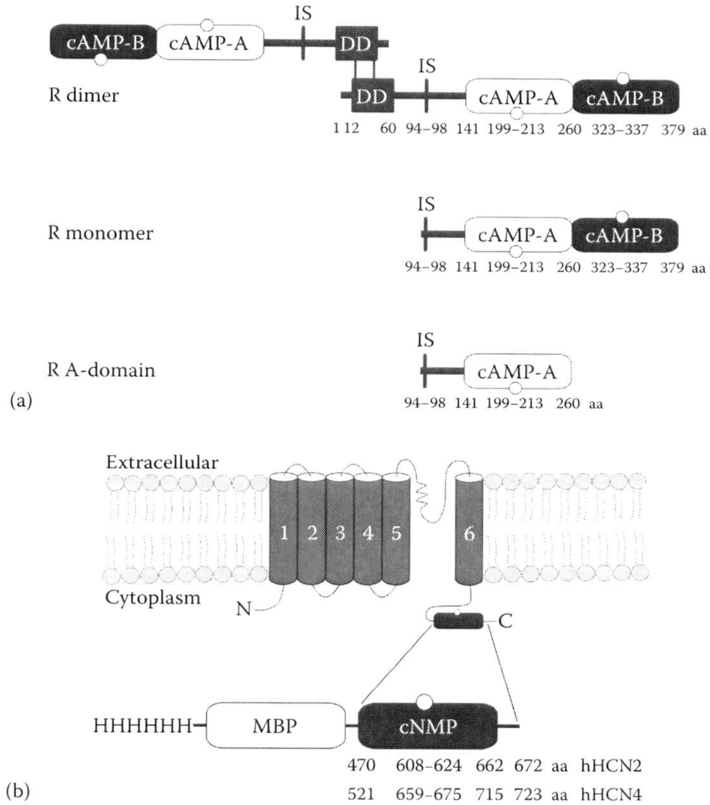

FIGURE 12.1 Linear organization of PKA R and HCN constructs. (a) The PKA R contains an N-terminal dimerization/docking (DD) domain followed by an inhibitory sequence (IS) for the interaction with the catalytic (C) subunit. At the C-terminus, two tandem CNB domains, cAMP-A and cAMP-B, are located. The mammalian type I R forms a dimer through two interchain disulfide bonds in the DD domains. An R monomer is constructed by deletion of the DD domain;[33] additionally, a truncation at the C-terminus yields a construct with only the inhibitory sequence and the CNB domain A (= R A-domain). (b) HCN channel monomers consist of six transmembrane domains with the pore region between domains 5 and 6. The CNB domain (cNMP) is localized at the intracellular C-terminus. (Figure modified according to Wahl-Schott, C., and Biel, M. *Cellular and Molecular Life Sciences* 66(3):470–494, 2009.) Recombinant HCN2 and HCN4 constructs contain the CNB domain with the C-linker fused to an N-terminal hexa His- and MBP-tag. The numbers indicate the amino acids (aa) outlining the respective domains; the circles represent the cNMP-binding sites.

obtain rate constants (k_{ass} and k_{diss}), R was injected at different concentrations in independent cycles. In general, an injection cycle consisted of the following steps:

1. *Association phase*: PKA R was injected for 200 s
2. *Dissociation phase*: At the end of the association phase, the flow was switched to running buffer for 200 s

3. *Regeneration*: The sensor chip surfaces were regenerated with either 0.2% (w/v) sodium dodecyl sulfate (SDS) or 3 M guanidinium HCl. After regeneration, water was injected for 1 min followed by buffer injections until the initial baseline level was reached

12.2.1.1.4 Data Analysis

Measured data were corrected by subtraction of the signal derived from the reference surface (surface 1) prior to evaluation. Kinetic analysis was performed with the BIAevaluation 4.1 software (Biacore, GE Healthcare). If possible, global fit analysis was employed, that is, one set of rate constants was applied to a set of kinetic data obtained using different concentrations of analyte. In principle, kinetic constants were calculated by linear regression of data using the pseudo first-order equation: $dR/dt = k_{ass}\, C\, R_{max} - (k_{ass}\, C + k_{diss})\, R_t$, where k_{ass} is the association rate constant, k_{diss} is the dissociation rate constant, C is the concentration of the injected analyte, and R is the response (at maximal response R_{max}, at time t, R_t). Affinity equilibrium binding constants were calculated from the equation $K_D = k_{diss}/k_{ass}$. For the transient kinetics of R A-domain on low density surfaces, where the equilibrium was reached nearly instantaneously, additional analyses according to Scatchard[39] were performed.

12.2.1.2 Results and Conclusions

To directly determine the binding of R to cAMP, RIα dimer was injected in concentrations ranging from 10 to 1000 nM (based on monomeric R) over a high-density 8-AHA-cAMP surface. A representative plot (700 nM R dimer) is shown in Figure 12.2a. The resulting binding curve displayed an initially linear association phase and a very slow dissociation phase. Analysis of the binding kinetics obtained at different concentrations using BIAevaluation 4.1 software did not fit any kinetic model using global fit. Therefore, rate constants were determined by fitting association and dissociation phases separately. Surprisingly, the observed dissociation rate was much slower than published previously based on ^{3}H-labeled cAMP binding assays.[40] Reattachment of an analyte (here, PKA R) is a common problem observed with molecules with high affinity and fast on-rate binding. Therefore, it was tested if this effect could be overcome by the addition of free cAMP to the dissociation phase. When adding 3 mM of free cAMP to the buffer in the dissociation phase, a 37-fold increased initial dissociation rate was observed (Figure 12.2a). This is probably because the rebinding and reattachment (see below) is reduced; however, the competition effects of soluble cAMP with immobilized analogue cannot be excluded.

Because the PKA R dimer has four CNB domains, one or more CNB domains may still be attached to the surface by binding to another adjacent 8-AHA-cAMP molecule before a complete dissociation is achieved. Thus, the R dimer *walks* over the surface without dissociating. Therefore, we used a truncated form of the PKA R with only two CNB domains (R monomer, Figure 12.1). Again, the dissociation was performed in the presence and absence of cAMP on a high-density 8-AHA-cAMP surface (Figure 12.2b). Similar to R dimer, a slightly enhanced dissociation rate constant in the presence of cAMP was observed. Because R monomer has still two CNB domains, reattachment effects are still possible. Interestingly, a difference in the dissociation behavior between R dimer and R monomer was detected during the

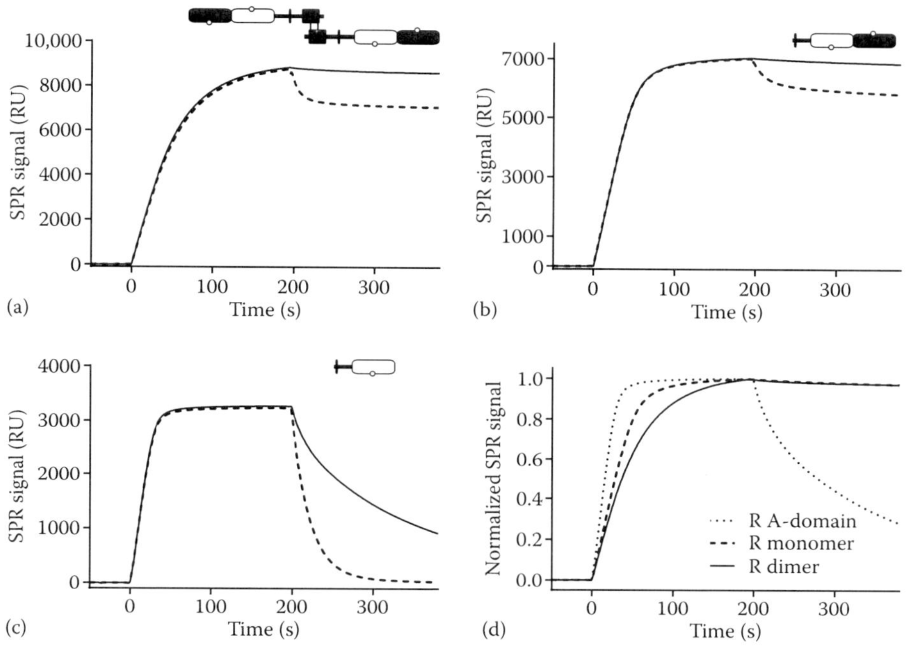

FIGURE 12.2 Reattachment and rebinding of cAMP-binding proteins. A total of 700 nM of cAMP-free R dimer (a), R monomer (b), and R A-domain (c), respectively, were injected over a high-density 8-AHA-cAMP sensor surface in the presence (dashed line) and absence of 3 mM soluble cAMP (solid line) during the dissociation phase, respectively. The slower dissociation in the absence of cAMP indicates reattachment events (solid lines). Rebinding can be overcome by adding cAMP during the dissociation phase, which binds to the dissociating R constructs and therefore prevents reassociation to the 8-AHA-cAMP surface (dashed lines). This becomes apparent, in particular, for the R A-domain (c). The overlay of the three constructs (without cAMP in the dissociation, d) clearly demonstrates the change in the association phase and a faster dissociation for R A-domain. The symbols represent the R constructs as defined in Figure 12.1.

regeneration phase. Although the R dimer was only completely dissociated from a high-density 8-AHA-cAMP sensor surface by injecting 3 M guanidinium-HCl twice for 1 min, a single 1-min injection was sufficient to regenerate R monomer.

To overcome the reattachment phenomenon completely, we further truncated the PKA R to a construct where only a single CNB domain remained (R A-domain). Surprisingly, the experiments in the presence and absence of cAMP still showed a major difference in the dissociation phases (Figure 12.2c). This demonstrates that on top of the reattachment, where several CNB domains act independently on the sensor surface, a rebinding phenomenon is occurring.[41] Due to the fast on-rate, the R tends to rebind instead of diffusing away from the sensor surface during the measurement. Again, competition effects of soluble cAMP with immobilized analogue cannot be excluded. Considering the already fast off-rate for R A-domain, the on-rate has to be extremely fast. Therefore, simplification of the biological system results in more consistent solid phase binding data.

According to the basic rate equation $dR/dt = k_{ass}\, C\, R_{max} - (k_{ass}\, C + k_{diss})\, R_t$, the association phase also includes dissociation events; therefore, the association should be affected by the rebinding phenomenon as well. Injection of free cAMP during the association phase is not an option as the high-affinity interaction between R and cAMP results in almost no binding to the surface at all (data not shown). Another strategy to reduce rebinding effects is the reduction of the surface density. Therefore, we produced a sensor chip with a gradient concentration of immobilized cAMP analogue. Three adjacent surfaces (flow cells 2–4) were covered with 0.035 mM (low density), 0.35 mM (medium density), and 3.5 mM 8-AHA-cAMP (high density), respectively. Again, we injected the PKA R constructs over these surfaces at a flow rate of 30 μL/min (Figure 12.3). At low surface densities, the shape of the association phase changed from a more linear to a curved form indicating a mass transfer limited interaction.[41,42] Mass transfer limitations occur when the rate constants are much faster than the transfer of analyte molecules between the zone of laminar flow and the unstirred zone at the sensor chip surface in a microfluidic system. In case of

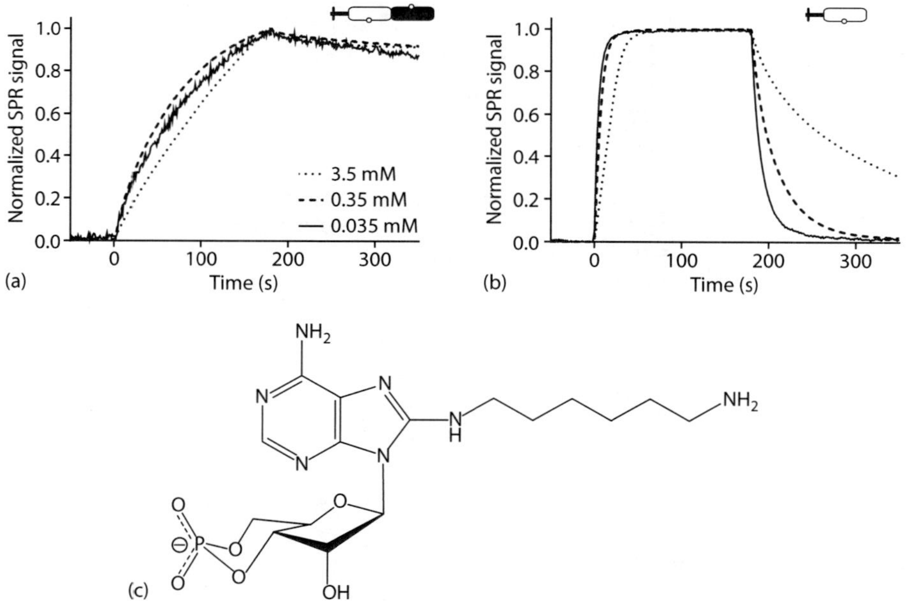

FIGURE 12.3 Association and dissociation of R constructs on 8-AHA-cAMP gradient surfaces. R monomer (500 nM, a) and R A-domain (700 nM, b) were injected over 8-AHA-cAMP (c) immobilized at different surface densities (high density, dotted line; medium density, dashed line; low density, solid line as indicated in the plot). For both constructs, the association rate is affected by the surface concentration. However, comparison of the dissociation phases of R monomer (a) and the single CNB domain construct (b) indicates that reattachment cannot be overcome if two (R monomer, a) or four cAMP-binding sites (R dimer, data not shown) are present. The symbols represent the R constructs as defined in Figure 12.1.

R monomer, the dissociation phase showed no difference on either of the surfaces with different densities of 8-AHA-cAMP (Figure 12.3a). To obtain consistent rate constants, several concentrations of R monomer were injected over the low-density surface (0.035 mM 8-AHA-cAMP). The association rate constant was $k_{ass} = 3.0 \times 10^4\ M^{-1}\ s^{-1}$ and the dissociation rate constant was $k_{diss} = 4.3 \times 10^{-3}\ s^{-1}$ (1:1 Langmuir one site association/dissociation equation [global fit]), resulting in an equilibrium dissociation constant (K_D) of 143 nM (see Table 12.1). When adding 3 mM cAMP to the dissociation phase, the k_{diss} was enhanced approximately twofold and a K_D of 213 nM was calculated.

Next, we injected 700 nM R A-domain simultaneously over the three gradient 8-AHA-cAMP surfaces. An overlay of the curves clearly demonstrated big differences in the kinetics (Figure 12.3b). Not only was the apparent rate of association changed but most notably the dissociation phase was accelerated up to 10-fold when comparing high-density and low-density surfaces. Again, to obtain accurate rate constants several concentrations of R A-domain (10–1000 nM) were injected over each 8-AHA-cAMP surface (Figure 12.4). In a 1:1 model, a 1:1 Langmuir (global fit) curve fit and the original data did not overlay very well when used on the high-density 8-AHA-cAMP sensor surface (3.5 mM; Figure 12.4a), which was reflected in a relative χ^2 value corresponding to more than 300% of the maximal binding response (R_{max}, see Table 12.2). The fit for R A-domain on medium surface densities

TABLE 12.1
Rate Constants and Thermodynamic Parameters for R Monomer/ 8-AHA-cAMP or cAMP Interaction Determined with SPR and ITC

	SPR		ITC
Ligand	**8-AHA-cAMP**		**cAMP**
Dissociation	**−cAMP**	**+cAMP**	–
k_{ass} [$M^{-1}\ s^{-1}$]($\times 10^6$)	0.03 ± 0.03	0.03 ± 0.03	–
k_{diss} [s^{-1}]($\times 10^{-3}$)	4.3 ± 2.6	6.4 ± 0.8[a]	–
K_D [nM]	143	213	6.7 ± 1.5
ΔH [kcal mol^{-1}]	–	–	−18.5 ± 0.8
ΔS [kcal $mol^{-1}\ K^{-1}$]	–	–	−0.026 ± 0.003
ΔG [kcal mol^{-1}]	–	–	−11.5 ± 0.9

Note: Data for SPR were derived from low-density surfaces (0.035 mM 8-AHA-cAMP) and fitted with Langmuir 1:1 one site exponential association/dissociation equations (global fit) using BIAevaluation software 4.1, data for dissociation + 3 mM cAMP were obtained from high-density surfaces (3.5 mM 8-AHA-cAMP). The dissociation phase was fitted with a single exponential equation using BIAevaluation software 4.1 or GraphPad Prism®. ITC data evaluation was performed with a single set of identical sites fit (binding sites $n = 2$) using MicroCal™ Origin®. For all values, the standard deviation of $n = 2$–6 experiments is indicated.

[a] Data derived from Langmuir 1:1 one site exponential dissociation (initial phase) equations using BIAevaluation software 4.1.

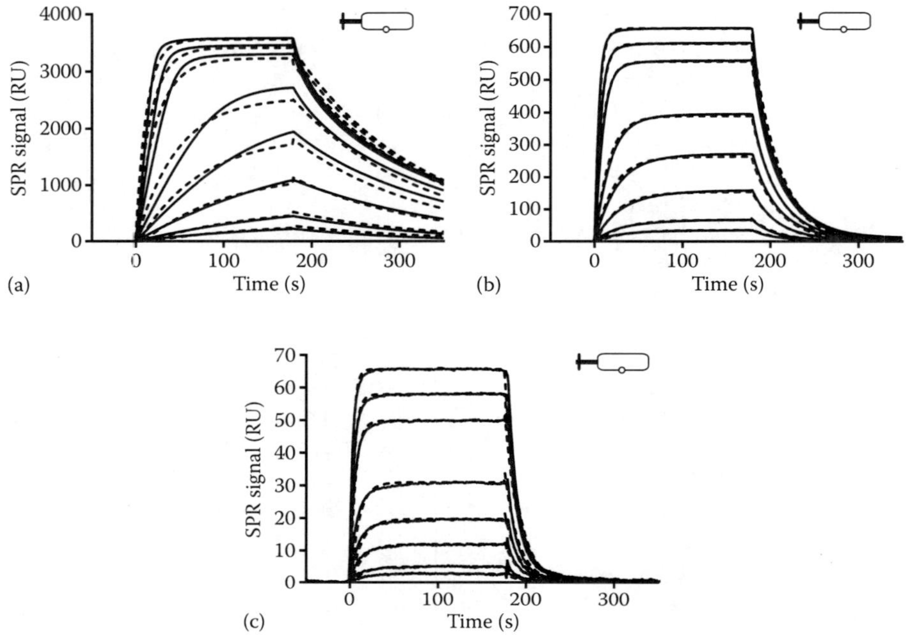

FIGURE 12.4 Apparent on-rates and off-rates are affected by the immobilization level of the coupled ligand. R A-domain was injected in a concentration range of 10 to 1000 nM over 8-AHA-cAMP at different surface densities (a: high density; b: medium density; c: low density). At high surface concentrations, the interaction is strongly mass transfer limited. Experimental data are shown as solid lines; dashed lines display the Langmuir 1:1 fit (global fit analysis using one site exponential association/dissociation equations). The respective rate and equilibrium binding constants are listed in Table 12.2. The symbols represent the R constructs as defined in Figure 12.1.

TABLE 12.2
Rate Constants for R A-Domain Determined with SPR on 8-AHA-cAMP Immobilized at Different Surface Densities

	High Density	Medium Density	Low Density
8-AHA-cAMP	**3.5 mM**	**0.35 mM**	**0.035 mM**
$k_{ass}[M^{-1}s^{-1}](\times 10^3)$	62 ± 15	150 ± 23	200 ± 45
$k_{diss}[s^{-1}](\times 10^{-3})$	7.3 ± 0.9	38 ± 8	70 ± 10
K_D [nM]	118	253	350
R_{max}/χ^2 [%]	308.4 ± 53.1	4.3 ± 0.3	1.0 ± 0.4

Note: The data were derived from the fits in Figure 12.4a through c, employing Langmuir 1:1 one site exponential association/dissociation equations using BIAevaluation 4.1. For all values, the standard deviation of $n = 3$ experiments is indicated.

(0.35 mM) was much better (relative χ^2 value = 4% of R_{max}) and rather indistinguishable from the reference subtracted measured data (Figure 12.4b). Further lowering the sensor surface density (0.035 mM) resulted in a fit describing the obtained data with even less deviation (Figure 12.4c) with a relative χ^2 of 1% of R_{max}. The resulting association rate constant was k_{ass} = 200 × 10^3 M^{-1} s^{-1} and the dissociation rate constant was k_{diss} = 70 × 10^{-3} s^{-1} with a K_D of 350 nM. Data for the dissociation phase derived from low-density sensor surfaces (Figure 12.4c) very much resembled the results acquired from experiments where cAMP was added to the dissociation phase (Figure 12.2c). Thus, lowering the surface density can help to overcome the rebinding phenomenon. Increasing the flow rate (e.g., up to 100 μL/min in the Biacore system) is another option to reduce mass transfer limitation and thus rebinding effects.[41]

12.2.1.3 Direct Binding Interaction with Protein Surfaces

An alternative strategy for the direct investigation of protein–cyclic nucleotide interaction is to immobilize proteins covalently to a chip surface and the interaction of cyclic nucleotides with the protein surface is determined.

12.2.1.3.1 Protein Coupling

HCN channel constructs were expressed and purified according to Möller et al.[11] and Lolicato et al.[43] The constructs described here consist of the C-linker and the CNB domain only (aa 470–672 for HCN2 and aa 521–723 for HCN4) fused to an N-terminal hexa His- and maltose binding protein (MBP) tag. These recombinant proteins were overexpressed in *E. coli* BL21 (DE3) Rosetta cells and purified through Co^{2+}-immobilized metal ion affinity chromatography (IMAC). For coupling to sensor chip surfaces, the fusion tag was cleaved off using rhinovirus 3C (HRV3C) protease and purified further through gel filtration.

The two isoforms, HCN2 and HCN4, were covalently coupled on the surface of a CM5 sensor chip (Biacore, GE Healthcare).

1. *Surface activation*: Using standard NHS/EDC coupling chemistry,[30] the surfaces were activated for 10 min
2. *Protein coupling*: Recombinant HCN2 and HCN4 fragments, both dissolved in 10 mM sodium acetate (pH 6.0), to a final concentration of 60 μg/mL, were injected for 20 min each on separate flow cells
3. *Surface deactivation*: Surfaces were deactivated with 1 M ethanolamine-HCl (pH 8.5), for 10 min

For reference, a flow cell was activated and deactivated subsequently without any protein coupled.

12.2.1.3.2 Direct Binding Interaction with Protein Surfaces

For the direct interaction of cyclic nucleotides with HCN channel proteins, concentration series of cyclic nucleotides were injected for 3 min over the protein surface. Because of the fast off-rates, no regeneration of the surface was necessary. The dissociation time was elongated up to 400 s. The data were analyzed with Biacore

T100 evaluation software 1.1.1 (Biacore, GE Healthcare) and GraphPad Prism plotting the RU values against the injection time.

12.2.1.4 Results and Conclusions

Both cAMP and cGMP in concentrations ranging from 1 μM to 2 mM were injected over HCN-coupled sensor chip surfaces. The resulting sensorgrams displayed transient binding kinetics (Figure 12.5). Following a fast association, the SPR signal reached a plateau indicating equilibrium between association and dissociation. The dissociation phase was initiated by switching to buffer after 3 min and the SPR signal almost immediately dropped back to the initial baseline level. This fast dissociation of the complex between the respectively coupled HCN constructs and the cyclic nucleotides was sufficient to regain a stable baseline (i.e., a constant SPR signal of 0 RU) and thus, no regeneration step was needed. Such a binding equilibrium can be analyzed according to Scatchard,[39] plotting the ratio of the equilibrium SPR signal and the analyte concentration against the equilibrium SPR signal. The slope of a linear fit is then equal to $-1/K_D$. The injection of low micromolar cGMP concentrations resulted in little to no SPR response (Figure 12.5b + d), indicating that the cGMP affinities of HCN2 and HCN4 are both in the higher micromolar range (K_D of

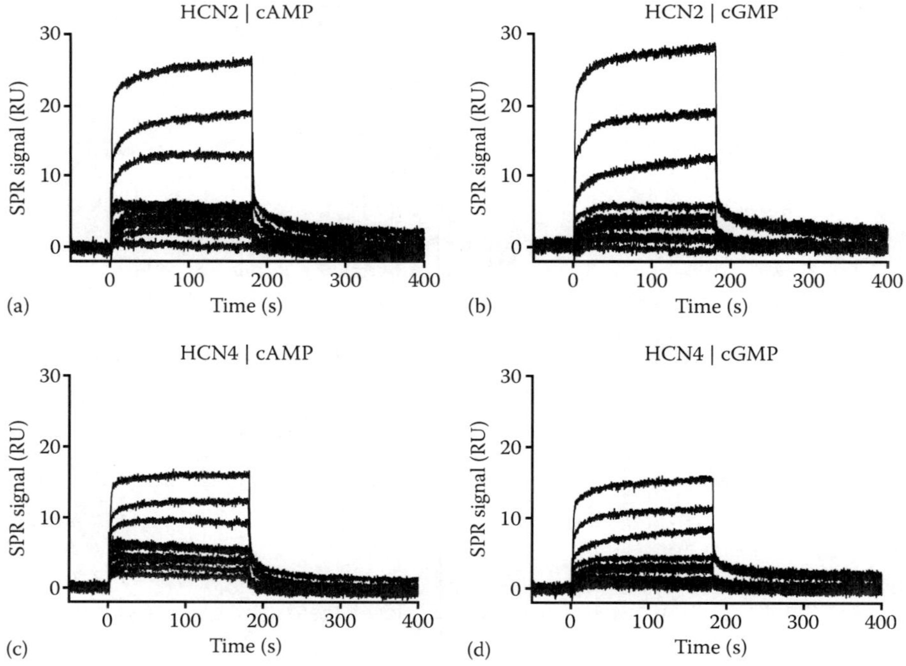

FIGURE 12.5 Direct binding of cyclic nucleotides to surface-coupled proteins. HCN2 (a + b) and HCN4 (c + d) were covalently coupled to a sensor chip surface. cAMP (a + c) and cGMP (b + d) were injected over the protein surfaces in a concentration ranging from 1 μM to 2 mM. Both nucleotides show transient kinetics characterized by fast on-rates and off-rates. The different maximum SPR signals between HCN2 and HCN4 surfaces indicate deviating surface densities.

approximately 10 μM). In contrast, the cAMP affinity of both proteins is one order of magnitude higher (K_D of approximately 1 μM).[11] The lower level of SPR signal on the HCN4 surface (Figure 12.5c and d) in comparison to the HCN2 surface implies that HCN4 was coupled at lower density. The sensorgrams clearly demonstrate that the coupling of protein to a sensor chip surface for cyclic nucleotide interaction analysis causes a significantly lower signal-to-noise ratio when compared with interactions with analogue surfaces (see Section 12.2.1.2; Figures 12.2–12.4). Because SPR biosensors are mass detectors, it is recommended to use the interaction partner with the bigger mass as an analyte. However, coupling of protein to a sensor chip surface allows us to measure the interaction with cyclic nucleotides or analogues, such as cAMP and cGMP, which cannot be covalently coupled with standard chemistry. Additionally, direct binding interactions with protein surfaces can be used to verify and complement results from analogue surface interactions. Moreover, artifacts caused by the reduction of the degrees of freedom can be minimized.

12.2.2 Fluorescence Polarization

Fluorescence polarization (FP) is an optical method to determine equilibrium binding constants for the interaction of two molecules. This technique is performed as a homogenous assay in microtiter plates. Interaction analyses with FP are rapid and inexpensive and can be scaled up for high-throughput screening.[26,44] In FP, the smaller interaction partner, as a rule of thumb, is labeled with a fluorophor (e.g., fluorescein). During incubation of the labeled sample with an interaction partner, a binding equilibrium is reached. Subsequently, the FP signal is determined with a fluorometer or a microtiter plate reader. Here, the fluorophor of the labeled interaction partner is excited with linear polarized light. In the time frame between fluorophor excitation and emission, the fluorescence lifetime, labeled molecules move freely in solution. Depending on their rotation, the fluorescence will be emitted in a plane of orientation deviating from the plane of the excitation light—the light is depolarized and the FP signal is low. In contrast, if the fluorescently labeled molecules are slowed down in their rotational speed over the fluorescence lifetime due to binding to a larger interaction partner, the emitted light is in the same plane of orientation as the excitation light resulting in a higher FP signal.[26,27,45,46] Determining the relative distribution between bound and unbound ligand allows the calculation of equilibrium binding constants.

Fluorescently labeled analogues such as 8-(2-[fluoresceinyl]aminoethylthio) adenosine-3′,5′-cyclic monophosphate (8-Fluo-cAMP)[47] and 8-(2-[fluoresceinyl]aminoethylthio)guanosine-3′,5′-cyclic monophosphate (8-Fluo-cGMP)[48] have been proven powerful tools for FP interaction analyses with CNB domain constructs.[11,26,49,50] Additionally, unlabeled cyclic nucleotides can be used as competitors for the fluorescent probe to determine half-maximal effective concentrations (EC_{50}). This indirect competition assay allows us to quantify the binding potential of such competitors without any specific changes in the assay design.[26,51]

In the following sections, we will describe a competitive binding assay (for a review, see Moll et al.[26]) for the interaction between HCN channel CNB domains and cyclic nucleotides.

12.2.2.1 Competition Assay

All cyclic nucleotides and analogues were purchased from BioLog Life Science Institute (Bremen, Germany) and were dissolved in 20 mM MOPS (pH 7), 150 mM NaCl, and 10% DMSO (v/v).

FP measurements were performed using a Fusion α-FP microtiter plate reader (PerkinElmer) with a read time of 2 s for each well, a 485 nm excitation filter, a 535 nm emission polarization filter, and a photomultiplier voltage of 1100 V. FP interaction analyses were carried out with all components in 20 mM MOPS (pH 7.0), 150 mM NaCl, and 0.005% (w/v) CHAPS in a 384-well microtiter (PerkinElmer, Optiplate, black) at room temperature. Data were analyzed with GraphPad Prism.

In competition experiments, both the protein concentration and the concentration of 8-Fluo-cAMP were fixed whereas the concentration of the competing cyclic nucleotide varied in concentration from the picomolar to the millimolar range. Plotting the FP signal against the logarithm of the CNA concentration and fitting the data with a sigmoidal dose–response curve allowed the determination of EC_{50} values. Measurements were repeated with different protein preparations and at least in duplicate for each cyclic nucleotide.

12.2.2.2 Results and Conclusions

FP can be used to monitor the interaction of CNB domains and fluorescently labeled analogues such as 8-Fluo-cAMP. Competitive assays can be performed to indirectly determine the apparent affinities of nonfluorescently labeled cyclic nucleotides and analogues.[26] For a competitive assay, both the fluorescent analogue concentration and the protein concentration are fixed at a level that results in 50% to 80% of the maximum FP signal of this interaction. This initial FP signal is then competed by increasing the concentration of the nonfluorescent analogue or cyclic nucleotide, respectively. Plotting the FP signal against the logarithm of the competitor concentration, the data can be fitted as a sigmoidal dose–response curve with a negative slope.

In the given example, we separately used cAMP and cGMP to compete with the interaction between 8-Fluo-cAMP and HCN2 or HCN4, respectively (Figure 12.6). For better graphical comparison, the FP signal was normalized by setting the initial FP signal to a value of 1.0 and the background FP signal to 0.0. Both cyclic nucleotides compete the binding of 8-Fluo-cAMP to the HCN CNB domain. However, cAMP is a more effective competitor than cGMP, which is demonstrated by an EC_{50} value for cAMP of approximately 2 μM and an EC_{50} value for cGMP of approximately 12 μM.[11] This is consistent with the results achieved from SPR experiments with protein-coupled surfaces (see Section 12.2.1.4), and clearly demonstrates how two independent methods can be used to validate data. FP data also indicate a higher cAMP selectivity of HCN2 (Figure 12.6a) in comparison to HCN4 (Figure 12.6b).

12.2.3 Isothermal Titration Calorimetry

In general, all methods described thus far require labeling or attachment to a solid phase, which may lead to artifacts caused by steric hindrance, loss of rotational freedom, or restrictions to certain conformations due to the modification introduced in the molecule of interest. In contrast, isothermal titration calorimetry (ITC) can

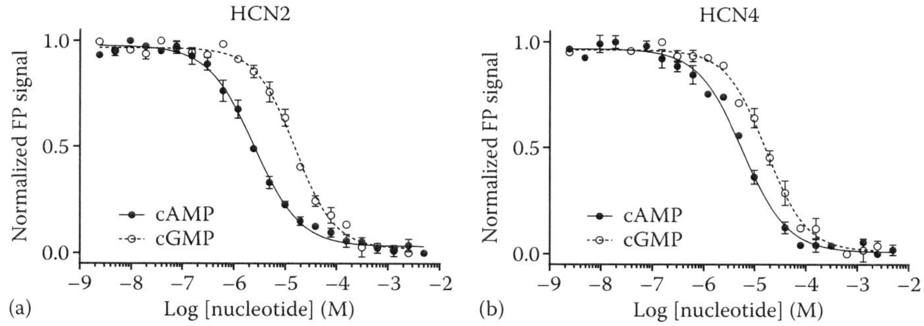

FIGURE 12.6 FP competition measurements of HCN channel constructs with cyclic nucleotides. Competition measurement of MBP-HCN2 (a) and MBP-HCN4 (b) with cAMP and cGMP, respectively. Normalized FP signal versus the logarithmic concentration of the respective cyclic nucleotide. Each point represents mean ± SEM from triplicate measurements. Data was fitted with a sigmoidal dose–response (variable slope) fit and EC_{50} values were determined using GraphPad Prism. Both HCN2 and HCN4 have a higher affinity for cAMP compared with cGMP.

be applied as a universal method for in-depth analysis of molecular interactions as long as the binding event generates or absorbs heat (which is generally the case).[52] ITC is truly label-free and has no limitations with regard to molecular weight and works in a direct assay format over a wide range of binding constants (in general, K_D values between 10 nM and 100 μM).[53,54] In a single ITC experiment, a complete set of thermodynamic parameters [change in enthalpy (ΔH), entropy (ΔS), and the Gibbs free energy (ΔG)] as well as the stoichiometry of the interaction (N) and the equilibrium binding constants (K_A, K_D) could be determined. However, the initial experiment will likely not be maximally informative and the assay conditions have to be optimized for each interaction. ITC experiments have to be planned carefully because this assay format requires rather large amounts of both interaction partners. The importance of the knowledge of the thermodynamic parameters ΔH and ΔS, which drive a molecular interaction, and their relevance in the improvement of lead compounds in drug development has been reviewed in an article by Ladbury.[53] However, the use of thermodynamic data to explain structural events associated with the interaction is complicated.

12.2.3.1 ITC Assay

In principle, the ITC experiment is performed by placing one interaction partner in the sample cell of the calorimeter at a constant temperature. The second, significantly more concentrated (a minimum of 20-fold to 40-fold molar excess with regard to the molar concentration of the binding sites) interaction partner is injected stepwise and accurately into the sample cell. The heat, released or absorbed in association with the binding process, is measured.

The interaction between R monomer and cAMP (Figure 12.7c), was analyzed with a VP-ITC microcalorimeter (MicroCal, GE Healthcare). The experiments were carried out in 20 mM MOPS (pH 7.0), 150 mM NaCl with 1 mM β-mercaptoethanol.

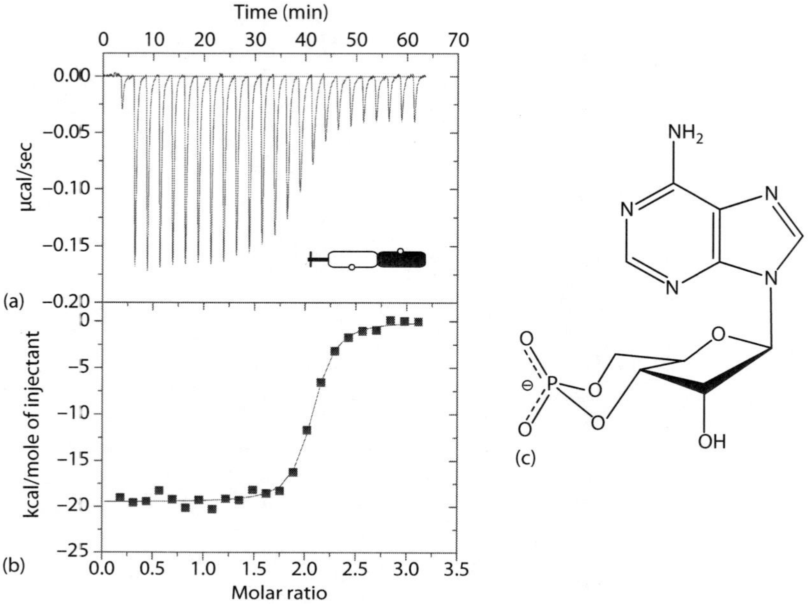

FIGURE 12.7 Analysis of cAMP binding to R monomer by ITC. (a) Titration of 1.1 μM R monomer (1.385 mL, cAMP free) with 40 μM cAMP (5 μL injection steps) in 20 mM MOPS, pH 7.0, 150 mM NaCl, 1 mM β-mercaptoethanol at 20.5°C (VP-ITC MicroCalorimeter; MicroCal, GE Healthcare), raw data. The heat released is plotted as a function of time. (b) Plot of heat exchange per mole of injectant (integrated areas under the respective peaks in panel a versus molar ratio of protein after subtraction of buffer control (data not shown). The best least squares fit, applying a model in which two cAMP per R monomer are bound (one binding site model, binding sites n = 2.02 ± 0.009), was performed with the software MicroCal Origin. (c) Chemical structure of cAMP. The symbol represents the R construct as defined in Figure 12.1.

cAMP was dissolved in the dialysis buffer of the purified protein to minimize artifacts arising from mismatched buffer components. All solutions were thoroughly degassed before use. The assay was performed as follows:

1. 1.1 μM cAMP-free R monomer was incubated at 20.5°C in the 1.385 mL cell of the apparatus
2. A 40 μM cAMP stock solution was injected in two steps of 1 μL (as described in the article by Mizoue and Tellinghuisen[55]) followed by 23 injections of 5 μL each. Each subsequent injection was performed automatically once the reaction returned to baseline levels (set to 150 s). The last injection leads to a calculated threefold final molar excess of cAMP to the cAMP binding sites in the cell

In a control experiment, 40 μM cAMP was injected in identical steps into buffer only and the resulting heat signal was subsequently subtracted from the raw data of the complex formation to correct for dilution and mixing effects. Data evaluation was performed with the software MicroCal Origin for ITC (MicroCal, GE Healthcare; see also Wiseman et al.[52]), including corrections for volume change during the titration (Figure 12.7).

12.2.3.2 Results and Conclusions

ITC is well suited for label-free analysis of the interaction between PKA R and cAMP. In the case of adding cAMP to R, heat is released until the protein is saturated, indicating that this interaction is exothermic. The heat produced is equal to the area under the peaks (Figure 12.7a). The change in enthalpy can be determined by plotting the heat produced against the molar ratio of ligand to protein (Figure 12.7b). Assuming that each R monomer can bind two molecules of cAMP, three different models were applied:

1. A single set of identical sites
2. Two sets of independent sites and
3. Sequential binding sites

Only the model with a single set of identical sites ($n = 2$; Figure 12.7b) yielded a satisfactory fit with a calculated ΔH of −18.5 kcal mol^{-1}, an ΔS of −0.026 kcal mol^{-1} K^{-1} and a ΔG of −11.5 kcal mol^{-1} (see Table 12.1). The matching model is consistent with the fact that in R monomer, cooperativity of the two CNB domains is lost.[33] Based on the basic thermodynamic principle, free energy can be calculated using equations: $\Delta G = \Delta H - T\Delta S = -RT\ln(1/K_D)$, where T is the absolute temperature and R is the universal gas constant. In the case of cAMP binding to R, the negative enthalpic contribution of −18.5 kcal mol^{-1} is favorable, whereas the positive entropic contribution ($-T\Delta S$) of +7 kcal mol^{-1} is unfavorable for the complex formation and reduces the affinity. The calculated K_D value is 6.7 nM. This is in good agreement with published FP data with a K_D value of 4 nM.[26]

12.3 CONCLUDING REMARKS

The identification and characterization of cyclic nucleotide–binding proteins is an important aspect of signal transduction research. The development and improvement of analogues has been a milestone in cyclic nucleotide research. Today, CNAs are essential tools for interaction analysis and can be used in a variety of applications. In this chapter, we have summarized the general concepts of SPR, FP, and ITC and described respective assays for interaction analyses of CNB domain constructs with cyclic nucleotides and analogues. Furthermore, interpretation of the results produced was given. In line with this, we critically discussed the pros and cons of each method. Additionally, ways to improve the consistency of data were shown. Besides the methodology discussed, other established BIA methods, such as stop-flow or rapid quench, can also be used to analyze protein–cyclic nucleotide interactions. More recently, another BIA method, microscale thermophoresis (MST) based on hydration-dependent movement along a temperature gradient, was developed. MST is performed in solution under both label and label-free formats, and will likely be useful for monitoring cyclic nucleotide interactions.

These different BIA approaches can be used complementarily because each methodology may associate with inherent obstacles and potential artifacts.

ACKNOWLEDGMENTS

We thank Frank Schwede (BioLog Life Science Institute, Bremen, Germany) for supplying the designed CNAs. Plasmids for expression of the HCN channel constructs were a kind gift of Anna Moroni (University of Milan, Milan, Italy). Robin Lorenz is supported by a PhD fellowship of the University of Kassel as a member of the graduate program Functionomics. This work was supported by the European Union (EU) FP7 collaborative project Affinomics (contract no. 241481) and Federal Ministry of Education and Research Project (FKZ 0316177F, No Pain). We acknowledge the Center for Interdisciplinary Nanostructure Science and Technology (CINSaT) of the University of Kassel for support of this work.

REFERENCES

1. Beavo, J.A., and Brunton, L.L. Cyclic nucleotide research—Still expanding after half a century. *Nature Reviews Molecular Cell Biology* 2002;3(9):710–718.
2. Berman, H.M., Ten Eyck, L.F., Goodsell, D.S., Haste, N.M., Kornev, A., and Taylor, S.S. The cAMP binding domain: An ancient signaling module. *Proceedings of the National Academy of Sciences of the United States of America* 2005;102(1):45–50.
3. Popovych, N., Tzeng, S.-R., Tonelli, M., Ebright, R.H., and Kalodimos, C.G. Structural basis for cAMP-mediated allosteric control of the catabolite activator protein. *Proceedings of the National Academy of Sciences of the United States of America* 2009;106(17):6927–6932.
4. Taylor, S.S., Buechler, J.A., and Yonemoto, W. cAMP-dependent protein kinase: Framework for a diverse family of regulatory enzymes. *Annual Review of Biochemistry* 1990;59:971–1005.
5. Su, Y., Dostmann, W.R., Herberg, F.W. et al. Regulatory subunit of protein kinase A: Structure of deletion mutant with cAMP binding domains. *Science* 1995;269(5225): 807–813.
6. Pfeifer, A., Ruth, P., Dostmann, W.R., Sausbier, M., Klatt, P., and Hofmann, F. Structure and function of cGMP-dependent protein kinases. *The Journal of Biological Chemistry* 1999;135:105–149.
7. Hofmann, F. The biology of cyclic GMP-dependent protein kinases. *The Journal of Biological Chemistry* 2005;280(1):1–4.
8. Rehmann, H., Prakash, B., Wolf, E. et al. Structure and regulation of the cAMP-binding domains of Epac2. *Nature Structural Biology* 2003;10(1):26–32.
9. Selvaratnam, R., Akimoto, M., Van Schouwen, B., and Melacini, G. cAMP-dependent allostery and dynamics in Epac: An NMR view. *Biochemical Society Transactions* 2012;40:219–223.
10. Benarroch, E.E. HCN channels: Function and clinical implications. *Neurology* 2013;80: 304–310.
11. Möller, S., Alfieri, A., Bertinetti, D. et al. Cyclic nucleotide mapping of hyperpolarization-activated cyclic nucleotide–gated (HCN) channels. *ACS Chemical Biology* 2014;9(5):1128–1137.
12. Kaupp, U.B., and Seifert, R. Cyclic nucleotide–gated ion channels. *Physiological Reviews* 2002;82(3):769–824.

13. Kannan, N., Wu, J., Anand, G.S. et al. Evolution of allostery in the cyclic nucleotide binding module. *Genome Biology* 2007;8(12):R264.
14. Diller, T.C., Madhusudan, Xuong, N.H., and Taylor, S.S. Molecular basis for regulatory subunit diversity in cAMP-dependent protein kinase: Crystal structure of the type II beta regulatory subunit. *Structure* 2001;9(1):73–82.
15. Schwede, F., Maronde, E., Genieser, H., and Jastorff, B. Cyclic nucleotide analogs as biochemical tools and prospective drugs. *Pharmacology & Therapeutics* 2000;87(2–3):199–226.
16. Gilman, A.G. A protein binding assay for adenosine 3′:5′-cyclic monophosphate. *Proceedings of the National Academy of Sciences of the United States of America* 1970;67(1):305–312.
17. Døskeland, S.O., and Ueland, P.M. A cAMP receptor from mouse liver cytosol whose binding capacity is enhanced by Mg++-ATP. *Biochemical and Biophysical Research Communications* 1975;66(2):606–613.
18. Døskeland, S.O., Ueland, P.M., and Haga, H.J. Factors affecting the binding of [3H] adenosine 3′:5′-cyclic monophosphate to protein kinase from bovine adrenal cortex. *Biochemical Journal* 1977;161(3):653–665.
19. Aye, T.T., Mohammed, S., van den Toorn, H.W.P. et al. Selectivity in enrichment of cAMP-dependent protein kinase regulatory subunits type I and type II and their interactors using modified cAMP affinity resins. *Molecular & Cellular Proteomics* 2009;8(5):1016–1028.
20. Bertinetti, D., Schweinsberg, S., Hanke, S.E. et al. Chemical tools selectively target components of the PKA system. *BMC Chemical Biology* 2009;9:3.
21. Hanke, S.E., Bertinetti, D., Badel, A., Schweinsberg, S., Genieser, H.-G., and Herberg, F.W. Cyclic nucleotides as affinity tools: Phosphorothioate cAMP analogues address specific PKA subproteomes. *New Biotechnology* 2011;28(4):294–301.
22. De Wit, R.J., Hekstra, D., Jastorff, B. et al. Inhibitory action of certain cyclophosphate derivatives of cAMP on cAMP-dependent protein kinases. *European Journal of Biochemistry/FEBS* 1984;142(2):255–260.
23. Dostmann, W.R. (RP)-cAMPS inhibits the cAMP-dependent protein kinase by blocking the cAMP-induced conformational transition. *FEBS Letters* 1995;375(3):231–234.
24. Valtcheva, N., Nestorov, P., Beck, A. et al. The commonly used cGMP-dependent protein kinase type I (cGKI) inhibitor Rp-8-Br-PET-cGMPS can activate cGKI in vitro and in intact cells. *The Journal of Biological Chemistry* 2009;284(1):556–562.
25. Moll, D., Prinz, A., Brendel, C.M. et al. Biochemical characterization and cellular imaging of a novel, membrane permeable fluorescent cAMP analog. *BMC Biochemistry* 2008;9:18.
26. Moll, D., Prinz, A., Gesellchen, F., Drewianka, S., Zimmermann, B., and Herberg, F.W. Biomolecular interaction analysis in functional proteomics. *Journal of Neural Transmission* 2006;113(8):1015–1032.
27. Moll, D., Zimmermann, B., Gesellchen, F., and Herberg, F.W. Current developments for the in vitro characterization of protein interactions. In: *Proteomics in Drug Research*, Wiley-CVH, Weinheim, Germany; 2006:159–172.
28. Stenberg, E., Persson, B., Roos, H., and Urbaniczky, C. Quantitative determination of surface concentration of protein with surface plasmon resonance using radiolabeled proteins. *Journal of Colloid and Interface Science* 1991;143:513–526.
29. Hahnefeld, C., Drewianka, S., and Herberg, F.W. Determination of kinetic data using surface plasmon resonance biosensors. *Methods in Molecular Medicine* 2004;94:299–320.
30. Johnsson, B., Löfås, S., and Lindquist, G. Immobilization of proteins to a carboxymethyldextran-modified gold surface for biospecific interaction analysis in surface plasmon resonance sensors. *Analytical Biochemistry* 1991;198(2):268–277.

31. Herberg, F.W., Maleszka, A., Eide, T., Vossebein, L., and Tasken, K. Analysis of A-kinase anchoring protein (AKAP) interaction with protein kinase A (PKA) regulatory subunits: PKA isoform specificity in AKAP binding. *Journal of Molecular Biology* 2000;298:329–339.
32. Sandberg, M., Taskén, K., Oyen, O., Hansson, V., and Jahnsen, T. Molecular cloning, cDNA structure and deduced amino acid sequence for a type I regulatory subunit of cAMP-dependent protein kinase from human testis. *Biochemical and Biophysical Research Communications* 1987;149:939–945.
33. Herberg, F.W., Dostmann, W.R., Zorn, M., Davis, S.J., and Taylor, S.S. Crosstalk between domains in the regulatory subunit of cAMP-dependent protein kinase: Influence of amino terminus on cAMP binding and holoenzyme formation. *Biochemistry* 1994;33(23):7485–7494.
34. Ringheim, G.E., and Taylor, S.S. Effects of cAMP-binding site mutations on intradomain cross-communication in the regulatory subunit of cAMP-dependent protein kinase I. *The Journal of Biological Chemistry* 1990;265(32):19472–19478.
35. Hahnefeld, C., Moll, D., Goette, M., and Herberg, F.W. Rearrangements in a hydrophobic core region mediate cAMP action in the regulatory subunit of PKA. *Biological Chemistry* 2005;386(7):623–631.
36. Wahl-Schott, C., and Biel, M. HCN channels: Structure, cellular regulation and physiological function. *Cellular and Molecular Life Sciences* 2009;66(3):470–494.
37. Cook, P.F., Neville, M.E., Vrana, K.E., Hartl, F.T., and Roskoski, R. Adenosine cyclic 3′,5′-monophosphate dependent protein kinase: Kinetic mechanism for the bovine skeletal muscle catalytic subunit. *Biochemistry* 1982;21(23):5794–5799.
38. Herberg, F.W., and Taylor, S.S. Physiological inhibitors of the catalytic subunit of cAMP-dependent protein kinase: Effect of MgATP on protein–protein interactions. *Biochemistry* 1993;32:14015–14022.
39. Scatchard, G. The attractions of proteins for small molecules and ions. *Annals of the New York Academy of Sciences* 1949;51:660–672.
40. Døskeland, S., and Ogreid, D. Characterization of the interchain and intrachain interactions between the binding sites of the free regulatory moiety of protein kinase I. *Journal of Biological Chemistry* 1984;259(4):2291–2301.
41. Herberg, F.W., and Zimmermann, B. Analysis of protein kinase interactions using biomolecular interaction analysis. In: *Protein Phosphorylation—A Practical Approach*, 2nd Ed. Oxford University Press; 1999:1–39.
42. Kortt, A.A., Gruen, L.C., and Oddie, G.W. Influence of mass transfer and surface ligand heterogeneity on quantitative BIAcore™ binding data. Analysis of the interaction of NC10 Fab with an anti-idiotype Fab. *Journal of Molecular Recognition* 1997;10: 148–158.
43. Lolicato, M., Nardini, M., Gazzarrini, S. et al. Tetramerization dynamics of C-terminal domain underlies isoform-specific cAMP gating in hyperpolarization-activated cyclic nucleotide–gated channels. *The Journal of Biological Chemistry* 2011;286(52): 44811–44820.
44. Owicki, J.C. Fluorescence polarization and anisotropy in high throughput screening: Perspectives and primer. *Journal of Biomolecular Screening* 2000;5(5):297–306.
45. Pope, A., Haupts, U., and Moore, K. Homogeneous fluorescence readouts for miniaturized high-throughput screening: Theory and practice. *Drug Discovery Today* 1999; 4(8):350–362.
46. Burke, T.J., Loniello, K.R., Beebe, J.A., and Ervin, K.M. Development and application of fluorescence polarization assays in drug discovery. *Combinatorial Chemistry & High Throughput Screening* 2003;6:183–194.

47. Schwede, F., Christensen, A., Liauw, S. et al. 8-Substituted cAMP analogues reveal marked differences in adaptability, hydrogen bonding, and charge accommodation between homologous binding sites (AI/AII and BI/BII) in cAMP kinase I and II. *Biochemistry* 2000;39(30):8803–8812.
48. Peña, I., and Domínguez, J.M. Thermally denatured BSA, a surrogate additive to replace BSA in buffers for high-throughput screening. *Journal of Biomolecular Screening* 2010; 15(10):1281–1286.
49. Guo, D.-C., Regalado, E., Casteel, D.E. et al. Recurrent gain-of-function mutation in PRKG1 causes thoracic aortic aneurysms and acute aortic dissections. *American Journal of Human Genetics* 2013;93(2):398–404.
50. Huang, G.Y., Kim, J.J., Reger, A.S. et al. Structural basis for cyclic-nucleotide selectivity and cGMP-selective activation of PKG I. *Structure* 2014;22(1):116–124.
51. Moll, D., Schweinsberg, S., Hammann, C., and Herberg, F.W. Comparative thermodynamic analysis of cyclic nucleotide binding to protein kinase A. *Biological Chemistry* 2007;388(2):163–172.
52. Wiseman, T., Williston, S., Brandts, J.F., and Lin, L.N. Rapid measurement of binding constants and heats of binding using a new titration calorimeter. *Analytical Biochemistry* 1989;179(1):131–137.
53. Ladbury, J.E. Calorimetry as a tool for understanding biomolecular interactions and an aid to drug design. *Biochemical Society Transactions* 2010;38(4):888–893.
54. Velázquez Campoy, A., and Freire, E. ITC in the post-genomic era...? Priceless. *Biophysical Chemistry* 2005;115(2–3):115–124.
55. Mizoue, L.S., and Tellinghuisen, J. The role of backlash in the "first injection anomaly" in isothermal titration calorimetry. *Analytical Biochemistry* 2004;326(1):125–127.

13 Dissecting the Physiological Functions of PKA Using Genetically Modified Mice

Brian W. Jones, Jennifer Deem, Linghai Yang, and G. Stanley McKnight

CONTENTS

13.1 INTRODUCTION

The protein kinase A (PKA) family includes the products of four regulatory (R) subunit genes and two catalytic (C) subunit genes in the mouse. In primates, a third C subunit has been postulated (Cγ) that seems to be the product of a retroposon. The mouse genome also contains a retroposon on the X chromosome derived from the *Prkaca* gene (Cα2 variant) but it is not expressed and contains mutations that would be expected to eliminate kinase activity.[1] Table 13.1 summarizes some of the key characteristics of the mouse PKA subunit genes and proteins. In general, the alpha isoforms of RI, RII, and C are expressed in all tissues whereas beta isoforms show more selective expression.

PKA assembles as a holoenzyme formed by a dimer of two R subunits that each binds a C subunit. Type I holoenzymes contain homodimers of RIα or RIβ subunits, although it is likely that heterodimers of RIα/RIβ can sometimes also form. Dissociation of type I holoenzymes occurs at a lower threshold of cAMP compared with type II holoenzymes containing homodimers of RIIα or RIIβ. The RIβ containing holoenzyme is more sensitive to cAMP-induced dissociation compared with RIα containing holoenzymes. One of the major differences between type I and type II holoenzymes is their subcellular localization and this is determined by interaction with A kinase anchoring proteins (AKAPs), which generally have a much greater affinity for the type II holoenzymes as discussed in the following section.

TABLE 13.1
Summary of PKA Subunit Properties

Gene	Splice/Promoter Variants	Protein Expressed	Tissue Expression	KO Phenotypes
Prkar1a	Two promoters identified	RIα	Ubiquitously expressed	Embryonic lethal E14
Prkar1b		RIβ	Neural selectivity in mice	LTP deficits in hippocampus
Prkar2a		RIIα	Ubiquitously expressed	No phenotypes noted
Prkar2b		RIIβ	Selective expression in WAT, BAT, and neurons	Lean, resistant to diet-induced obesity (DIO), hyperlocomoter activity
Prkaca	Alternative promotors give rise to two variants	Cα1, Cα2	Cα1 expressed ubiquitously, Cα2 expressed in male postmeiotic germ cells	Cα1: perinatal lethality, infertility Cα2: male infertility
Prkacb	Multiple promoters and splice variants	Cβ1, Cβ2, Cβ4	Cβ1 expressed ubiquitously, other isoforms show neural or lymphoid expression	Cβ1: synaptic plasticity defects Cβ (all isoforms): resistant to DIO

13.2 USING SUBUNIT-SPECIFIC KO MICE

To study the physiological functions of PKA, it was essential to develop mouse models with loss-of-function mutations in each of the PKA subunits. We recognized that there could be significant overlap in function of the various R and C subunits and that compensation would occur, which would at least partially rescue discrete physiological functions. Nevertheless, the analysis of these PKA subunit KO mouse lines has already produced a broad series of insights into the role of individual PKA holoenzymes in physiology, which in turn led to more specific conditional approaches described in later sections of this chapter. The genes encoding the individual subunits of PKA have been knocked out in the mouse by conventional mouse genetic techniques that produce whole animal disruptions of each subunit throughout development and in the adult. All these KO mice are available through the Mutant Mouse Regional Resource Centers coordinated by Jackson Laboratories. One exception is the RIβ KO line, which was the first line constructed; we unfortunately lost this line before it could be cryopreserved in a repository.

Because multiple PKA subunits are expressed in all mouse cells and are likely to play a modulatory role in virtually every intracellular signaling pathway, investigators interested in studying a specific physiological process may need to investigate several KO mouse lines to determine which subunits are most relevant to their studies. Two of the regulatory subunits, RIα and RIIβ, have also been targeted using Cre recombinase and loxP technology to obtain conditional mutations in specific tissues and these are also available to investigators. Conditional mutants have advantages especially when the whole animal KO is lethal, as is the case for RIα. However, the conditional mutants necessitate a careful understanding of the limitations of Cre recombinase techniques and may not be necessary for an initial examination of function in a specific tissue. A brief discussion of the various mutant lines available is given in the following section along with examples of phenotypes that have already been studied and a discussion of some of the advantages and limitations of these approaches for studying complex physiological processes.

13.2.1 *Prkar1a* (RIα) KO Mice

The *Prkarla* gene was disrupted by inserting a neomycin resistance cassette into the third exon effectively preventing expression of RIα protein. Homozygote disruption of RIα expression results in defects in mesoderm differentiation and migration resulting in growth retardation and embryonic lethality. The embryos fail to develop a heart tube at E8.5 and are resorbed by E10.5.[2] These early defects are likely caused by excess activity of the uninhibited C subunits because the severe effects of the RIα KO can be partially rescued by a heterozygous KO of the *Prkaca* (Cα) gene and an even greater rescue is seen with a homozygous Cα KO background (Figure 13.1). The double knockout animals still do not survive to birth. This result suggests that, at least during embryogenesis, the other R subunits are not able to compensate and maintain cAMP control over expressed C subunits.

The heterozygous RIα KO also has defects as adults and demonstrates an increase in tumor formation and decreased male fertility.[3,4] A similar disease state known as Carney complex has been identified in humans who have inherited loss-of-function mutations in one allele of the RIα gene. Cardiac myxomas occur commonly in Carney complex patients along with other noncardiac tumors, whereas the mice were more likely to develop hepatocellular carcinomas and cardiac tumors were not observed. This autosomal dominant multiple neoplasia syndrome may be directly related to the loss of RIα-dependent inhibition of C subunit. Work from the Kirschner laboratory has shown that Schwann cell tumors develop when RIα is specifically knocked out in that cell type using Cre-Lox technology. Heterozygosity at either the Cα or Cβ gene can suppress tumorigenesis.[5] Recently, it was shown that a mammary-specific null mutation of RIα induced spontaneous breast tumors and this correlated with increased PKA catalytic activity.[6]

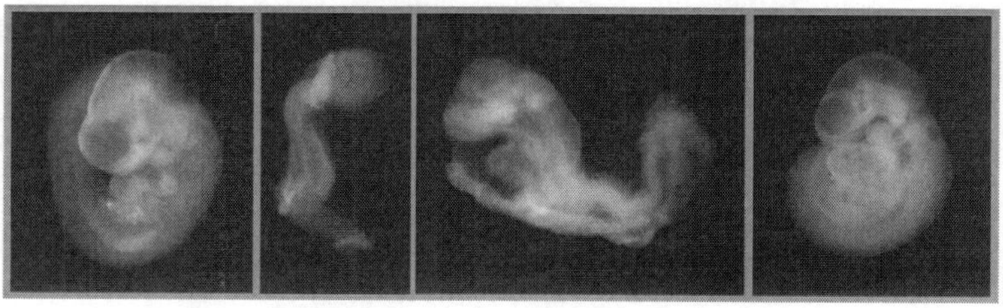

FIGURE 13.1 **(See color insert.)** E9.5 embryos of R1α or Cα mutant mice. Loss of RIα results in early embryonic lethality that is partially rescued if Cα is also reduced or deleted.

The evidence shown in previous sections strongly suggests that unregulated C subunit activity can be oncogenic and recent clinical studies support this idea. A chromosomal translocation that fuses the DNAJB1 promoter and first exon to the *PRKACA* gene in humans results in fibrolamellar hepatocellular carcinoma associated with increased expression of an active chimeric C subunit in the tumor cells.[7] Thirty-five percent of adrenal tumors associated with Cushing's syndrome were recently reported to contain a specific mutation (Leu206Arg) in the *PRKACA* gene that interferes with RIα binding and results in increased phosphorylation of PKA targets.[8]

In addition to the overall increased tumor formation in both humans and mice with only one functional RIα gene, sperm defects and reduced male fertility are also observed in both species. The sperm defects in male mice can be rescued on a heterozygous Cα2 KO background in which there is a 50% reduction in expression of the male germ cell-specific Cα2 protein.[4] In summary, unregulated C subunit is apparently the primary mechanism underlying most, if not all, RIα null phenotypes.

13.2.2 *Prkar1b* (RIβ) KO Mice

The RIβ gene was disrupted by inserting a neomycin resistance gene directly into exon 3 and this resulted in a complete loss of RIβ protein. RIβ KO mice are healthy and fertile but exhibit defects in synaptic plasticity in both the hippocampus[9] and visual cortex.[10] Long-term depression (LTD) is decreased in the Schaffer collateral/CA1 synapse and in the perforant path/dentate granule cell synapse. The mossy fiber/CA3 pathway is known to depend on PKA activity in the presynaptic dentate granule neuron.[11] The RIβ KO mice have decreased long-term potentiation (LTP) when measured after a high-frequency train stimulus (HFS), but forskolin-induced LTP remains intact. This is interesting because PKA activity is required for both forskolin forms of LTP, yet RIβ KO only affects the HFS-induced component, implying that another form of the kinase (i.e., type II) is required for forskolin-induced LTP. Despite the deficits in mossy fiber/CA3 LTP and Schaffer collateral/CA1 LTD, no deficits in behavior are evident. The behaviors tested include exploratory behavior in open field, contextual learning in either a Barnes maze or Morris water maze, and contextual and cued fear conditioning. Deficits in more subtle forms of learning and memory would not have been detected by these tests.

RIβ is expressed throughout the nervous system in mice. RIβ KO mice have diminished tissue-injury evoked (nociceptive) pain responses whereas their response to acute peripheral nerve injury remained intact.[12] This suggested that C-fiber activation after thermal, mechanical, or chemical insult does not depend on RIβ-PKA, but those aspects of the inflammatory response and sensitization of primary afferent nociceptors are RIβ-dependent. In the RIβ KO mice, there is compensation by RIα at the protein level without increased expression of RIα mRNA. One of the differences between RIβ-containing versus RIα-containing holoenzymes is the increased sensitivity of the RIβ holoenzyme to cAMP-evoked activation. Perhaps this sensitivity to cAMP is a major factor in the responses that are altered in the RIβ KO mice. In addition to changes in sensitivity to cAMP, the quaternary structure of the holoenzymes formed with either RIα or RIβ are quite different and this may lead to changes in PKA interactions with other binding partners.[13]

13.2.3 *Prkaca* (Cα) KO Mice

Due to its ubiquitous expression and involvement in many critical physiological systems, one might have predicted that deletion of the Cα catalytic subunit of PKA would be lethal. Nevertheless, *Prkaca* null mice can survive to adulthood, although they are severely runted and many die just after birth.[14] Compensation in the form of increased expression of Cβ is seen, but the levels of total PKA activity decline dramatically in most tissues. Analysis of spermatogenesis in adult male Cα KO reveals that although sperm develop normally, they are incapable of forward motility. We later discovered that an alternative promotor/splice form of Cα called Cα2 is specifically expressed in sperm and is absolutely required for cAMP-stimulated sperm motility and fertility.[15,16] Although the severe runting of Cα KO mice precluded their usefulness in studying specific physiological roles of PKA, the Cα heterozygotes were helpful in demonstrating that phenotypes associated with deficiencies in RIα are due to excess C subunits.

13.2.4 Cα2 Isoform-Specific KO

As alluded to previously, we discovered that an alternative promoter/exon in the first intron of the *Prkaca* gene is activated in midpachytene spermatogenesis and that this novel transcript splices into exon 2 of the Cα gene[15] (Figure 13.2). A novel sperm C subunit called Cs had been previously reported in ovine sperm[17] and we demonstrated that mice also express this slightly smaller Cα subunit that contains seven distinct, nonmyristylated, N-terminal amino acids. The protein product is called Cα2 and is the same as Cs described by the Witman laboratory.

We took advantage of this unique Cα2 first exon and mutated the ATG in the mouse so that no functional protein could be made, but the *Prkaca* promoter and other structural elements remained intact. Mice carrying this Cα2-specific deletion are normal except that the males are infertile and sperm are unresponsive to bicarbonate-induced increases in sperm motility. Other aspects of capacitation including increased tyrosine phosphorylation are also absent.[16]

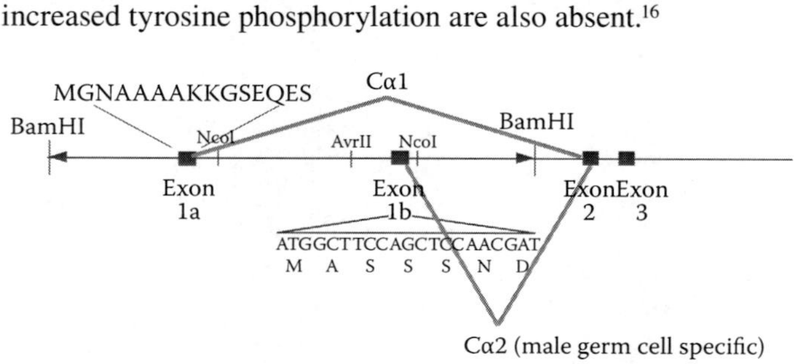

FIGURE 13.2 Diagram of Cα gene. Alternative splicing from exon 1a or 1b distinguishes Cα isoforms.

13.2.5 *Prkacb* (Cβ) KO Mice

Low stringency screening of cDNA libraries with a probe for the Cα subunit revealed a closely related Cβ mRNA.[18] In mice, the Cβ mRNA is most abundant in the brain but present in all tissues examined. A KO of the Cβ gene was first constructed by deleting part of exon 1 and inserting the neomycin resistance gene.[19] After this first KO was made, we discovered that the Cβ gene has at least three other promoter start sites that give rise to four different Cβ variants (Cβ1, Cβ2, Cβ3, and Cβ4).[20] A more detailed study of all of the Cβ variants has been published.[21] The original KO disrupted Cβ1 and left Cβ2, Cβ3, and Cβ4 intact. The Cβ1 mice are healthy and have no major behavioral defects,[22] but electrophysiological recording in the hippocampus revealed defects in both late phase LTP and LTD in the Schaffer collateral/CA1 pathway.[19] The mossy fiber/CA3 pathway is also defective in the Cβ1 KO mice with a substantial decrease in HFS-induced LTP.[22] Subsequently, a Cβ KO mouse was constructed that removed all Cβ isoforms by replacing the common exon 2. Compensation occurred in the form of increased expression of Cα, especially in the brain.[23] The full Cβ KO mice are partially resistant to diet-induced obesity and the subsequent development of insulin resistance.[24]

13.2.6 Role of PKA in Neural Tube Development

The availability of both Cα and Cβ1 KO mice allowed us to examine mice with a combined Cα and Cβ1 deficiency. No embryos homozygous for both the Cα and Cβ1 disruption were observed, suggesting that they fail very early in development. However, we observed mice born with spina bifida and genotyping revealed that they were Cα heterozygotes and Cβ1 KO. The reduced PKA activity seen in mice with loss of any three of the four C alleles ($C\alpha^{-/-}/C\beta1^{-/+}$ or $C\alpha^{+/-}/C\beta1^{-/-}$) results in early neural tube defects and expansion of the Sonic hedgehog (Shh) signaling response in the thoracic to sacral regions of the neural tube.[25] PKA is known to be a negative regulator of Shh signaling and this is likely due to PKA-dependent phosphorylation of the Gli family of transcription factors. The neural tube phenotype caused by PKA deficiency is not dependent on Shh itself but requires the presence of Gli2 activity. Deletion of Gli2 rescues the phenotype of the PKA-deficient neural tube, indicating that PKA normally exerts negative regulatory control of Gli2 transcriptional activity (Huang and McKnight, unpublished observations). A recent study has shown that PKA localizes to the basal body of the primary cilium where it inhibits Gli2 activation and controls the activity of the Shh pathway.[26]

13.2.7 *Prkar2a* (RIIα) KO Mice

The RIIα subunit of PKA is expressed in all tissues examined in the mouse and is thought to play a major role in anchoring PKA to AKAPs in specific cell types. An RIIα KO mouse line was constructed by replacing the first exon of the *Prkar2a* gene with a neomycin resistance cassette and this completely prevented the expression of RIIα protein. The mice are healthy and fertile with no obvious behavioral defects. We examined the potentiation of calcium currents in skeletal muscle because muscle expresses both RIIα and RIα and anchored PKA has been shown to be necessary for potentiation of the calcium channel by trains of brief depolarizations. RIIα KO mice show normal voltage-dependent potentiation, but this potentiation is more sensitive to disruption by peptides that interfere with AKAP/R subunit interactions. Compensation occurs in skeletal muscle in the form of increased levels of RIα, but the C subunit continues to colocalize with L-type Ca^{2+} channels on the T tubules. Although the binding affinity of RIα for AKAPs is much less that for RIIα, it seems that the RIα subunit is able to compensate and localize sufficient PKA near the Ca^{2+} channels to retain potentiation.[27]

RIIα protein is highly expressed in mouse sperm and PKA is localized to the midpiece and distal region of the principal piece of the flagellum. RIIα KO sperm develop normally and there is no alteration in capacitation-induced motility. Male fertility is unaffected in the RIIα KO, indicating that compensation by other R subunits is sufficient to regulate PKA activity and sperm motility. A large compensatory increase in RIα is observed in the RIIα KO sperm and immunohistochemistry demonstrated that most of the RIα and C subunit in the RIIα KO sperm are localized to the cytoplasmic droplet and little PKA is anchored along the flagellum.[28] Because we have demonstrated with the Cα2 KO mouse that PKA activity is absolutely required for sperm motility and male fertility,[16] we conclude that either anchoring of PKA along the flagellum must not be essential for its physiological function or the small amounts of RIα-PKA that are anchored must be sufficient.

13.2.8 *Prkar2b* (RIIβ) KO Mice

The RIIβ subunit of PKA is selectively expressed in neurons, white and brown fat cells, pituitary, thyroid, Sertoli cells in the testis, and granulosa cells in the ovary. The RIIβ KO was made by replacing most of the first exon with the neomycin resistance gene. The homozygote RIIβ KO mice appear normal and healthy but weigh slightly less than littermate WT mice. Analysis of organ weights revealed that the reason the RIIβ KO mice weighs less than WT littermates is largely due to a substantial loss of triglyceride storage in white adipose tissue (WAT). Because RIIβ is the major regulatory subunit in WAT and also brown adipose tissue (BAT), our initial thinking was that defects in those tissues might be contributing to the lean phenotype.[29,30] However, we were also aware of the high expression of RIIβ in neurons and the potential that a PKA mutation in neurons could be affecting energy homeostasis, giving rise to the lean phenotype. In addition to their lean phenotype, RIIβ KO mice display a dramatic increase in nocturnal locomotor activity. This suggests the possibility that a portion of the lean phenotype could be related to the exercise-induced increase in energy expenditure. Further dissection of the role of individual tissues and cell types in the lean and hyperactive phenotypes depended on the generation of conditional mutants of RIIβ.

13.2.8.1 Generation of RIIβ Lox-Stop Mice

To determine the tissue-specific function of RIIβ-PKA, we generated RIIβ lox-stop mice in which a lox-stop cassette was inserted into the 5′-UTR region of the RIIβ (*Prkar2b*) gene (Figure 13.3). RIIβ expression is disrupted but can be re-expressed

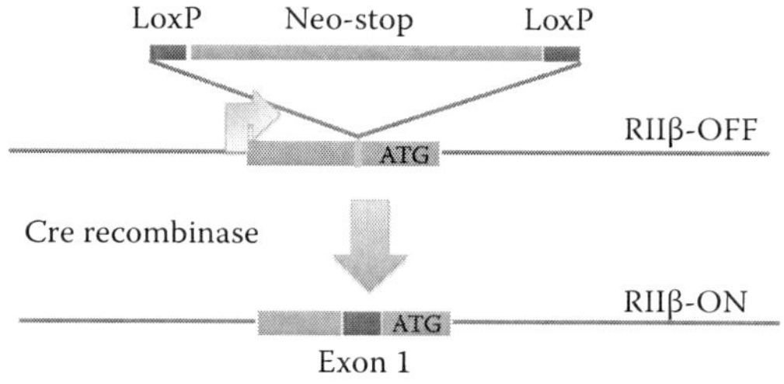

FIGURE 13.3 Diagram of RIIβ lox-stop strategy. The inserted neomycin cassette prevents the expression of wild-type RIIβ.

by Cre-dependent removal of the lox-stop cassette in RIIβ lox-stop mice.[31,32] We generated a series of mice with tissue-specific RIIβ re-expression by crossing the RIIβ lox-stop mice to multiple Cre transgenic lines to determine which tissues were most important in the genesis of specific phenotypes.

13.2.8.2 Dopamine Receptor-2–Expressing Medium Spiny Neurons Control the Hyperactivity Phenotype of RIIβ KO Mice

RIIβ KO mice exhibit a twofold to threefold increase in nocturnal locomotor activity that has been hypothesized to be caused by PKA deficiency in the striatum.[33,34] The basal resting energy expenditure, as measured by oxygen consumption per gram of lean body mass, is comparable between RIIβ KO and wild-type control mice,[32] but a significant increase in nocturnal energy expenditure is observed in the mice that exhibited hyperlocomotor activity. To determine if increased activity-induced energy expenditure is the major cause of leanness, we crossed the RIIβ lox-stop mice to striatum-specific DARPP32-Cre, D1R-Cre, and dopamine receptor-2 (D2R)-Cre transgenic mice. We found that DARPP32-Cre activates RIIβ re-expression specifically in the medium spiny neurons (MSNs) of the striatum and reverses the hyperactivity phenotype but not the leanness of RIIβ KO mice.[32] In contrast, D1R-Cre induces RIIβ re-expression in multiple brain regions including D1R-expressing MSNs, but does not reverse the hyperactivity. When a D2R-Cre is used to re-express RIIβ in the D2R neurons of the striatum, the hyperactivity is reversed but the animals remain lean. These results indicate that increased activity and leanness are two independent phenotypes and RIIβ-PKA deficiency in D2R neurons leads to hyperactivity. We speculate that RIIβ deficiency in D2R neurons might impair their neuronal activity and promote locomotion through the indirect pathway.[32]

13.2.8.3 Lean Phenotype of RIIβ KO Mice Depends on GABAergic Hypothalamic Neurons

Despite the increased basal lipolysis in fat tissues, the leanness of RIIβ KO mice is rescued by synapsin-Cre– or nestin-Cre–induced RIIβ re-expression in neurons, but not by aP2-Cre-induced adipocyte-specific RIIβ re-expression.[32] By crossing the RIIβ lox-stop mice to RIP2-Cre mice, as well as the use of AAV-Cre virus-induced RIIβ re-expression, we determined that hypothalamic deficiency of RIIβ-PKA is the major cause of the leanness seen in RIIβ KO mice. Moreover, selective RIIβ re-expression in GABAergic neurons driven by Vgat-Cre[35] reverses the leanness of RIIβ KO mice. Taken together, these results suggest that RIIβ-PKA deficiency in hypothalamic GABAergic neurons leads to decreased adiposity. Further studies indicate that RIIβ deficiency leads to enhanced leptin sensitivity in the hypothalamus (L. Yang and G.S. McKnight, unpublished data), suggesting that the leanness of RIIβ KO mice is due to changes in signal transduction pathways in specific neurons that control the magnitude and duration of the response to leptin.

13.2.8.4 RIIβ KO Mice Exhibit Deficits in Response to Dopaminergic Drugs

One of the brain regions that expresses the highest level of RIIβ is the striatum and this region undergoes the most dramatic decrease in total PKA activity in the whole animal RIIβ KO. Only about 25% of the total striatal PKA remains in the RIIβ KO and the mice display locomotor and gene expression phenotypes.[36] In addition to the hyperlocomotor behavior that depends on RIIβ-PKA mutation in the D2R neurons, the mice are also resistant to the effects of the D2R antagonist haloperidol. In WT mice, haloperidol induces c-Fos and neurotensin in the D2R neurons of the dorsal lateral striatum. These mice also display cataleptic behavior (inhibition of voluntary movement) in response to haloperidol. The RIIβ KO mice are resistant to both the behavioral and gene expression changes associated with haloperidol.[37] In contrast, the RIIβ KO mice are hypersensitive to the locomotor stimulation seen with chronic low doses of amphetamine,[36] exhibit decreased sensitivity to the sedative properties of ethanol, and an increase in voluntary consumption of ethanol-containing solutions.[38]

13.2.9 Advantages and Limitations in the Use of PKA Subunit KO Mice

13.2.9.1 Advantages

The role of PKA in complex physiological systems that control body weight regulation, reproduction, development, or aging must be studied in the context of the whole animal. The use of genetically manipulated mice is invaluable to such studies and allows investigators to directly test whether a specific gene product is required or involved in any specific physiological function. Conditional strategies that lead to either tissue-specific KO—or conversely, tissue-specific re-expression of a gene product in a null background—allow the identification of the major cell types involved in discrete physiological regulation. Genetic dissection of the mechanisms involved in KO phenotypes can be employed by crossing mutant mice to look for enhancer interactions or suppressor interactions. Because we have been studying a discrete family of isoforms of the R and C subunits of PKA, it has been illuminating in some cases to examine double-mutant mice. For example, the developmental abnormalities in RIα KO embryos are partially rescued on either a Cα heterozygote or Cα KO background and the increased tumor formation in the RIα heterozygote is substantially rescued on a Cα heterozygote background.[39] This suppressor interaction argues that the phenotypes of the RIα mutants are due to an excess of unregulated Cα subunit. An enhancer interaction is seen when combining a Cβ KO with a Cα heterozygote: either mutant animal alone is healthy and fertile but a combined Cβ KO/Cα het develops neural cord defects and is born with spina bifida.[25]

13.2.9.2 Limitations

Null mutations in specific PKA subunits result in multiple biochemical changes including decreased total PKA activity, increased basal PKA activity, compensatory increases in other PKA subunits, and altered subcellular localization of PKA holoenzymes. These changes could be variable in different cell types depending on the relative complement of R and C subunits of PKA and the type of AKAPs in the cell. Observed physiological phenotypes show that the animal cannot fully compensate for the loss of a specific subunit that is essential in a physiological pathway, but the lack of an effect of a specific PKA subunit KO on a physiological process does not demonstrate that the specific PKA subunit is normally extraneous.

13.3 DISSECTION OF PKA FUNCTIONS IN VIVO USING CONDITIONAL DOMINANT NEGATIVE, CONSTITUTIVELY ACTIVE, OR CHEMICAL GENETIC APPROACHES

To directly regulate PKA activity in cells in vivo, we created conditional alleles of RIα and Cα that contained mutations, which we have shown in cell culture to either inhibit or enhance PKA activity. These conditional alleles were either silenced by the presence of a lox-STOP-lox sequence in an intron or by the presence of a Cre-lox–dependent switchable coding sequence. The goal of these protocols is to modulate PKA activity in specific cell types in vivo.

13.3.1 Conditional RIαB Dominant Negative Mice

As stated previously, RIα is ubiquitously expressed in a wide range of cell types and tissues. Thus, a mutant RIα lacking cAMP binding capability but retaining the ability to bind to PKA catalytic subunits would serve as an ideal endogenous inhibitor of PKA in multiple cell types or tissues. RIαB is a mutant RIα isoform with a G324D amino acid substitution in cAMP binding site B and was initially characterized in cAMP-resistant S49 mouse lymphoma cells.[40,41] This mutation decreases the cAMP affinity of RIα more than 100-fold,[42] so that the mutant RIαB remains bound to PKA catalytic subunit even at relatively high concentrations of cAMP and PKA activation is suppressed. We generated a *knock-in* mouse line in which one RIα (*Prkar1a*) allele is modified to encode a silent, but Cre-inducible, RIαB[43] (Figure 13.4). By crossing this mouse line to a series of Cre transgenic mice, we characterized the distinct physiological functions of PKA in different cells and tissues.

13.3.2 RIαB Expression in Embryonic Stem Cells and Gene Expression

In mouse embryonic stem (ES) cells that are heterozygous for the targeted RIαB lox-stop allele, wild-type RIα is expressed from only one allele and the expression of the mutant RIαB is disrupted by the lox-stop cassette inserted in the intron upstream of the mutant allele.[43] We showed that the loss of an RIα allele does not cause significant reduction of total PKA activity in RIαB-OFF ES cells compared with wild-type ES cells, indicating that a single RIα allele is sufficient to maintain wild-type PKA

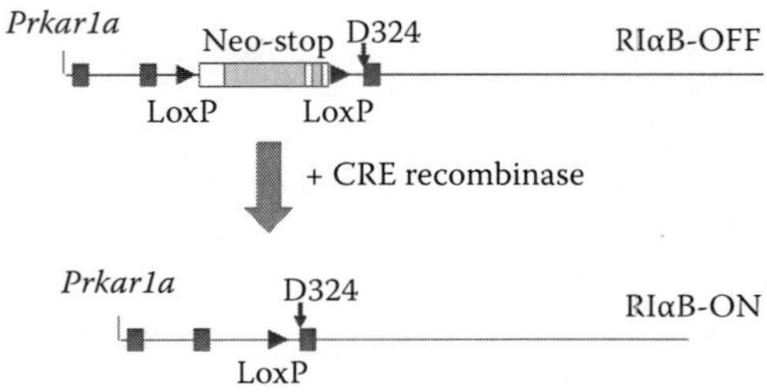

FIGURE 13.4 Diagram of the mutant RIαB allele in both OFF and ON configurations. The inserted neomycin cassette prevents the expression of dominant negative RIα mutant.

levels.[43] Transfection of the RIαB ES cells with a plasmid encoding Cre recombinase leads to the removal of the lox-stop cassette and expression of the RIαB allele. The expression of RIαB causes a 60% reduction in both basal and cAMP-stimulated PKA activity in RIαB-ON ES cells compared with either wild-type or RIαB-OFF ES cells. Notably, this reduction of PKA activity does not impair proliferation or survival of the ES cells. However, cAMP-inducible gene expression is greatly attenuated in RIαB-ON ES cells. To determine the effect of RIαB expression on gene expression, we transfected a cAMP-responsive element (CRE)-luciferase reporter construct into ES cells. Forskolin-induced transcription of the luciferase reporter is inhibited by 90% in RIαB-ON ES cells compared with wild-type or RIαB-OFF ES cells. This dramatic suppression of gene expression by RIαB protein expressed from a single endogenous allele in ES cells supports its application in vivo as a dominant inhibitor of PKA.

13.3.3 RIαB Expression in Liver and Its Effects on Glucose Homeostasis

The cAMP-PKA signaling pathway regulates glucose homeostasis partly through its regulation of glycogen metabolism and gluconeogenesis in the liver in response to fasting signals such as glucagon and epinephrine. PKA phosphorylation of CREB stimulates the expression of peroxisome proliferator-activated receptor-γ coactivator 1 (PGC1-α), which in turn activates the expression of several gluconeogenic enzymes including phosphoenolpyruvate carboxykinase (PEPCK) and glucose-6-phosphatase (G6Pase).[44,45]

Mice with liver-specific expression of RIαB are obtained by crossing the RIαB lox-stop mouse line to an albumin (alb)-Cre transgenic line.[46] Both basal and cAMP-stimulated PKA activities are dramatically decreased in liver extracts from RIαB/alb-Cre mice compared with wild-type or RIαB-OFF mice. Despite the significant inhibition of PKA activity, fasting-induced transcription of PGC1-α, PEPCK, and G6Pase in liver are essentially normal in RIαB/Alb-Cre mice. These mice have a slightly lower blood glucose level after 24 h of fasting compared with wild-type and RIαB controls, but the difference is not statistically significant. Surprisingly, we found that RIαB/Alb-Cre mice have enhanced glucose disposal during glucose tolerance tests, but their acute response to an insulin challenge seems normal. One possible explanation for the normal expression of gluconeogenic genes is that the inhibition of PKA activity by RIαB is not complete and the remaining kinase activity is sufficient to induce gluconeogenic gene expression. The incomplete inhibition of PKA activity might be because RIαB is expressed from a single allele with the other allele expressing wild-type RIα and the presence of RIIα PKA in hepatocytes. We took a further step to examine the phenotypes of RIαB/Alb-Cre with an RIIα knockout background and found greater glucose disposal ability in these mice compared with RIαB/Alb-Cre with a wild-type background. PKA activity is inhibited to a greater extent in the liver of RIIα KO/RIαB/Alb-Cre mice compared with RIαB/Alb-Cre mice with a wild-type background, but further study is needed to determine if the greater suppression of PKA activity impairs gluconeogenic gene expression in the liver. Also, it would be worthwhile to examine whether these mice with suppressed liver PKA activity are resistant to diet-induced insulin resistance and diabetes because a recent study suggested that metformin, the most frequently prescribed drug for type 2 diabetes, suppressed liver cAMP-PKA signaling to antagonize the gluconeogenic effect of glucagon and thus lower blood glucose.[47]

13.3.4 RIαB Expression in Enteric Neurons

PKA has been suggested to play important roles in the enteric nervous system (ENS).[48] RIαB expression was activated in the ENS by crossing the RIαB lox-stop mice to a proteolipid protein (PLP)-Cre or Hox11L1-Cre transgenic line. Both Cre lines activate RIαB expression in enteric neurons and result in significant inhibition of PKA activity and CREB phosphorylation.[49] The RIαB/PLP-Cre and RIαB/Hox11L1-Cre mice exhibit a similar phenotype resembling intestinal pseudo-obstructive disease in humans and the mice die shortly after weaning.[49] Severe distension of proximal intestine and impaired intestine motility are observed, although the mice seem to have normal histological structures of the intestinal nervous system and muscle.[49] These results suggest that the pseudo-obstruction is not attributable to a major disruption of ENS development, but rather inhibition of PKA may interfere with neurotransmission in the mature ENS. This mouse model suggests that PKA could be a target for the development of pharmacological therapies for chronic intestinal pseudo-obstruction.

13.3.5 RIαB Expression in Kidney Leads to Diabetes Insipidus

Water channel aquaporin 2 (AQP2) is highly expressed in the kidney collecting duct and is essential for vasopressin-regulated water reabsorption.[50] PKA phosphorylation of AQP2 at serine 256 is required for vasopressin-induced AQP2 trafficking from vesicles to the apical membrane of the principal cells in the collecting duct,[51] a key process for water reabsorption to produce concentrated urine.[52] However, cAMP-independent signaling pathways may also contribute to the effect of vasopressin[51] and direct evidence for PKA involvement in the physiological regulation of AQP2 in vivo was lacking. We selectively expressed RIαB in the collecting duct by crossing the RIαB lox-stop mice to a Sim1-Cre transgenic mouse line.[53,54] Sim1-Cre coexpresses with AQP2 in the inner medullar collecting duct principal cells using tdTomato as a Cre reporter.[55] The RIαB/Sim1-Cre mice exhibit severe polydipsia and polyuria characteristic of diabetes insipidus. Basal and cAMP-stimulated PKA activities are significantly attenuated in the inner medullar extracts from RIαB/Sim1-Cre mice and phosphorylation of PKA substrates is inhibited. As expected, we observed decreased phosphorylation of AQP2 at serine 256 in RIαB/Sim1-Cre mice and severely impaired accumulation of AQP2 protein in the apical membrane of principal cells in basal condition when mice have free access to drinking water. ddAVP administration or dehydration partially rescues the defects. Unexpectedly, total AQP2 protein level is also greatly reduced in the kidney of RIαB/Sim1-Cre mice compared with wild-type or RIαB control mice, although the AQP2 mRNA levels are comparable between the groups. Our results suggest that sustained PKA activity is indispensable for AQP2 protein expression, potentially through phosphorylation-dependent posttranslational effects. PKA is also important for the apical membrane insertion of AQP2 and whole body water homeostasis.

13.3.6 RIαB Expression in Striatal Neurons

The striatum regulates many fundamental behaviors such as feeding, motor, and motivational behaviors. Cyclic AMP-PKA signaling in the striatum has been suggested to be involved in the dopaminergic regulation of multiple physiological and psychological behaviors.[56] Mice with striatum-specific RIαB expression were obtained by crossing the RIαB lox-stop mice to a DARPP32-Cre transgenic line. RIαB/DARPP32-Cre mice exhibit hypophagia, growth retardation, and impaired striatum-dependent behaviors such as locomotion and rotarod performance.[57] RIαB expression in the striatum produces a small inhibition of total PKA activity, but is much more effective in reducing the activity of type I PKA. Type I PKA is localized in the cell body whereas the majority of PKA in striatal neurons is in the form of RIIβ-PKA and is localized in dendrites. These results support the concept that different subtypes of PKA have specialized physiological functions that are likely attributed to their subcellular localization, especially in highly polarized cells such as neurons.

13.3.7 Advantages and Limitations in the Use of Dominant Negative RIα

13.3.7.1 Advantages

In RIαB/Cre mice, the expression level of RIαB is under the control of endogenous *Prkar1a* gene control elements. This physiological level of expression may avoid problems caused by protein overexpression with other techniques such as virus-induced expression. Due to the broad expression pattern of RIα, the RIαB lox-stop mice can be useful to study PKA signaling in many different cellular and physiological processes in vivo using the Cre/loxP system. Furthermore, because RIαB is expressed from a single allele in the presence of wild-type RIα and other regulatory subunits of PKA, RIαB expression often results in only partial inhibition of PKA. This may be especially important when studying PKA signaling in tissues where PKA activity is crucial for survival and complete disruption of PKA may lead to embryonic lethality. RIαB mice may be useful to dissect the physiological functions of type I versus type II PKA in some cell types due to the distinct subcellular distribution of RI and RII subunits.

13.3.7.2 Limitations

RIαB lox-stop mice are heterozygous for RIα and males exhibit sperm abnormalities and subsequent infertility. Both sexes have late onset development of soft tissue sarcomas, hemangiosarcomas, chondrosarcomas, and hepatocellular carcinomas.[3] These phenotypes make it complicated when RIαB/Cre mice are used to study PKA functions in related cell types and tissues. RIαB/Cre mice can only be bred from male Cre mice and female RIαB mice because of the infertility of male RIαB mice. As mentioned previously, partial PKA inhibition by RIαB in some cell types (such as hepatocytes) could result in no obvious phenotype, which may lead to underestimating the role of PKA in some responses. Because of the broad expression pattern of RIα, the specificity of RIαB expression in RIαB/Cre mice is critically dependent on the expression pattern of Cre. This may lead to unexpected phenotypes if the Cre expression is not highly restricted. The extent to which PKA activity is inhibited by Cre-induced RIαB expression in different cell types or tissues may vary depending on the level of endogenous RIα protein relative to other PKA subtypes in the cell. Thus, the RIαB/Cre method may not be suitable for studying cell types in which the level of endogenous RIα protein is very low compared with other PKA regulatory subunits.

13.3.8 Expression of a Conditional Allele of the Cα Gene with Constitutive Activity

The R subunits of PKA normally bind tightly to the Cα subunit and inhibit its activity. Overexpression of the C subunit in cell culture has been shown to lead to a compensatory increase in RIα, which effectively keeps C inhibited in a holoenzyme complex. We performed a directed mutagenesis study on the mouse Cα subunit of PKA to identify mutations that rendered the C subunit resistant to inhibition by R subunits.[58] Several mutations that produce partial constitutive activity were identified and, when combined, they render the C subunit completely resistant to inhibition by RIα or RIIα. This constitutively active Cα has been used extensively in cell culture and, more recently, we constructed a conditional allele of *Prkaca*, CαR, that allows tissue-specific expression in vivo.[59] The CαR allele has a mutation that changes Trp196 to Arg and this causes a fivefold decrease in its ability to bind and be inhibited by R subunits (Figure 13.5).

The CαR-OFF allele is activated to CαR-ON specifically in hepatocytes by crossing CαR with albumin-Cre transgenic mice. The CαR/alb-Cre mice exhibit defects in blood glucose regulation have fasting hyperglycemia and are glucose-intolerant. In addition, levels of the metabolic regulator, fructose-2,6-bisphosphate (F-2,6-P_2) are decreased in the fed state, as might be expected under conditions of increased constitutive C activity because PKA decreases the synthesis of F-2,6-P_2. Messenger RNA levels for gluconeogenic genes including PEPCK, PGC-1, and G-6-P are not affected, but the mRNA for glucokinase is greatly reduced in the fed state.[59]

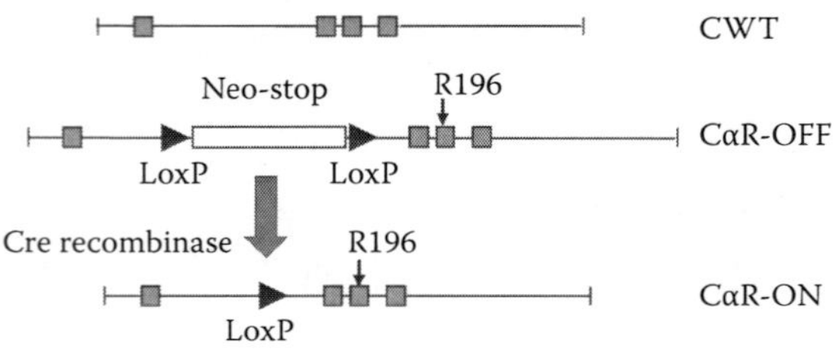

FIGURE 13.5 Diagram of the CαR allele in both OFF and ON states. The inserted neomycin cassette prevents the expression of constitutively active Cα mutant.

13.3.8.1 Limitations and Comments

The CαR mouse line can be useful in studies on the in vivo actions of PKA in specific cell types, but there are limitations that must be considered. The CαR mouse is heterozygote for the Cα allele and has a decrease in total levels of Cα in the whole animal. Although we have not yet detected physiological phenotypes in the heterozygote, it is important to test for this possibility. The overall level of increased PKA activity is also difficult to determine. Although we see that half of the mRNA expressed in the CαR-ON cells are mutants, as expected, the increase in basal activity is much smaller than expected and the total PKA activity is more similar to CαR-OFF cells or tissues. Studies have shown that the free C subunit is unstable in cells due to protein degradation and that C only becomes stable when bound to R subunit in a holoenzyme. We conclude that the constitutively active C subunit turns over rapidly in the cell, although it clearly phosphorylates targets and causes biological effects.

13.3.9 Chemical Genetic Approach to Allow Tissue-Specific Inhibition of PKA In Vivo

In collaboration with the Shokat laboratory, we identified *gatekeeper* amino acid mutations in the Cα and Cβ subunits that allow novel inhibitors to gain access to the ATP binding pocket and inhibit C subunit activity.[60] This $C\alpha_{M120A}$ (Met to Ala at position 120) mutation was engineered into the mouse genome as a conditional allele (Figure 13.6). In the absence of Cre recombinase expression, the allele makes WT

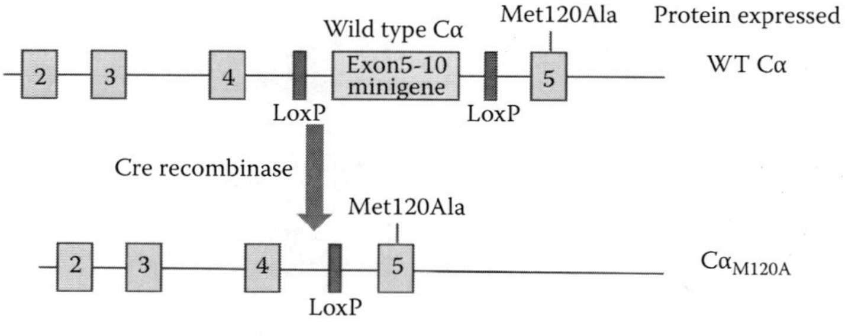

FIGURE 13.6 Diagram of the $C\alpha_{M120A}$ allele and exon 5–10 minigene. Not shown: wild-type exons 6–10.

Cα using the exon 5 to 10 minigene. After excision of the minigene by Cre recombination of the loxP sites, the product is the $C\alpha_{M120A}$ protein, which functions normally but can be inhibited by the ATP analogue, 1NM-PP1, a pyrazolo[3,4-d]pyrimidine inhibitor that does not inhibit WT Cα.[60,61] The usefulness of this approach is demonstrated by examining sperm isolated from mice expressing the mutant $C\alpha_{M120A}$ in all tissues. The activation of sperm motility and capacitation by bicarbonate is blocked by 1NM-PP1 in a reversible manner demonstrating the essential role of PKA and specifically Cα in this process.[61]

13.3.9.1 Limitations

Most tissues express both Cα and Cβ whereas our mutation only makes Cα sensitive to inhibition by the drug 1NM-PP1. Cβ can compensate for the function of Cα. This is a limitation especially when examining neural tissues where Cβ is a major fraction of the total C subunit protein. We expected that we could overcome this problem by crossing the $C\alpha_{M120A}$ mice onto a Cβ KO background, but the overall expression of WT Cα from the minigene is reduced to 50% of normal. On a Cβ KO background, this additional deficiency of Cα leads to developmental disorders in the neural tube and spina bifida in the newborn mice.[25] This limits the usefulness of this $C\alpha_{M120A}$ mouse line to tissues in which Cβ is not able to compensate for acute inhibition of $C\alpha_{M120A}$ by 1NM-PP1 and its analogs.

13.4 ANCHORING PROTEIN MUTATIONS THAT AFFECT PKA FUNCTION OR LOCALIZATION

Early studies into the subcellular localization of PKA subunits found that RI subunits are typically isolated in the cytosolic fraction whereas RII subunits are found in the particulate associated with organelles or membranes. Subcellular localization is mediated by AKAPs, a diverse family of phylogenetically unrelated proteins characterized by the presence of an approximately 30 amino acid amphipathic helix that permits docking to RI or RII.[62] The helix is a dynamic secondary structure capable of interacting with the antiparallel four-helix bundle, known as the docking and dimerization domain (D/D), created by homodimerization of the regulatory subunits of PKA. The D/D domain creates a fold that is conserved in all regulatory subunits with sequence differences determining an AKAP's ability to interact with RI, RII, or both. In many cases, the amphipathic helix and the D/D *fold* interact with nanomolar affinity.

The secondary structure and relative location of hydrophobic or charged side groups in the amphipathic helix is essential, even though various AKAPs show little identity in the primary sequence of their AKAP domain. Due to this lack of conserved sequence or common ancestry, discovering novel AKAPs has depended on identifying protein bands by far Western blot using purified, labeled RII subunits as a probe.[63] Recently, Burgers et al. developed a bioinformatic tool, THAHIT, that can map RII-binding domains on known AKAPs or be used in databases searches to identify novel AKAPs.[64]

AKAPs traditionally received names corresponding to their apparent molecular weight following SDS-PAGE, for example, AKAP150 runs near 150 kDa. This scheme is a source of confusion because several AKAPs exhibit multiple splice variants or exhibit different molecular weights across species. Some AKAPs were previously uncharacterized gene products, whereas other members of the family, such as microtubule-associated protein 2 (MAP2), had already been studied in other roles and subsequently found to contain an RII-binding domain. We refer to AKAPs by their gene names to avoid confusion.

AKAPs are multifunctional scaffolding proteins that bind to PKA as well as multiple components of signaling networks including other kinases, phosphatases, phosphodiesterases, cyclase, and substrates. The AKAPs are typically anchored to membranes or organelles in the cell. Although AKAPs are typically discussed as presenting PKA in proximity to its phosphorylation targets, it is equally important to consider how AKAPs might increase activity by localizing PKA near sites of cAMP production or, alternatively, decrease activity by sequestering PKA away from some subcellular regions. The loss of an AKAP, unlike the loss of a particular subunit of PKA, is expected to result in the regional delocalization of PKA rather than a change in the overall amount of PKA in the cell. Although named for the ability to anchor PKA, many AKAPs must also be viewed as multiprotein scaffolds; any manipulations to an AKAP may likely affect much more than just PKA localization.

A common mechanism to disrupt PKA anchoring by AKAPs is to introduce the peptide Ht31, which corresponds to the PKA-binding domain of AKAP13 and competes with endogenous AKAPs for PKA.[62] This approach has the advantage of not disrupting other protein–protein interactions on the AKAP. As a peptide, Ht31 does not normally cross the plasma membrane unless stearated or myristylated. Alternatively, Ht31 is routinely injected into cells or expressed via plasmid or viral vectors. Other AKAP-competing peptides have since been developed that show preference toward either RI or RII subunits.[65–67] A disadvantage to this method is that, even by using RI- or RII-selective peptides, it does not selectively target only one AKAP present in the cell. Thus, it is not possible to determine which AKAP is involved in the observed effect.

13.4.1 AKAP1 KO Mice

AKAP1 (also called S-AKAP84, D-AKAP1, AKAP121, and AKAP149) has as many as 15 splice variants that code proteins. A common N-terminal exon encodes a mitochondrial targeting sequence and PKA binding domain.[68] This exon makes up the majority of the protein (572 amino acids out of 903 total), and we targeted this exon for deletion in an AKAP1 knockout. Although most AKAPs show almost no affinity for RI subunits of PKA, AKAP1 binds both RI and RII with high affinity.[69]

The role of AKAP1 localization to mitochondria was recently reviewed by Merrill and Strack.[70] In addition to binding PKA, AKAP1 also directly interacts with a host of other molecules including phosphatases, phosphodiesterase PDE4a, E3 ubiquitin ligase Siah2, dynamin-related protein Drp1, and RNA. This large multimolecular complex is anchored to the outer mitochondrial membrane and functions to inhibit Drp1 from severing mitochondria during certain physiological and pathophysiological states. For example, loss of AKAP1 in hippocampal neurons can lead to mitochondrial fission and eventually apoptosis.[71,72]

13.4.1.1 Oocyte Maturation Defects

Mammalian oocyte maturation is arrested during meiosis in the final stages of follicular development. The luteinizing hormone surge corresponds to a decrease in cAMP and PKA activity that relieves the meiotic block and allows the oocyte to complete maturation. A key observation during this time is that PKA localizes to mitochondria and we identified AKAP1 as the protein required for PKA localization.[73] This discovery began as a serendipitous observation that although AKAP1 heterozygous females and homozygous knockout males are fertile, homozygous knockout females show approximately 10-fold reduced fertility. Further examination revealed that this subfertility is due to a failure of oocytes to resume meiosis.

AKAP1 is not the only AKAP expressed in oocytes or that is involved in oocyte maturation. When Ht31 is injected into immature wild-type oocytes, it promotes maturation, even if the oocytes are incubated in IBMX to maximize cAMP levels and suppress maturation.[73] This indicates that PKA anchoring is essential for meiotic arrest whereas anchoring to mitochondria promotes meiosis.

13.4.2 AKAP5

AKAP5 was originally characterized after copurification with brain RIIβ from bovine brain extracts and is also known as AKAP75 (bovine), AKAP79 (human), and AKAP 150 (rat/murine).[74,75] Much of the research on AKAP5 has focused on its role in synaptic plasticity because of its enrichment within dendrites and the postsynaptic density. AKAP5 also performs roles in peripheral nerve cell activation, cardiac calcium handling, and vascular smooth muscle function.[76,77]

AKAP5 contains a series of domains essential for interactions with membrane-associated proteins, PKC, PKA, and downstream substrates of these kinases. The mature protein contains three basic targeting subdomains, a membrane-associated guanylate kinase (MAGUK) protein-binding domain, a PxIxIT-type docking motif, an amphipathic helix, and a leucine zipper domain (Figure 13.7). The complement of binding partners for AKAP5 depends on cell type and activity. Three basic domains at the N-terminus facilitate AKAP5 interactions with phospholipids, F-actin, adenylyl cyclases, and cadherins.[78,79] Additionally, the first basic domain contains an anchor site for PKC and the potassium channel, KCNQ2/3. AKAP5 is reversibly palmitoylated and this regulates its localization to dendritic spines.[80] The PxIxIT domain allows interaction with calcineurin (CaN or protein phosphatase 2B).

13.4.2.1 AKAP5 Mutant Mice

AKAP5 is encoded by a single exon in mice. Two AKAP5 KO mice have been produced. To create the first, a neomycin resistance cassette was inserted to replace

FIGURE 13.7 Domain map of AKAP5. AKAP5 binds multiple proteins and interacts with membranes. aHe, the alpha helix responsible for binding to PKA RII subunits.

most of the AKAP5 coding sequence.[81,82] A second AKAP5 KO was developed by flanking the coding sequence with loxP sites allowing a conditional KO in the presence of Cre recombinase.[83]

Deleting the entire protein could result in the delocalization of many binding partners in addition to PKA. Thus, we also hoped to create a mutant AKAP5 lacking only its ability to bind PKA. Fortunately, the PKA binding domain is near the extreme C-terminus of the protein, so we targeted the C-terminal region of AKAP5 for truncation by inserting a stop codon at leucine 710. This produces a truncated protein, named AKAP5D36, lacking the last 36 amino acids that contains the amphipathic helix necessary for R subunit binding. However, this region also contains a leucine zipper motif that can associate with L-type calcium channels.[84] Both the complete KO and the AKAP5D36 mutant mouse lines are fertile and demonstrate no obvious phenotypical differences from wild-type mice.

13.4.2.2 Role in Phosphorylation of Calcium Channels

Hippocampal extracts from AKAP5 KO mice display reduced phosphorylation of $Ca_V1.2$ at serine 1928 following β-agonist treatment.[81] AKAP5 within the dendritic spines and cell body of CA1 neurons of the hippocampus is associated with the β2-adrenergic receptor (β2-AR), $Ca_V1.2$, adenylyl cyclases, and the phosphatase PP2A in a large signaling complex.[85] AKAP5-mediated calcium signaling is required for calcium-dependent nuclear factor of activated T-cells (NFAT)–mediated transcription in neurons.[84,86] Similarly, in cardiac extracts, AKAP5 coimmunoprecipitates with $Ca_V1.2$ and the loss of AKAP5 results in the loss of β-agonist enhancement necessary for calcium-induced calcium release.[87] Although neurons and cardiomyocytes are functionally and structurally dissimilar, both are excitable and require tight control of calcium for downstream signaling.

Calcium sparklet activity, a localized elevation in intracellular calcium, may be a solitary event from a single L-type calcium channel or the sustained activity of clustered L-type calcium channels. This activity is dependent on PKC modulation of the L-type calcium channel and is lost in the AKAP5 KO leading to defective regulation of vascular smooth muscle by angiotensin II through a PKC-dependent pathway.[88]

13.4.2.3 Regulation of TRPV Channels

TRPV1, also known as the capsaicin receptor, detects multiple noxious stimuli, including heat, depending on the sensitivity of the channel in nociceptive neurons. PKC-dependent phosphorylation alters the threshold for TRPV activation, increasing its sensitivity, whereas PKA reduces the channel's ability to be desensitized. Enhanced G protein–coupled receptor signaling by inflammatory mediators such as prostaglandin E_2 (PGE_2) leads to inflammatory thermal hyperalgesia. Studies utilizing the AKAP5D36 mouse demonstrated the necessity of anchoring of PKA by AKAP5,[89] whereas studies utilizing the AKAP5 KO determined the role of anchored PKC[90] in TRPV channel modulation. More recent studies into the anchoring of AKAP5 to the membrane by the phospholipid phosphatidylinositol-4,5-bisphosphate (PIP2) have demonstrated additional layers of modulation of the TRPV channel. High intracellular levels of PIP2 levels reduce TRPV–AKAP5 interaction, whereas degradation of PIP2 increases the association and as a result enhances the sensitization of the channel by increasing levels of anchored PKA near the channel.[78]

In arterial myocytes, TRPV4 and AKAP5 interact in a dynamic fashion leading to cellular relaxation. PKC is activated by Gq-coupled receptors leading to an increase in local calcium and diacylglycerol. Once released from AKAP5, PKC modulates the TRPV4 channel, leading to local calcium-induced calcium release (CICR) and activation of the hyperpolarizing Ca^{2+}-activated potassium (BK) channels, relaxing the myocyte. Importantly, signaling through angiotensin II receptors leads to an increase in the proximity of AKAP5 and TRPV4.[76] AKAP5 is also important for PKA-mediated potentiation of the calcium channel $Ca_V1.2$ in arterial myocytes downstream of angiotensin II signaling. The total calcium flux through a TRPV channel is nearly 100 times greater than a $Ca_V1.2$ channel. However, the $Ca_V1.2$ channel calcium sparklet rate is much greater, leading to greater increase in local and subsequently global calcium concentration and muscle contraction.

13.4.2.4 Behavioral and Electrophysiological Phenotypes

AKAP5 has been found to interact, either directly or by mutual interaction, with a number of ion channels including NMDA and AMPA receptors,[91] KCNQ2/3,[92] Kv4.2,[93] TRPV1,[89] and $Ca_V1.2$.[81] AKAP5 coordinates PKA and CaN near their mutual target, the AMPA receptor. Phosphorylation of Ser845 on the receptor promotes its trafficking to the membrane and its activity within a dendritic spine.[91] AMPA-Rs are responsible for the majority of fast excitatory synaptic transmissions. Delocalization of PKA in AKAP5 mutant mice can lead to changes in LTP and LTD in CA1 hippocampal neurons and leads to defects in learning and memory.[82,83]

13.4.3 AKAP7 KO

The *Akap7* gene codes for at least four alternatively spliced proteins (also called AKAP15/18). The shortest isoforms, AKAP7α and AKAP7β, are only 81 or 104 amino acids, respectively, arise from the same promoter, and differ only by the inclusion of a 69-base pair segment inserted between the first and last exon of AKAP7α. The longest isoforms, AKAP7γ and AKAP7δ, are 314 and 353 amino acids, respectively. Their difference depends on an alternative start site for AKAP7γ downstream of the AKAP7δ initiation codon. Some species, such as rats, lack this alternative start site and thus make AKAP7δ but not AKAP7γ.

13.4.3.1 Discriminating between Isoforms

The only coding sequence shared by all four isoforms is the terminal exon. This region is important for several reasons. First, it contains the amphipathic helix that binds to PKA regulatory subunits;[94,95] that is, the anchoring domain that makes AKAP7 an AKAP. Second, it contains a modified leucine zipper domain, just C-terminal to the PKA binding domain that is essential for interacting with certain L-type Na+ or Ca2+ channels.[96] Third, to date, the only commercially available antibodies that reliably detect AKAP7 protein recognize epitopes in this region.[97]

Some work has been done in overexpression systems to identify differences in subcellular localization of the various isoforms. The short isoforms have one myristoylated and two palmitoylated residues in the extreme N-terminus that are necessary and sufficient for anchoring the protein to the plasma membrane.[95] One report indicates that AKAP7α and AKAP7β traffic to either the basolateral or apical membrane, respectively, in polarized MDCK cells (a model of transporting epithelium).[98] This was observed by overexpressing GFP-tagged forms of either protein. Whether this mirrors localization in vivo has not been tested. Not only is this difficult to study due to the lack of antibodies that would distinguish AKAP7β from other isoforms, but AKAP7β expression levels are extremely low. Almost no signal was detectable by RT-PCR using primers specifically designed to detect the "β" exon and only one band at approximately 15 kDa is observed on immunoblot.[97] This is in contrast to the report by Fraser et al., which was the first to detect the AKAP7β isoform.[95]

Long isoforms of AKAP7 do not contain sites for lipid modification—and indeed, are present in the soluble fraction of cells—but do contain a nuclear localization sequence that is necessary and sufficient for transporting AKAP7γ/δ to the nucleus under certain conditions.[99] AKAP7 bound to PKA seems to be excluded from the nucleus, although whether this is due to PKA covering the NLS or some other reason is not known. The role of AKAP7 in the nucleus is also entirely unclear. The central domain of AKAP7γ/δ shares structural and sequence homology with RNA-interacting proteins and contains an AMP-binding site, although no phosphoesterase activity could be detected.[100,101] Recent work showed that the central domain of AKAP7γ/δ can rapidly degrade trimers of 2′,5′-oligoadenylate (2-5A), strands of which are responsible for activating RNase L as part of an antiviral response.[101]

Besides the discussed differences in sequence, structure, and subcellular localization, there is also a dramatic difference in tissue expression of the short versus long isoforms. The long isoforms were detected in every tissue that we examined, albeit at very low levels.[97] This does not mean, however, that AKAP7γ/δ are expressed in every cell. We used RiboTag analysis to identify transcripts in dentate granule cells of the mouse dentate gyrus (DG). The RiboTag transgenic mouse expresses a modified gene for Rpl22, a 60S ribosomal protein, in a Cre-dependent manner.[102] The modified Rpl22 contains three hemagglutinin tags, allowing purification of the poly-ribosome and its associated transcripts from one cell type in a complex tissue by immunoprecipitation with antihemagglutinin antibodies. RiboTag analysis of mRNA transcripts extracted from the DG of the hippocampus revealed that AKAP7γ expression in the DG granule cells is considerably lower than in the total hippocampus, suggesting that DG neurons do not express the long isoforms (unpublished data). In contrast, AKAP7α is easily detected only in brain lysates and is faintly detected in lung and to a much lesser degree in other tissues. AKAP7α is expressed in every major brain region, but it is restricted to certain neurons within different regions. For example, AKAP7α is highly expressed in DG neurons but no other cell type in the hippocampus—that is, exactly the opposite expression pattern of AKAP7γ/δ.

13.4.3.2 AKAP7 in Cardiac Signaling

AKAP7 was first reported as a 15-kDa protein that copurified and coimmunopreciptated with L-type calcium channels from rabbit skeletal muscle and later rat heart.[103,104] Subsequent analysis using AKAP7KO mice and newly available antibodies revealed that, in fact, the short isoform of the protein is not expressed in cardiomyocytes and expression is skeletal muscle was extremely low.[97] These more recent results suggest that the discovery of AKAP7 was rather serendipitous because it was apparently a contaminant that did not originate in the cells intended for study.

The long isoforms of AKAP7 are detected in cardiomyocytes. Some studies suggest a role for AKAP7δ in regulating phospholamban phosphorylation in rat cardiomyocytes.[77] This role is cast in doubt, however, by the discovery that AKAP7KO show normal calcium handling and phospholamban phosphorylation. It is also difficult to reconcile AKAP7's very broad—almost universal—distribution with it having a very specific interaction with a protein such as phospholamban with rather restricted expression.

In summary, there seems to be no expression of the short isoforms of AKAP7 in cardiomyocytes or indeed in much of the heart. The long isoforms are present, but to date there is no certain requirement for their presence for any physiological function in heart.

13.4.3.3 Challenges

AKAP7 presents several challenges that are often encountered when using knockout techniques. As noted, the *Akap7* gene has a total of 10 exons (although only two to eight are used in any one transcript). The gene spans over 147 kb of DNA, with introns between 4 and 61 kb, thereby limiting the ability to target more than one exon for deletion via standard homologous recombination techniques (Figure 13.8). The remaining exons could, in theory, still be transcribed and translated, although of course the hope, which must ultimately be tested when making a knockout, is that any residual transcripts would be destroyed through nonsense-mediated decay.

In the case of *Akap7*, we designed primers to distinguish between the long versus short isoforms. Unfortunately, because the AKAP7α exon is so short and much of its 5′-UTR inhospitable to PCR primers, we placed the 3′ primer in the shared exon. Thus, we cannot specifically detect levels of exon α alone. Our primers for the long isoforms detect an approximately twofold decrease in mRNA from knockout tissue. Thus, knockout animals continue to transcribe *Akap7* long isoforms despite disruption of the 3′-UTR. The fact that we detect less mRNA might suggest that it is less stable and targeted for degradation.

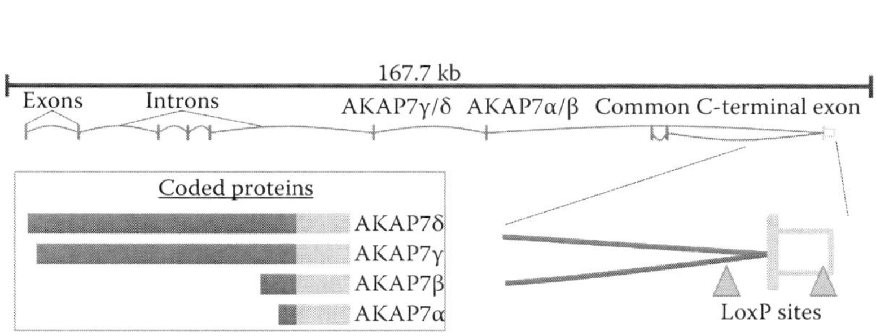

FIGURE 13.8 Gene structure and knockout strategy for AKAP7. Deletion of the terminal exon affects all AKAP7 isoforms.

The poor availability of antibodies to detect and distinguish between AKAP7 isoforms has created many challenges in the study of AKAP7. The only reliable antibodies for immunoblot or immunofluorescence microscopy recognize the one exon that all isoforms share in common. Thus, whereas protein levels of the various isoforms can be distinguished by immunoblot on the basis of size, no such discrimination is possible when staining tissue or cells for microscopy; that is, there is no way to know whether the detected signal is a long or short isoform. The mere presence of long isoform mRNA in KO animals may have little or no biological relevance unless it leads to translation into protein.

13.5 CONCLUDING REMARKS

The techniques for targeted manipulation of the mouse genome have been improving rapidly since the initial creation of transgenic mice in 1981 and the development of targeted mutagenesis using ES cells in 1989. We have discussed the application of these techniques in studies on the physiological functions of PKA. Studies that rely on targeted mutations in specific PKA subunits are subject to the limitation that compensation by other subunits can occur over the course of development. Efforts to overcome this limitation have involved studying mice carrying mutations in multiple PKA subunits, but this can lead to severe physiological defects or nonviable animals. More recent work has begun to use conditional knockouts and site-directed mutations in specific protein domains, but compensatory changes that partially restore function are still almost certain to occur. Thus, an especially important area to develop will be the creation of mutations with effects that can be rapidly turned on and off in vivo to allow investigators to circumvent the limitations inherent to a family of interacting subunits. We are optimistic that such tools will help to further unravel the mechanisms by which PKA regulates normal physiology.

REFERENCES

1. Cummings, D. E., Edelhoff, S., Disteche, C. M., and McKnight, G. S. 1994. Cloning of a mouse protein kinase A catalytic subunit pseudogene and chromosomal mapping of C subunit isoforms. *Mammalian Genome: Official Journal of the International Mammalian Genome Society* 5: 701–706.
2. Amieux, P. S., Howe, D. G., Knickerbocker, H. et al. 2002. Increased basal cAMP-dependent protein kinase activity inhibits the formation of mesoderm-derived structures in the developing mouse embryo. *The Journal of Biological Chemistry* 277: 27294–27304.
3. Veugelers, M., Wilkes, D., Burton, K. et al. 2004. Comparative PRKAR1A genotype-phenotype analyses in humans with Carney complex and prkar1a haploinsufficient mice. *Proceedings of the National Academy of Sciences of the United States of America* 101: 14222–14227.
4. Burton, K. A., McDermott, D. A., Wilkes, D. et al. 2006. Haploinsufficiency at the protein kinase A RI alpha gene locus leads to fertility defects in male mice and men. *Molecular Endocrinology* 20: 2504–2513.
5. Yin, Z., Pringle, D. R., Jones, G. N., Kelly, K. M., and Kirschner, L. S. 2011. Differential role of PKA catalytic subunits in mediating phenotypes caused by knockout of the Carney complex gene Prkar1a. *Molecular Endocrinology* 25: 1786–1793.

6. Beristain, A. G., Molyneux, S. D., Joshi, P. A. et al. 2014. PKA signaling drives mammary tumorigenesis through Src. *Oncogene*, epub ahead of print on March 2014 .
7. Honeyman, J. N., Simon, E. P., Robine, N. et al. 2014. Detection of a recurrent DNAJB1-PRKACA chimeric transcript in fibrolamellar hepatocellular carcinoma. *Science* 343: 1010–1014.
8. Goh, G., Scholl, U. I., Healy, J. M. et al. 2014. Recurrent activating mutation in PRKACA in cortisol-producing adrenal tumors. *Nature Genetics* 46: 613–617.
9. Brandon, E. P., Zhuo, M., Huang, Y. Y. et al. 1995. Hippocampal long-term depression and depotentiation are defective in mice carrying a targeted disruption of the gene encoding the RI beta subunit of cAMP-dependent protein kinase. *Proceedings of the National Academy of Sciences of the United States of America* 92: 8851–8855.
10. Hensch, T. K., Gordon, J. A., Brandon, E. P. et al. 1998. Comparison of plasticity in vivo and in vitro in the developing visual cortex of normal and protein kinase A RIbeta-deficient mice. *The Journal of Neuroscience: The Official Journal of the Society for Neuroscience* 18: 2108–2117.
11. Weisskopf, M. G., Castillo, P. E., Zalutsky, R. A., and Nicoll, R. A. 1994. Mediation of hippocampal mossy fiber long-term potentiation by cyclic AMP. *Science* 265: 1878–1882.
12. Malmberg, A. B., Brandon, E. P., Idzerda, R. L. et al. 1997. Diminished inflammation and nociceptive pain with preservation of neuropathic pain in mice with a targeted mutation of the type I regulatory subunit of cAMP-dependent protein kinase. *The Journal of Neuroscience: The Official Journal of the Society for Neuroscience* 17: 7462–7470.
13. Taylor, S. S., Ilouz, R., Zhang, P., and Kornev, A. P. 2012. Assembly of allosteric macromolecular switches: Lessons from PKA. *Nature Reviews. Molecular Cell Biology* 13: 646–658.
14. Skalhegg, B. S., Huang, Y., Su, T. et al. 2002. Mutation of the C alpha subunit of PKA leads to growth retardation and sperm dysfunction. *Molecular Endocrinology* 16: 630–639.
15. Desseyn, J. L., Burton, K. A., and McKnight, G. S. 2000. Expression of a nonmyristylated variant of the catalytic subunit of protein kinase A during male germ-cell development. *Proceedings of the National Academy of Sciences of the United States of America* 97: 6433–6438.
16. Nolan, M. A., Babcock, D. F., Wennemuth, G. et al. 2004. Sperm-specific protein kinase A catalytic subunit Calpha2 orchestrates cAMP signaling for male fertility. *Proceedings of the National Academy of Sciences of the United States of America* 101: 13483–13488.
17. San Agustin, J. T., Leszyk, J. D., Nuwaysir, L. M., and Witman, G. B. 1998. The catalytic subunit of the cAMP-dependent protein kinase of ovine sperm flagella has a unique amino-terminal sequence. *The Journal of Biological Chemistry* 273: 24874–24883.
18. Uhler, M. D., Chrivia, J. C., and McKnight, G. S. 1986. Evidence for a second isoform of the catalytic subunit of cAMP-dependent protein kinase. *The Journal of Biological Chemistry* 261: 15360–15363.
19. Qi, M., Zhuo, M., Skalhegg, B. S. et al. 1996. Impaired hippocampal plasticity in mice lacking the Cbeta1 catalytic subunit of cAMP-dependent protein kinase. *Proceedings of the National Academy of Sciences of the United States of America* 93: 1571–1576.
20. Guthrie, C. R., Skalhegg, B. S., and McKnight, G. S. 1997. Two novel brain-specific splice variants of the murine C beta gene of cAMP-dependent protein kinase. *The Journal of Biological Chemistry* 272: 29560–29565.
21. Funderud, A., Henanger, H. H., Hafte, T. T. et al. 2006. Identification, cloning and characterization of a novel 47 kDa murine PKA C subunit homologous to human and bovine C beta2. *BMC Biochemistry* 7: 20.
22. Huang, Y. Y., Kandel, E. R., Varshavsky, L. et al. 1995. A genetic test of the effects of mutations in PKA on mossy fiber LTP and its relation to spatial and contextual learning. *Cell* 83: 1211–1222.

23. Howe, D. G., Wiley, J. C., and McKnight, G. S. 2002. Molecular and behavioral effects of a null mutation in all PKA C beta isoforms. *Molecular and Cellular Neurosciences* 20: 515–524.
24. Enns, L. C., Morton, J. F., Mangalindan, R. S. et al. 2009. Attenuation of age-related metabolic dysfunction in mice with a targeted disruption of the C beta subunit of protein kinase A. *The Journals of Gerontology. Series A, Biological Sciences and Medical Sciences* 64: 1221–1231.
25. Huang, Y., Roelink, H., and McKnight, G. S. 2002. Protein kinase A deficiency causes axially localized neural tube defects in mice. *The Journal of Biological Chemistry* 277: 19889–19896.
26. Tuson, M., He, M., and Anderson, K. V. 2011. Protein kinase A acts at the basal body of the primary cilium to prevent Gli2 activation and ventralization of the mouse neural tube. *Development (Cambridge, England)* 138: 4921–4930.
27. Burton, K. A., Johnson, B. D., Hausken, Z. E. et al. 1997. Type II regulatory subunits are not required for the anchoring-dependent modulation of Ca2+ channel activity by cAMP-dependent protein kinase. *Proceedings of the National Academy of Sciences of the United States of America* 94: 11067–11072.
28. Burton, K. A., Treash-Osio, B., Muller, C. H., Dunphy, E. L., and McKnight, G. S. 1999. Deletion of type IIalpha regulatory subunit delocalizes protein kinase A in mouse sperm without affecting motility or fertilization. *The Journal of Biological Chemistry* 274: 24131–24136.
29. Cummings, D. E., Brandon, E. P., Planas, J. V. et al. 1996. Genetically lean mice result from targeted disruption of the RII beta subunit of protein kinase A. *Nature* 382: 622–626.
30. Planas, J. V., Cummings, D. E., Idzerda, R. L., and McKnight, G. S. 1999. Mutation of the RIIbeta subunit of protein kinase A differentially affects lipolysis but not gene induction in white adipose tissue. *The Journal of Biological Chemistry* 274: 36281–36287.
31. Czyzyk, T. A., Sikorski, M. A., Yang, L., and McKnight, G. S. 2008. Disruption of the RIIbeta subunit of PKA reverses the obesity syndrome of Agouti lethal yellow mice. *Proceedings of the National Academy of Sciences of the United States of America* 105: 276–281.
32. Zheng, R., Yang, L., Sikorski, M. A. et al. 2013. Deficiency of the RIIbeta subunit of PKA affects locomotor activity and energy homeostasis in distinct neuronal populations. *Proceedings of the National Academy of Sciences of the United States of America* 110: E1631–E1640.
33. Newhall, K. J., Cummings, D. E., Nolan, M. A., and McKnight, G. S. 2005. Deletion of the RIIbeta-subunit of protein kinase A decreases body weight and increases energy expenditure in the obese, leptin-deficient ob/ob mouse. *Molecular Endocrinology* 19: 982–991.
34. Nolan, M. A., Sikorski, M. A., and McKnight, G. S. 2004. The role of uncoupling protein 1 in the metabolism and adiposity of RII beta-protein kinase A-deficient mice. *Molecular Endocrinology* 18: 2302–2311.
35. Vong, L., Ye, C., Yang, Z. et al. 2011. Leptin action on GABAergic neurons prevents obesity and reduces inhibitory tone to POMC neurons. *Neuron* 71: 142–154.
36. Brandon, E. P., Logue, S. F., Adams, M. R. et al. 1998. Defective motor behavior and neural gene expression in RIIbeta-protein kinase A mutant mice. *The Journal of Neuroscience: The Official Journal of the Society for Neuroscience* 18: 3639–3649.
37. Adams, M. R., Brandon, E. P., Chartoff, E. H. et al. 1997. Loss of haloperidol induced gene expression and catalepsy in protein kinase A-deficient mice. *Proceedings of the National Academy of Sciences of the United States of America* 94: 12157–12161.

38. Thiele, T. E., Willis, B., Stadler, J. et al. 2000. High ethanol consumption and low sensitivity to ethanol-induced sedation in protein kinase A-mutant mice. *The Journal of Neuroscience: The Official Journal of the Society for Neuroscience* 20: RC75.
39. Charitakis, K., Wilkes, D., Kim, L. et al. 2008. Molecular mechanisms of PRKAR1A-dependent cardiac tumorigenesis. *Circulation* 118: S-885.
40. Steinberg, R. A. 1984. Fine-structure mapping of charge-shift mutations in regulatory subunit of type I cyclic AMP-dependent protein kinase. *Molecular and Cellular Biology* 4: 1086–1095.
41. Clegg, C. H., Correll, L. A., Cadd, G. G., and McKnight, G. S. 1987. Inhibition of intracellular cAMP-dependent protein kinase using mutant genes of the regulatory type I subunit. *The Journal of Biological Chemistry* 262: 13111–13119.
42. Woodford, T. A., Correll, L. A., McKnight, G. S., and Corbin, J. D. 1989. Expression and characterization of mutant forms of the type I regulatory subunit of cAMP-dependent protein kinase. The effect of defective cAMP binding on holoenzyme activation. *The Journal of Biological Chemistry* 264: 13321–13328.
43. Willis, B. S., Niswender, C. M., Su, T., Amieux, P. S., and McKnight, G. S. 2011. Cell-type specific expression of a dominant negative PKA mutation in mice. *PLoS One* 6: e18772.
44. Yoon, J. C., Puigserver, P., Chen, G. et al. 2001. Control of hepatic gluconeogenesis through the transcriptional coactivator PGC-1. *Nature* 413: 131–138.
45. Puigserver, P., and Spiegelman, B. M. 2003. Peroxisome proliferator-activated receptor-gamma coactivator 1 alpha (PGC-1 alpha): Transcriptional coactivator and metabolic regulator. *Endocrine Reviews* 24: 78–90.
46. Postic, C., Shiota, M., Niswender, K. D. et al. 1999. Dual roles for glucokinase in glucose homeostasis as determined by liver and pancreatic beta cell-specific gene knockouts using Cre recombinase. *The Journal of Biological Chemistry* 274: 305–315.
47. Miller, R. A., Chu, Q., Xie, J. et al. 2013. Biguanides suppress hepatic glucagon signalling by decreasing production of cyclic AMP. *Nature* 494: 256–260.
48. Barlow, A., de Graaff, E., and Pachnis, V. 2003. Enteric nervous system progenitors are coordinately controlled by the G protein-coupled receptor EDNRB and the receptor tyrosine kinase RET. *Neuron* 40: 905–916.
49. Howe, D. G., Clarke, C. M., Yan, H. et al. 2006. Inhibition of protein kinase A in murine enteric neurons causes lethal intestinal pseudo-obstruction. *Journal of Neurobiology* 66: 256–272.
50. Dibas, A. I., Mia, A. J., and Yorio, T. 1998. Aquaporins (water channels): Role in vasopressin-activated water transport. *Proceedings of the Society for Experimental Biology and Medicine* 219: 183–199.
51. Brown, D. 2003. The ins and outs of aquaporin-2 trafficking. *American Journal of Physiology. Renal Physiology* 284: F893–F901.
52. McDill, B. W., Li, S. Z., Kovach, P. A., Ding, L., and Chen, F. 2006. Congenital progressive hydronephrosis (cph) is caused by an S256L mutation in aquaporin-2 that affects its phosphorylation and apical membrane accumulation. *Proceedings of the National Academy of Sciences of the United States of America* 103: 6952–6957.
53. Balthasar, N., Dalgaard, L. T., Lee, C. E. et al. 2005. Divergence of melanocortin pathways in the control of food intake and energy expenditure. *Cell* 123: 493–505.
54. Gilbert, M. L., Yang, L., Su, T., and McKnight, G. S. 2015. Expression of a dominant negative Protein Kinase A (PKA) mutation in kidney elicits a diabetes insipidus phenotype. *American Journal of Physiology-Renal Physiology*, epub ahead of print on January 2015.
55. Madisen, L., Zwingman, T. A., Sunkin, S. M. et al. 2010. A robust and high-throughput Cre reporting and characterization system for the whole mouse brain. *Nature Neuroscience* 13: 133–140.

56. Svenningsson, P., Nishi, A., Fisone, G. et al. 2004. DARPP-32: An integrator of neurotransmission. *Annual Review of Pharmacology and Toxicology* 44: 269–296.
57. Yang, L., Gilbert, M. L., Zheng, R., and McKnight, G. S. 2014. Selective expression of a dominant-negative type Ialpha PKA regulatory subunit in striatal medium spiny neurons impairs gene expression and leads to reduced feeding and locomotor activity. *The Journal of Neuroscience: The Official Journal of the Society for Neuroscience* 34: 4896–4904.
58. Orellana, S. A., and McKnight, G. S. 1992. Mutations in the catalytic subunit of cAMP-dependent protein kinase result in unregulated biological activity. *Proceedings of the National Academy of Sciences of the United States of America* 89: 4726–4730.
59. Niswender, C. M., Willis, B. S., Wallen, A. et al. 2005. Cre recombinase-dependent expression of a constitutively active mutant allele of the catalytic subunit of protein kinase A. *Genesis* 43: 109–119.
60. Niswender, C. M., Ishihara, R. W., Judge, L. M. et al. 2002. Protein engineering of protein kinase A catalytic subunits results in the acquisition of novel inhibitor sensitivity. *The Journal of Biological Chemistry* 277: 28916–28922.
61. Morgan, D. J., Weisenhaus, M., Shum, S. et al. 2008. Tissue-specific PKA inhibition using a chemical genetic approach and its application to studies on sperm capacitation. *Proceedings of the National Academy of Sciences of the United States of America* 105: 20740–20745.
62. Carr, D. W., Hausken, Z. E., Fraser, I. D., Stofko-Hahn, R. E., and Scott, J. D. 1992. Association of the type II cAMP-dependent protein kinase with a human thyroid RII-anchoring protein. Cloning and characterization of the RII-binding domain. *The Journal of Biological Chemistry* 267: 13376–13382.
63. Carr, D. W., and Scott, J. D. 1992. Blotting and band-shifting: Techniques for studying protein–protein interactions. *Trends in Biochemical Sciences* 17: 246–249.
64. Burgers, P. P., van der Heyden, M. A., Kok, B., Heck, A. J., and Scholten, A. 2015. A systematic evaluation of protein kinase A–A-kinase anchoring protein interaction motifs. *Biochemistry* 54: 11–21.
65. Alto, N. M., Soderling, S. H., Hoshi, N. et al. 2003. Bioinformatic design of A-kinase anchoring protein-in silico: A potent and selective peptide antagonist of type II protein kinase A anchoring. *Proceedings of the National Academy of Sciences of the United States of America* 100: 4445–4450.
66. Carlson, C. R., Lygren, B., Berge, T. et al. 2006. Delineation of type I protein kinase A-selective signaling events using an RI anchoring disruptor. *The Journal of Biological Chemistry* 281: 21535–21545.
67. Gold, M. G., Lygren, B., Dokurno, P. et al. 2006. Molecular basis of AKAP specificity for PKA regulatory subunits. *Molecular Cell* 24: 383–395.
68. Huang, L. J., Wang, L., Ma, Y. et al. 1999. NH2-Terminal targeting motifs direct dual specificity A-kinase-anchoring protein 1 (D-AKAP1) to either mitochondria or endoplasmic reticulum. *The Journal of Cell Biology* 145: 951–959.
69. Huang, L. J., Durick, K., Weiner, J. A., Chun, J., and Taylor, S. S. 1997. Identification of a novel protein kinase A anchoring protein that binds both type I and type II regulatory subunits. *The Journal of Biological Chemistry* 272: 8057–8064.
70. Merrill, R. A., and Strack, S. 2014. Mitochondria: A kinase anchoring protein 1, a signaling platform for mitochondrial form and function. *The International Journal of Biochemistry & Cell Biology* 48: 92–96.
71. Cribbs, J. T., and Strack, S. 2007. Reversible phosphorylation of Drp1 by cyclic AMP-dependent protein kinase and calcineurin regulates mitochondrial fission and cell death. *EMBO Reports* 8: 939–944.
72. Merrill, R. A., Dagda, R. K., Dickey, A. S. et al. 2011. Mechanism of neuroprotective mitochondrial remodeling by PKA/AKAP1. *PLoS Biology* 9: e1000612.

73. Newhall, K. J., Criniti, A. R., Cheah, C. S. et al. 2006. Dynamic anchoring of PKA is essential during oocyte maturation. *Current Biology: CB* 16: 321–327.
74. Carr, D. W., Stofko-Hahn, R. E., Fraser, I. D., Cone, R. D., and Scott, J. D. 1992. Localization of the cAMP-dependent protein kinase to the postsynaptic densities by A-kinase anchoring proteins. Characterization of AKAP 79. *The Journal of Biological Chemistry* 267: 16816–16823.
75. Sarkar, D., Erlichman, J., and Rubin, C. S. 1984. Identification of a calmodulin-binding protein that co-purifies with the regulatory subunit of brain protein kinase II. *The Journal of Biological Chemistry* 259: 9840–9846.
76. Mercado, J., Baylie, R., Navedo, M. F. et al. 2014. Local control of TRPV4 channels by AKAP150-targeted PKC in arterial smooth muscle. *The Journal of General Physiology* 143: 559–575.
77. Lygren, B., Carlson, C. R., Santamaria, K. et al. 2007. AKAP complex regulates Ca2+ re-uptake into heart sarcoplasmic reticulum. *EMBO Reports* 8: 1061–1067.
78. Jeske, N. A., Por, E. D., Belugin, S. et al. 2011. A-kinase anchoring protein 150 mediates transient receptor potential family V type 1 sensitivity to phosphatidylinositol-4,5-bisphosphate. *The Journal of Neuroscience: The Official Journal of the Society for Neuroscience* 31: 8681–8688.
79. Gorski, J. A., Gomez, L. L., Scott, J. D., and Dell'Acqua, M. L. 2005. Association of an A-kinase-anchoring protein signaling scaffold with cadherin adhesion molecules in neurons and epithelial cells. *Molecular Biology of the Cell* 16: 3574–3590.
80. Keith, D. J., Sanderson, J. L., Gibson, E. S. et al. 2012. Palmitoylation of A-kinase anchoring protein 79/150 regulates dendritic endosomal targeting and synaptic plasticity mechanisms. *The Journal of Neuroscience: The Official Journal of the Society for Neuroscience* 32: 7119–7136.
81. Hall, D. D., Davare, M. A., Shi, M. et al. 2007. Critical role of cAMP-dependent protein kinase anchoring to the L-type calcium channel Cav1.2 via A-kinase anchor protein 150 in neurons. *Biochemistry* 46: 1635–1646.
82. Weisenhaus, M., Allen, M. L., Yang, L. et al. 2010. Mutations in AKAP5 disrupt dendritic signaling complexes and lead to electrophysiological and behavioral phenotypes in mice. *PLoS One* 5: e10325.
83. Tunquist, B. J., Hoshi, N., Guire, E. S. et al. 2008. Loss of AKAP150 perturbs distinct neuronal processes in mice. *Proceedings of the National Academy of Sciences of the United States of America* 105: 12557–12562.
84. Oliveria, S. F., Dell'Acqua, M. L., and Sather, W. A. 2007. AKAP79/150 anchoring of calcineurin controls neuronal L-type Ca2+ channel activity and nuclear signaling. *Neuron* 55: 261–275.
85. Davare, M. A., Avdonin, V., Hall, D. D. et al. 2001. A beta2 adrenergic receptor signaling complex assembled with the Ca2+ channel Cav1.2. *Science* 293: 98–101.
86. Murphy, J. G., Sanderson, J. L., Gorski, J. A. et al. 2014. AKAP-anchored PKA maintains neuronal L-type calcium channel activity and NFAT transcriptional signaling. *Cell Reports* 7: 1577–1588.
87. Nichols, C. B., Rossow, C. F., Navedo, M. F. et al. 2010. Sympathetic stimulation of adult cardiomyocytes requires association of AKAP5 with a sub-population of L-type calcium channels. *Circulation Research* 107: 747–756.
88. Navedo, M. F., Nieves-Cintron, M., Amberg, G. C. et al. 2008. AKAP150 is required for stuttering persistent Ca2+ sparklets and angiotensin II-induced hypertension. *Circulation Research* 102: e1–e11.
89. Schnizler, K., Shutov, L. P., Van Kanegan, M. J. et al. 2008. Protein kinase A anchoring via AKAP150 is essential for TRPV1 modulation by forskolin and prostaglandin E2 in mouse sensory neurons. *The Journal of Neuroscience: The Official Journal of the Society for Neuroscience* 28: 4904–4917.

90. Jeske, N. A., Patwardhan, A. M., Ruparel, N. B. et al. 2009. A-kinase anchoring protein 150 controls protein kinase C-mediated phosphorylation and sensitization of TRPV1. *Pain* 146: 301–307.
91. Tavalin, S. J., Colledge, M., Hell, J. W. et al. 2002. Regulation of GluR1 by the A-kinase anchoring protein 79 (AKAP79) signaling complex shares properties with long-term depression. *The Journal of Neuroscience: The Official Journal of the Society for Neuroscience* 22: 3044–3051.
92. Bal, M., Zhang, J., Hernandez, C. C., Zaika, O., and Shapiro, M. S. 2010. Ca2+/calmodulin disrupts AKAP79/150 interactions with KCNQ (M-Type) K+ channels. *The Journal of Neuroscience: The Official Journal of the Society for Neuroscience* 30: 2311–2323.
93. Lin, L., Sun, W., Kung, F., Dell'Acqua, M. L., and Hoffman, D. A. 2011. AKAP79/150 impacts intrinsic excitability of hippocampal neurons through phospho-regulation of A-type K+ channel trafficking. *The Journal of Neuroscience: The Official Journal of the Society for Neuroscience* 31: 1323–1332.
94. Gray, P. C., Johnson, B. D., Westenbroek, R. E. et al. 1998. Primary structure and function of an A kinase anchoring protein associated with calcium channels. *Neuron* 20: 1017–1026.
95. Fraser, I. D., Tavalin, S. J., Lester, L. B. et al. 1998. A novel lipid-anchored A-kinase anchoring protein facilitates cAMP-responsive membrane events. *The EMBO Journal* 17: 2261–2272.
96. Hulme, J. T., Ahn, M., Hauschka, S. D., Scheuer, T., and Catterall, W. A. 2002. A novel leucine zipper targets AKAP15 and cyclic AMP-dependent protein kinase to the C terminus of the skeletal muscle Ca2+ channel and modulates its function. *The Journal of Biological Chemistry* 277: 4079–4087.
97. Jones, B. W., Brunet, S., Gilbert, M. L. et al. 2012. Cardiomyocytes from AKAP7 knockout mice respond normally to adrenergic stimulation. *Proceedings of the National Academy of Sciences of the United States of America* 109: 17099–17104.
98. Trotter, K. W., Fraser, I. D., Scott, G. K. et al. 1999. Alternative splicing regulates the subcellular localization of A-kinase anchoring protein 18 isoforms. *The Journal of Cell Biology* 147: 1481–1492.
99. Brown, R. L., August, S. L., Williams, C. J., and Moss, S. B. 2003. AKAP7gamma is a nuclear RI-binding AKAP. *Biochemical and Biophysical Research Communications* 306: 394–401.
100. Gold, M. G., Smith, F. D., Scott, J. D., and Barford, D. 2008. AKAP18 contains a phosphoesterase domain that binds AMP. *Journal of Molecular Biology* 375: 1329–1343.
101. Gusho, E., Zhang, R., Jha, B. K. et al. 2014. Murine AKAP7 has a 2′,5′-phosphodiesterase domain that can complement an inactive murine coronavirus ns2 gene. *MBio* 5: e01312–e01314.
102. Sanz, E., Yang, L., Su, T. et al. 2009. Cell-type-specific isolation of ribosome-associated mRNA from complex tissues. *Proceedings of the National Academy of Sciences of the United States of America* 106: 13939–13944.
103. Gray, P. C., Tibbs, V. C., Catterall, W. A., and Murphy, B. J. 1997. Identification of a 15-kDa cAMP-dependent protein kinase-anchoring protein associated with skeletal muscle L-type calcium channels. *The Journal of Biological Chemistry* 272: 6297–6302.
104. Hulme, J. T., Lin, T. W., Westenbroek, R. E., Scheuer, T., and Catterall, W. A. 2003. Beta-adrenergic regulation requires direct anchoring of PKA to cardiac CaV1.2 channels via a leucine zipper interaction with A kinase-anchoring protein 15. *Proceedings of the National Academy of Sciences of the United States of America* 100: 13093–13098.

Index

D

E

F

G

H